Strategic
Technology
Management

McGraw-Hill Engineering and Technology Management Series
Michael K. Badawy, Ph.D., Editor in Chief

Strategic Technology Management

Frederick Betz

McGraw-Hill, Inc.

New York   San Francisco   Washington, D.C.   Auckland   Bogotá
Caracas   Lisbon   London   Madrid   Mexico City   Milan
Montreal   New Delhi   San Juan   Singapore
Sydney   Tokyo   Toronto

Library of Congress Cataloging-in-Publication Data

Betz, Frederick.
    Strategic technology management / Frederick Betz.
        p.   cm. — (McGraw-Hill engineering and technology management
series)
    Includes bibliographical references and index.
    ISBN 0-07-005137-2
    1. Technology—Management.   2. Technological innovations.
    3. Strategic planning. I. Title. II. Series.
    T49.5.B45   1993
    658.4′062—dc20 Allen County Public Library   92-35846
                    900 Webster Street   CIP
                    PO Box 2270
                    Fort Wayne, IN 46801-2270

1 2 3 4 5 6 7 8 9 0 DOC/DOC 9 9 8 7 6 5 4 3

ISBN   07-005137-2

*The sponsoring editor for this book was Robert Hauserman, the editing
supervisor was Jane Palmieri, and the production supervisor was
Suzanne W. Babeuf. It was set in Century Schoolbook by McGraw-Hill's
Professional Book Group composition unit.*

Printed and bound by R. R. Donnelley & Sons Company.

This book is printed on acid-free paper.

To my mother, Sarah Kaiser Betz, and my father, Frederick William Betz

Contents

Series Introduction

Technology is a key resource of profound importance for corporate profitability and growth. It also has enormous significance for the well-being of national economies as well as international competitiveness. Effective management of technology links engineering, science, and management disciplines to address the issues involved in the planning, development, and implementation of technological capabilities to shape and accomplish the strategic and operational objectives of an organization.

Management of technology involves the handling of technical activities in a broad spectrum of functional areas including basic research; applied research; development; design; construction, manufacturing, or operations; testing; maintenance; and technology transfer. In this sense, the concept of technology management is quite broad, since it covers not only R&D but also the management of product and process technologies. Viewed from that perspective, the management of technology is actually the practice of integrating technology strategy with business strategy in the company. This integration requires the deliberate coordination of the research, production, and service functions with the marketing, finance, and human resource functions of the firm.

That task calls for new managerial skills, techniques, styles, and ways of thinking. Providing executives, managers, and technical professionals with a systematic source of information to enable them to develop their knowledge and skills in managing technology is the challenge undertaken by this book series. The series will embody concise and practical treatments of specific topics within the broad area of engineering and technology management. The primary aim of the series is to provide a set of principles, concepts, tools, and techniques for those who wish to enhance their managerial skills and realize their potentials.

The series will provide readers with the information they must have and the skills they must acquire in order to sharpen their managerial performance and advance their careers. Authors contributing to the series are carefully selected for their expertise and experience. Although the series books will vary in subject matter as well as approach, one major feature will be common to all of them: a blend of practical applications and hands-on techniques supported by sound research and relevant theory.

The target audience for the series is quite broad. It includes engineers, scientists, and other technical professionals making the transition to management; entrepreneurs; technical managers and supervisors; upper-level executives; directors of engineering; people in R&D and other technology-related activities; corporate technical development managers and executives; continuing management education specialists; and students in engineering and technology management programs and related fields.

We hope that this series will become a primary source of information on the management of technology for practitioners, researchers, consultants, and students, and that it will help them become better managers and pursue the most rewarding professional careers.

DR. MICHAEL K. BADAWY
Professor of Management of Technology
The R. B. Pamplin College of Business
Virginia Polytechnic Institute and State University
Falls Church, Virginia

Preface

Technology in a firm is the knowledge of the productive capabilities of the firm's businesses. This book covers the formulation of a technology strategy in order to anticipate, create, and use technology for economic advantage. It focuses on the problem of managing strategic technologies—rapidly changing core technologies essential to the future competitive position of a corporation.

This book is intended for professionals who manage technical change in organizations—those responsible for the research and development (R&D), manufacturing, and marketing of new products or services and for the improvement of manufacturing capability. Examples of types of problems addressed are

1. How to identify where the next breakthroughs are coming in rapidly changing technologies.

2. How to manage innovation in the product-development cycle for rapid responsiveness.

3. How to focus technology strategy in a company that makes different kinds of products for different markets.

4. How to link R&D with corporate strategy.

5. How to acquire externally developed technology from universities, suppliers, and other partners.

Although technology is widely recognized to be essential to competitiveness, it has been one of the most difficult activities with which management has to deal. There are many reasons why some corporations have failed to exploit technological advantages and over time lost positions to competitors (even though earlier they had significant market shares, large R&D budgets, trained and skilled technical personnel, and large and active marketing and sales forces).

For example, some corporations have had incomplete planning processes at the corporate level, focusing only on financial control and failing to address and utilize the technical competencies within the firm for strategic advantage. Some corporations have physically separated corporate research laboratories from strategic business units and then have failed to effectively integrate research with product development. Some corporations have suffered from a "not invented here" syndrome and have failed to actively seek out and exploit relevant and important technologies developed elsewhere. Some corporations have used techniques for technology forecasting and planning but only in a naive and unsophisticated manner that resulted as much in misguidance as in correct insight about the future. Some corporations have successfully introduced new high-tech products only later to lose them to competitors, who further innovated and improved on the manufacture of the products.

Since there are so many ways to fail in commercially exploiting innovation, it is important to carefully attend to it. It is useful both to assemble a complete set of technology forecasting, planning, and implementation tools and to understand enough about the nature of technology to use those tools properly. This book is special for two reasons— (1) because it does include tools for forecasting, planning, and technology transfer all in one place, and (2) because it provides an in-depth examination of the nature of technological invention so that one can understand how to manage technical activities.

Many techniques are assembled here—for forecasting the rates of technology change, for forecasting the directions of technology change, for forecasting the impact of technological change on the structures of markets and industries, and for facilitating the implementation of new technology into products, services, and manufacturing. The book approaches technology strategy at both procedural and methodological levels. Management procedures are necessary to understand how to manage technology, but methodological approaches to technology are also necessary in order to understand what is being managed.

We know that there have been two approaches to management theory: (1) a superficial approach that thought general management principles could be used to manage any business without a profound knowledge of that business, and (2) a substantial approach that thought both general management principles and an understanding of the specific business to be managed were both necessary for good management. This book takes the second approach—asserting that for good management, the general principles of management must be adapted and refined to the special conditions of the process being managed (which here is technological change).

Accordingly, you may find this is one of the few books that reviews the essence of technological change—invention and engineering design. Invention and design are the activities in which the human spirit interacts with nature in order to affect the destiny of nature itself, and since humanity is also a part of nature, this makes a curious interaction. *Invention* is the original creation of a logic to manipulate nature for a purpose. *Design* is the subsequent use of that logic in products, processes, or services. How invention and design are possible provides the basis for planning research for technological progress.

This book is also intended as a text for courses in the management of technology and in engineering management—taught either in business schools or in engineering schools. It is particularly appropriate for a course on forecasting and planning technology. The text extensively uses *actual* case studies to illustrate theory and methodology.

This book is organized for a semester course of 15 sessions. Logically, one forecasts, plans, and then implements technologies. For pedagogical purposes, however, the order of presentation is reversed—beginning with planning and implementing technology before studying the techniques of technology forecasting and research planning (since the latter get rather esoteric):

Planning technological change in diversified corporations (Chaps. 1, 2, 3, and 4)

Understanding invention—the heart of technological change (Chap. 5)

Implementing new technology in new products or services or manufacturing (Chaps. 6 and 7)

Forecasting the economic impacts of technological change (Chaps. 8, 9, and 10)

Forecasting technological change and planning research (Chaps. 11, 12, 13, 14, and 15)

In the nineteenth and twentieth centuries, the world was organized into industrialized countries and nonindustrialized countries. As the twenty-first century begins, this older pattern is rapidly fading. In Europe, the Americas, Asia, Africa—on all continents—all nations are industrializing (some effectively and some ineffectively). Increasingly, the competitive differences between firms (and between countries) have come to depend on their abilities to acquire, develop, and focus new technologies on market need. Today there is no permanent technology advantage for any firm. There are only temporary lead times in technology. This makes managing strategic technologies essential for long-term survival.

Frederick Betz

Technology and Economic Goals

We know that technological change is a major factor in long-term commercial failure or success. New technologies create new markets or substitute in existing markets by obsoleting the affected current technologies and any of the products, services, or production processes in which these technologies are embedded. We also know that in a firm there are two principal economic goals for managing strategic technologies: (1) to innovate new markets or (2) to dominate and keep existing markets. In this book we will examine the problems of anticipating, planning, and implementing technological change for commercial advantage.

Technology forecasting is difficult, for the future is never predictable. Yet clear trends of technical change often can be identified. Certainly, goals for improving existing technologies can be formulated, and technically focused research can be planned, funded, and managed.

Forecasting market development and competitive conditions under technological change is also difficult. However, general patterns can be identified in most histories of new markets created by new technologies and in the competitive conditions as these markets evolved. These patterns can assist the formulation of marketing and competitive strategies as technologies change.

Technology planning and implementation are even more difficult activities than forecasting, for then resources are staked and futures risked. Yet even here, practice and studies have formulated and identified useful approaches. There are techniques and procedures which facilitate proper market attention for research efforts and which facilitate the proper transfer of technical advances into competitive products, production, and services.

An important problem in technology strategy arises from the fact that most large firms are diversified, and since different businesses use many different technologies, the number of relevant technologies to a diversified firm can be large. Hence technology strategy in a diversified firm can be complicated. Despite these complications, however, planning even for diversified firms still should be based on all the strategic bases of the business enterprise: technology, market, capital, production, and organization.

In the 1980s, business history highlighted the lesson that trying to manage modern corporations with only a financial strategy certainly did not guarantee long-term success. In the United States and Europe, many corporate conglomerates showed a common pattern of eventually having a market value of the whole corporation greater when taken apart than when kept together (e.g., as demonstrated in the corporate demise of RCA in 1985). The exceptions to this pattern have been firms, such as 3M, that developed core technological competencies—in addition to financial and marketing competencies.

> *Strategic technologies are the rapidly changing core technical competencies that provide competitive edges to the businesses of the corporation.*

My purpose in this book is to describe methods for managing core technical competencies of a firm.

ILLUSTRATION
Development of the Commodore 64
Personal Computer

The scope of our focus on anticipating technological change and on planning and implementing technological change can be illustrated by an interesting case of a new product development, the Commodore 64 personal computer (in the fledgling personal computer industry in the early 1980s, when it rocked and knocked out some of its major competitors). Moreover, it is a sobering illustration that a single innovative product success does not guarantee long-term commercial success. Also, the case highlights many important issues about technology strategy during the early phases of a new high-tech industry.

In 1981–1982, the Commodore company (then under Jack Tramiel as president) designed and produced a new personal computer called the *Commodore 64.* Although now long obsolete as a product, it was then a hot and very competitively priced product for the then-emergent home computer industry. In fact, it pushed the prestigious Texas Instruments Company out of the home computer market while also contributing to the death of a new and high-flying company, Atari. (In the

1990s, Atari did still exist in name—but as a wholly new company and, ironically, under the ownership and control of Tramiel, who had earlier helped kill it.)

We remind ourselves that the strategic technology underlying the personal computer was the semiconductor integrated circuit (IC) chip, which, although only invented in 1959, had by 1980 already passed through three generations of rapid technological change—small-scale integration, middle-scale integration (MSI), and large-scale integration (LSI). LSI was a set of production processes enabling the fabrication of IC chips with *thousands* of transistors on a single chip. This transistor density had enabled computer designers to fabricate on a single chip, the heart of a computer, its the central processing unit (called then a *microprocessor*). The microprocessor chip made possible the new personal computer industry, and the first relatively complete and assembled personal computer to be marketed was the Apple in 1976.

By 1981, personal computer companies were emerging as a new industry. Then the personal computer market served two distinct and new markets, a business market for word processing, spreadsheets, and accounting and a home market for games. Both markets were served by the same computers. Apple was the leader, closely followed by Tandy, Texas Instruments, Atari, and Commodore. The prices of the computers ranged from $1500 to $2500.

The home market was, of course, very price sensitive to the $1000 barrier. Tramiel at Commodore first produced a personal computer called *PET,* which competed poorly against the then leader Apple. In addition, Tramiel watched the rapid sales of a new personal computer product from England, the Sinclair, then selling for $100. The Sinclair, however, did not even have a real keyboard. Therefore, at Commodore, Tramiel stopped producing his PET computer and introduced a new, cheaper model called the *VIC-20,* priced around $500 (a thousand dollars cheaper than the Apple). This was Commodore's first really big seller, but the VIC-20 was limited in speed and memory compared with its competitors, the Apple II model and the Atari 800. All computers then contained 8-bit word-length microprocessors (except the Texas Instruments 99/4a, which did use a 16-bit microprocessor, but its designers had not utilized it to show any performance advantage over competitor's products).

Also back in 1981, it was the video game market that was the hot and rapidly increasing market for personal computers. There were a lot of kids who loved the new video games and a lot of parents willing to spend $100 on a kid, but few parents then were willing to spend more. To get performance up and price down in this game market, the product required increased speed for graphics and sound. And the

Atari 800 pioneered the technical solution to speeding up graphics performance by incorporating a special video-display chip to assist the microprocessor.

This is the background, and the Commodore 64 story began in 1981 at a small chip manufacturing company, MOS Technology, then located in West Chester, Pennsylvania. In January, a group of semiconductor engineers decided to design two special chips for graphics and for sound. Albert J. Charpentier led the group, and he was head of MOS's Large-Scale Integration (LSI) Group. These new chips would be available to whomever then wanted to make "the world's best video game" (Perry and Wallich, 1985, p. 48).

What Charpentier's group wanted to do was to follow-up on Atari's technical solution by designing a newer and more advanced set of graphics and sound chips—to provide game capability as fast as the chips in the Atari 800. Charpentier proposed this project to his boss, Charles Winterble, who then was director of worldwide engineering for Commodore. Winterble approved the project.

It is interesting to note why Commodore owned MOS Technology. MOS Technology had been a new start-up company in the early 1970s when chip technology advanced from a density of hundreds of transistors on a chip in MSI technology to the thousands of transistors on a chip in LSI technology. MOS Technology had designed and produced the then popular microprocessor 6502, which was the 8-bit microprocessor that powered both the original Apple computers and Commodore's PET computer.

By the middle of the 1970s, however, several Japanese firms (such as Hitachi, Toshiba, and others) mastered LSI technology and produced a 16K memory chip, which was then the most dense memory chip available. Memory chips sold as the largest volume of the IC chip market. In 1975, Japanese chip producers grabbed a significant world market share (20 percent) of the 16K market. In addition, they aggressively lowered the price of memory chips and raised production quality. This hurt many of the new small chip producers, such as MOS—many of which had been started on the new LSI technology. So even when MOS was only a few years old, the entry of larger and better-financed Japanese competitors into its market dealt it a death blow.

In 1976, therefore, MOS was on the financial ropes, and Commodore, one of its customers for the 6502 microprocessor chip, picked up the company—"at 10 cents on the dollar" (Perry and Wallich, 1985, p. 48). This was how Charpentier of MOS came to work for Winterble of Commodore (thanks to the changing conditions of international competition in strategic technologies).

This also illustrates what happens to many small high-tech firms as (1) radical technological innovation allows new small firms to start up, but (2) then later many established competitors from other industries enter and drive down prices and profitability. (Later I will provide a fuller description of this kind of scenario and discuss how the concepts of the technology life cycle and of the industrial economic-value chain can be used to help plan technology strategy for such scenarios in times of rapid technological change.)

When Charpentier's group decided to make graphics and sound chips, their first step in the design process was to find out what the current high-quality chip products could do. This kind of practice in design is called *competitive benchmarking*. The group spent the first 2 weeks finding and looking at comparable graphics and sound chips industry-wide. They also looked at the graphics capabilities of the most advanced game and personal computers available (Perry and Wallich, 1985, p. 49):

> "We looked heavily into the Mattel Intellivision," recalls Winterble. "We also examined the Texas Instruments 99/4a and the Atari 800. We tried to get a feel for what these companies could do in the future by extrapolating from their current technology. That made it clear what the graphics capabilities of our machine had to be."

Competitive benchmarking for new product design requires not only that one find and look at the best and most advanced competing products but also that one *extrapolate* the future technical capability of competitors. One should design new products to beat not only the current products but also the *future* products of competitors.

The design of products to beat future products of competitors is the fundamental reason why technology forecasting *is vital to new product development.*

Next, the design team at MOS borrowed all the ideas from competitors' products that they liked. One can borrow a competitor's ideas (as long as they are not patented, copyrighted, or obtained wrongly as trade secrets). In designing new products, technology should be approached as a *cumulative as well as anticipatory state.* This means that in the design and production of new products, the technical goal should be to include at least all the best current practice and the highest current performance and then add to them.

Yet it is not enough to just try to include all the best features in a product because doing so might make the product too expensive for a target market.

Innovative product design should balance "best practice" with a competitive advantage in the proper ratio of performance to cost, and this is the fundamental reason why technology strategy is vital to new product development.

Next, the design team looked at *constraints* that would limit how much of the best practice and how much of the highest performance and improvements they could squeeze into the product design. In the case of design of the new chips, the most important constraint was that the line width of the etching structures in the silicon for inserting the transistors was then limited to no smaller than 5 μm (five millionths of a meter). This constraint limited the number of transistors and parts and hence the number and complexity of circuits and finally the number of features and speed of performance.

The fundamental challenges in new product design are always the tradeoffs that need to be made between desirable performance and technical and cost constraints.

The fundamental reasons that technology implementation is important to new product development are to provide improved product performance at a lower price.

The design team then had two things: First, they had to construct the *wish list* of features and performance criteria they would like to see on the chips, and second, they had to *limit* the numbers of circuits and features they could put on one chip with 5-μm technology. Then came the next step in the design process: "Then we prioritized the wish list from what must be in there to what ought to be in there to what we'd like to have..." (Perry and Wallich, 1985, p. 49).

This kind of logic in product design is basic. First, there should be a decision as to what the customer needs that would be better than what competitors now or could (by extrapolation) in the future supply. Second, there should be a decision to prioritize desirable features into "must," "ought," and "like to have." If a design cannot accomplish both the "must" and "ought" features, it will likely be a technical failure. If the design also cannot accomplish many of the "like-to-have" features, it probably will not be a commercial success.

Accordingly, we can see (1) that a new technology must be correctly anticipated and (2) that the proper tradeoff between performance and cost must be made in the initial phases of product design.

Commercial success critically depends on appropriate design decisions made between corporate research and product development.

With the prioritized wish list of required and desirable features, Charpentier's group then worked on laying out the circuits and transistors for the chips. Then several more product design decisions had to be made (such as sophistication versus simplicity, time to completion versus complexity, etc.): "Because design time rather than silicon was at a premium, the chips were laid out simply rather than compactly. 'We did it in a very modular fashion...'" (Perry and Wallich, 1985, p. 49).

The actual design process consists of many details, decisions, and tradeoffs. Scientific and generic engineering principles can help guide some of these. Other decisions are based on practical engineering experience and practical product and production experience. Some decisions simply are judgments about what the customer should want, and some are about technical pride. Design decisions are a mixture.

Design is the first step in new product development. Next comes prototype production and testing. Fortunately, MOS had a chip-fabrication line that could be interrupted to produce samples of the new design. This was necessary for debugging and refining the design until it worked: "David A. Ziembicke, then a production engineer at Commodore, recalls that typical fabrication times were a few weeks and that in an emergency the captive fabrication facility could turn designs around in as little as four days" (Perry and Wallich, 1985, p. 49). The cooperation between production and product design here facilitated the exceptionally fast design to product time of 9 months. Thus, by November of 1981, the chips were completed, and samples were produced and tested.

All designs require debugging and refining, and reducing redesign time is one of the major keys to competitiveness. This is another important point about competitiveness that we will explore later in managing the product-development cycle.

> For economic success, the links between technology strategy and product development must also include production strategy.

Next came a very significant change in this case—a changed vision of the product. In November of 1981, Charpentier and Winterble reported the success of the project to the CEO of Commodore, Jack Tramiel. While listening to their sales plans for the chips, Tramiel had in the back of his mind the recent success of Commodore's new personal computer product, the VIC-20. Therefore, when Charpentier and Winterble showed Tramiel the new chips, Tramiel imagined a new personal computer product for Commodore that would use the competitive advantage of the new technology and provide a product successor to the popular VIC-20. Tramiel saw that the new chips could be used in a new product as good as the best on the market but at a fraction of the price. He also was anticipating other new technological products soon to be introduced, new 64K random access memory (RAM) chips: "'Jack [Tramiel] made the bet that by the time we were ready to produce a product, 64K RAMs would be cheap enough for us to use,' Charpentier said" (Perry and Wallich, 1985, p. 51).

This is an example of the integration of technology strategy with business strategy.

Tramiel decided to use Charpentier's graphics and sound chips not to sell to competitors but to produce a new product called the *Commodore 64 personal computer* with a major competitive advantage in price: "When the design of the Commodore 64 began, the overriding goals were simplicity and low cost. The initial cost of the Commodore 64 was targeted at $130; it turned out to be $135" (Perry and Wallich, 1985, p. 51).

Technological Innovation and Technology Strategy

Let us pause in this illustration to briefly review the key points generally known about the relationships between technology strategy and technological innovation.

Technological innovation is usually defined as the invention, development, and introduction into the marketplace of new products, processes, or services that embody new technology. Technological innovation does begin with invention, an idea of a way to do things. In U.S. law, if the idea is new, useful, and nonobvious, it is a patentable invention. If it is not patentable, then the invention becomes either the generically available technology of an industry or the proprietary working knowledge of a firm.

It is well known that invention may be motivated either primarily by a desire to advance technique or primarily to satisfy a specific market need. The first motivation has been called *technology push* and the second *market pull*. And in the literature on innovation, there have been many arguments as to which provides the most frequent sources of innovation. However, whether motivated by market pull or technology push, commercially successful innovation should (1) match technological superiority to market application and (2) provide competitively priced, high-quality goods.

It is also well known that technology can be used for competitive advantage through either of the two general competitive strategies which

Eric Porter has particularly expounded: (1) product differentiation and/or (2) low-cost and high-quality production leadership (Porter, 1985).

Within the current concepts of total quality control for business success, there are several kinds of quality that must be addressed (Godfrey and Kolssar, 1988):

1. The quality of how well a product performs its central function.
2. The overall quality of a product design in terms of its suitability to an application.
3. The quality of a production process that reproduces quantities of a product rapidly, without defects, and at low cost.
4. The quality of attending to and responding to customer needs in durability and maintenance of the product.

Thus any given technological innovation may affect one or more of the senses of product quality: performance, applicability, production, and durability.

We also know that technology strategy is a complex problem partly because several technologies are usually fundamental to any business, partly because several businesses are involved in a diversified firm, and partly because innovation processes are tricky to manage.

Finally, we know that there are five basic management principles that underpin all technological innovation processes:

1. Technology should be conceived of as a competitive factor within a business system.
2. New technological potential should be forecast.
3. Technology forecasts should be implemented through planned technology strategies.
4. New products using the new technology must be marketed with special attention to the problems of new markets.
5. Technology strategy and business strategy must be closely integrated.

These have been elaborated by different authors. Michael Porter (1985) emphasized that the firm is a value-adding business system in which technology is viewed as one of the competitive factors. Erich Jansch (1967), Brian Twiss (1968), Joseph Martino (1983), and others have argued the importance of and possibility of anticipating technical change. Lowell Steele (1989) and Arthur Rubenstein (1989) emphasized the importance of managerial perspectives in planning and organizing technology strategy. Roland Schmitt (1985) and William Davidow (1986) emphasized that high-tech products provide special marketing problems over older-technology markets. In 1987, I emphasized that corporate strategy that ignores or fails to integrate technology strategy may fail to position the corporation for long-term survival.

Using technology as a competitive factor must begin with anticipating technological change. Techniques for the business organization to systematically anticipate relevant technological change have been loosely called *technology forecasting*. However, this term may invite misunderstanding, for technology is never "forecasted" in the sense of prediction. Technological change can only be anticipated or planned—hopefully seen coming some time in the future.

> *The usefulness of technology forecasting techniques lies not in the notion of prediction, but in their value to business planning for systematically searching out potential and relevant technical change.*

Technology forecasting needs methods for (1) understanding the scientific and engineering basis upon which technological advances may be possible and (2) estimating the direction, magnitude, and rate of possible technological change.

The next step after technology forecasting is formulating technology strategy. Even when properly anticipated, technological change needs to be created for a given business enterprise in an appropriate form and in a timely manner.

> *The formulation of technology strategy is an organizational commitment to deliberately pursue a direction of technical change.*

ILLUSTRATION (*Continued*)
Exploiting the Product Innovation
of the Commodore 64

The competitive advantage of the Commodore 64 was its price. At the time, it did have roughly comparable or superior performance to competing products—but at one-fourth the cost of production.

Tramiel had set performance and price goals for a new product, the Commodore 64, that would use Charpentier's new chips. The next problem, however, was to design the computer. This new design was relatively straightforward, since it was a redesign of existing personal computer designs. The system architect for the design was Robert Yannes.

In every new product, however, no matter how clear the design, bugs always occur—in both design and production. Unfortunately, these small bugs also can have major competitive impacts. Despite the cleverness of the new Commodore 64 and its very low price, the product quickly gained a reputation for poor quality that eventually hurt Commodore's business reputation.

For example, one of the bugs was in Charpentier's graphics chip design, which required a "patch" in order to allow the computer to display both black-and-white and color information without conflict. This problem was quickly fixed by Charpentier and later corrected in a second model of the graphics chip. Another bug occurred in the new design from using a read-only memory (ROM) chip from the previous VIC-20. Used in the new product system of the Commodore 64, it proved sensitive to spurious voltage spikes that occurred when control over the bus changed. Unfortunately, the result was to produce a visible "sparkle"— small spots of light dancing randomly on the display screen. This defect was found after 3 weeks of hard work and corrected.

There were other problems that came both from design choices and from production. For example, cheap components were purchased that degraded performance. Assembly errors also were made that affected performance. Finally, there was also a very major problem with the disk drive, which operated too slowly. This problem came from (1) a marketing-influenced decision for compatibility with the VIC-20 and (2) a deliberate decision by Commodore to scrimp on software development.

In the end, all the bugs, problems, and poor market and production decisions added up to a public perception of a poor-quality product: "The one major flaw of the C-64 is not in the machine itself, but in its disk drive. With a reasonably fast disk drive and an adequate disk-operating system (DOS), the C-64 could compete in the business market with the Apple....With the present disk drive, though, it is hard pressed to lose its image as a toy" (Perry and Wallich, 1985, p. 51).

One sees in this example that the design of a product is only part of the competitive problem. Bugs will always occur, and wrong decisions will always get made. It is important to correct problems rapidly, improve production quality, and produce a new model that corrects wrong decisions. Otherwise, the competitive advantage of the new technology can be lost quickly for other than technical reasons.

In the product-development cycle, rapid product improvement is fundamental to maintaining market share after a successful innovation.

Still, with its relatively low price, the Commodore 64 was a commercial hit for a time. The Commodore 64 was single-handedly responsible for driving Texas Instruments out of the home personal computer market and a major contributor to the bankruptcy of Atari.

Even early in a new technology industry, price still importantly counts as a competitive factor.

To see this, we should review the fact that the price war in home personal computers had begun in 1980, even in the infancy of the new industry. Tramiel had produced the Commodore PET computer in 1979 as a competitor to Apple and Tandy, and it did only so-so. During 1980, Tramiel had watched a very cheap personal computer, made by Sinclair in England, with only a toy keyboard but priced at $100, sell well. In contrast, personal computers with usable keyboards and peripherals then sold for from $1000 to $2000.

Responding, Tramiel introduced the VIC-20, priced at $250 dollars, early in 1981 and watched it sell well. It sold despite the fact that it had limited performance compared with the Apple, Tandy, Atari, and Texas Instruments machines. Texas Instruments then responded to Tramiel's aggressively priced VIC-20 by cutting the price of its 99/4a to an even lower $199. It also sold well, and Texas Instruments launched a crash program for a new, cheaper product (the TI/8).

Thus, by early January of 1982, when Texas Instruments personnel went to Las Vegas for the winter consumer electronics trade show, they felt pretty good because they did not see anything yet with a 16-bit microprocessor. They did see the new Commodore 64 at that show, but did not then appreciate its potential price significance. However, the Atari people caught the significance: "All we saw at our [Commodore's] booth were Atari people with their mouths dropping open, saying, `How can you do that for $595?'" (Perry and Wallich, 1985, p. 51).

After that show, Commodore turned on the competitive heat. Tramiel cut the VIC-20 price to $125. Texas Instruments had to match that price, but their profit margin on the 99/4a fell to zero. In April, Tramiel struck again, lowering the VIC-20 price to $99, at which Commodore could still make a little money. Texas Instruments again had to match that price but their 99/4a lost money. All through 1982 and 1983, Texas Instruments' home computer division lost money, much money. Atari also lost money.

Texas Instruments was putting its hope on a new model, the TI 99/2. Finally, in 1983, Texas Instruments took the new product to the summer consumer electronics trade show, a full year after introduction of the Commodore 64. Of course, by then, Tramiel had already aggressively lowered the price of the Commodore 64 (from its initial price of $595 of the previous year).

Texas Instruments saw then that their new product could still not match the rapidly declining price of the Commodore 64. After the show,

Fred Bucy called a meeting of the Texas Instruments group and said, "I don't think this product can make any money. Does anyone disagree?" (Nocera, 1984, p. 65). No one disagreed. The new product model of the TI 99/8 was dead before production. And with it died Texas Instruments' home computer business. Texas Instruments then announced their withdrawal from the home computer market.

The great irony, of course, is that the technical and scientific capability of Texas Instruments was then, and still is, far greater than the capability of Commodore. Atari, even with its still technically superior home computer, could not meet Commodore's price, and the corporate owner closed Atari, selling off its inventory to Tramiel.

> *Commercially successful technology strategy is not the same as technical capability.*

This illustration of the Commodore 64 product innovation shows the major commercial impact that an appropriately timed innovation in the product-development process of one competitor can have on competitors whose product-development cycles lag seriously in timeliness and effectiveness.

Planned Technological Change

I can now state the single most important concept which you will study in this book—*to understand why and how technological change is possible as a planned activity and hence manageable.*

Planned technological change is the deliberate development of new or improved functionality. *Functionality* is the ability to achieve a goal. Planning technological progress is possible because (1) all technologies are functionally defined physically structured systems, (2) technological progress can be planned by intending to improve the technical performance of a system, and (3) improvements can occur through expanding the functional logic of the system and/or developing structural alternatives for the physical basis of the system.

Radically new and basic technological inventions cannot be planned, and hence radical innovation cannot be managed directly. Yet all original inventions are useless until improved. All inventions must be refined. Incremental innovation can be managed directly. Thus the management of strategic technologies is the planned improvement of the technologies necessary and critical to long-term business success.

ILLUSTRATION (*Concluded*)
The Temporal Nature of the
Commodore 64 Product

We saw how the product-development cycle at that point in time at
Commodore created an enormous competitive success. Business competition continues, however, and although Commodore was very successful at that time, it subsequently lost its competitive advantage.
There are several questions that will be enlightening to discuss about
this case:

1. How did Charpentier anticipate the possible technological change?
2. How did Charpentier plan and create technological change?
3. How did Tramiel, Winterble, and Charpentier plan a product to use
   the new technology?
4. How was the product/technology strategy integrated with Commodore's business strategy?
5. Why did the product, after such dramatic initial success, eventually
   fail to build the business position of Commodore?

Charpentier had two different sources of information in mind when his
group decided to make a new generation of graphics and sound chips.
The first source was a *market need,* since the home electronics game
market was exploding in 1980 and 1981. The second source was the *technical advances* in chip density in 1980–1981 that were moving toward
very-large-scale integration (VLSI), which Charpentier understood as
providing the capability of creating a chip with 5-μm line-width technology. Together these provided a *market* and a *technical basis* for anticipating new technology. These two kinds of information sources for
anticipating new technology are called *market pull* and *technology push.*

Implementation of technological change occurs within discrete technical projects that are planned, budgeted, and managed. In the second
question, we note that Charpentier first obtained approval from
Winterble, Commodore's director of engineering, for the project, with
an agreed-on project budget and deadlines. Technical goals, market
goals, and budget and schedule goals were set. Technology strategy
must be translatable into research and development (R&D) projects.

Determining what kind of a product in which to implement new technology requires business criteria as well as technical criteria. For the
third question, we note that the final product use of the new technology
involved not just engineering personnel (Charpentier and Winterble)
but the CEO of Commodore, Tramiel. The product use of the technology was redefined by Tramiel, from sale of the new chip to internal use

of the chip for a new personal computer model. Tramiel's business strategy focused on the opportunities of the personal computer market as providing a larger market and higher profitability than as a chip supplier.

Integration of technology strategy and business strategy requires a focus on both the competitive advantages of a new-technology product and a marketing and pricing strategy that exploits those advantages. About the fourth question, we note that Tramiel's business judgment focused on the market size and price sensitivity of the home computer market. Tramiel correctly judged that at that time price was more immediately important than technical capability. If the Commodore 64 could deliver the same or slightly better technical performance than competing home personal computers but at a fraction of the cost, it could knock out competition. And it did.

Product performance must include not only technical performance but also quality. Any new product by itself will never ensure the continuing business success of any company. Product improvement must be continual. For the last question, we note that Tramiel did not apparently appreciate the impact that customer perception of quality would have on the long-term competitive position of Commodore. Poor decisions were made in design, production, procurement, and software development that together created a public perception of Commodore's poor quality and disappointing performance. Quality was never emphasized, nor was the product improved in a timely manner for better performance. The C-64 did not ensure Commodore's future because it quickly became obsolete.

Every product, however, provides a company with an opportunity to impress customers with the "quality" of the company.

In the long term, a company survives through a perception of quality, particularly as technological change slows and when competition forces prices down toward 'commodity' type pricing.

Risks in Technological Innovation

The technical risks of meeting business goals in a timely and practical manner are important:

Can a technical goal be achieved?

Can it be achieved on time for a product introduction?

Can it be achieved in a reasonable budget?

If achieved, can it be produced in a product at a cost that will allow the product to be sold and be profitable?

For example, Lowell Steele (1989, pp. 52–66) has listed some of the hard-won lessons about risks and success in innovation:

1. Unless a new technology offers really new functionality or significantly superior performance, it will not succeed as an innovation.

2. Technologies are systems, and until the system as a whole is complete and competent enough for an application, a new technology cannot succeed.

3. A new technology diffuses into use for several reasons, and the rate of diffusion varies. Accordingly, the rate at which a new market grows depends on many factors.

4. Although the customer determines ultimate success or failure of an innovation, the criteria on which a customer judges may be multidimensional, and customer judgment may change over time.

5. Successful products in a new technology depend as importantly on standards and infrastructure as on performance.

6. Because of a variety of factors, both technical and economic, a radically new technology usually requires a new business organization for successful innovation.

The first lesson is the key to commercial success in innovating a new technology. Technology essentially is about function and performance. If one is not offering really new functionality or the potential for intrinsically superior performance, then success in the marketplace against established technologies is not likely.

The second lesson is important because the fact that any technology is a system allows one to plan both incremental innovations in the technology and (infrequently) next generations of the technology. The systems perspective on technology is one of the central concepts we will exploit in this book.

The third, fourth, and fifth lessons are important because the economic benefits of technological change are found in the marketplace. And the ability to plan the when, why, and how of technological penetration into the marketplace is the key to making or losing money in new technologies.

The sixth lesson, about the importance of organizational factors in creating risks in innovation, is also a major source of risk in technological innovation. When innovating new technology, it is necessary to have very close coordination of the technical and business functions of a firm. Yet, as anyone with the experience of trying to get research, manufacturing, marketing, and finance personnel working closely together will have experienced, such coordination is difficult to achieve in practice.

So what makes the topic of strategic technology so difficult? The problems of technology strategy arise from the complexities of modern technological systems and from modern organizational structures of diversified firms. The personnel responsible for the technical functions of a firm and those responsible for the business functions have been trained very differently. Also, they have experienced different aspects of the business system. Accordingly, they actually perceive the world very differently. They even almost live in two different worlds—physical and economic.

Technical personnel predominantly see and live in a world of physical nature. Since the training of technical personnel is in the physical sciences, they come to appreciate that the most important thing about the world is that it is made up of matter and energy and forces—physical stuff. Business-functional personnel live in a world of social contracts and money. The training of business personnel responsible for marketing sales and accounting and production often is in economics and business. They appreciate primarily how much the world consists of human cooperation and competition—in which money measures, facilitates, and gauges human interaction.

Which perception is correct—the world as physics or the world as money? Of course, both are correct, for the world has both a physical nature and a social nature. However, personnel trained primarily in one view have difficulty appreciating the alternative view.

The logical forms of thinking in which the two major groups of corporate personnel (technical and functional) have been trained to perceive and analyze the world differ significantly and impede their communication and cooperation. For example, personnel trained in science and engineering are taught to think with such concepts as "design," "analysis," "synthesis," "problem solving," "technical service," "research and development," etc. In contrast, personnel trained in economics and business are taught to think with such concepts as "strategy and planning," "capitalization and budgeting," "organization," "staffing," "control," "evaluation and profitability," "expected utility and discounted futures," "supply-demand curves," etc.

To reduce the risks in innovation, it takes special attention and effort to integrate the two strategic concerns of business and technology. The technical concerns of the engineers must be translatable and integratable into the strategic concerns of management. The business concerns of managers must be capable of handling technical risk and exploiting technical progress.

Achieving communication and cooperation between technical and business functions and integrating business and technology strategies are not easy. Moreover, this does not involve simply teaching some management principles to technical personnel or, conversely, teaching a little technology to financial personnel. Both approaches have been tried in some engineering or business schools. The problem goes deeper. It has to do with having a refined perception of, sensitivity for, and commitment to the role of technology in the enterprise system.

> *The problem in managing strategic technologies is to* adapt and refine *the logic of management so as to effectively deal with the logic of technology.*

The way to deal with the risks of technological innovation is to develop a corporate ability to strategically manage its core technology competencies in the enterprises of its businesses. To build such a capability, a corporation must be clear about the different logics in the innovation processes that create short- and long-term competitive advantages.

ILLUSTRATION
When the United States Innovated and
Subsequently Lost the Mass Memory
(DRAM) Chip Market

As an illustration of the importance of the ability to manage strategic technologies (not only at a firm level but also at a national level), we can review the dramatic changes in market positioning in the world's semiconductor industry that occurred in the 1980s. Figure 1.1a shows the market shares of U.S. and Japanese firms in semiconductor production during that time. In total semiconductor chip production, the U.S. world market share fell over the course of the decade from 60 to 38 percent, while Japan surged ahead into market dominance.

In particular, Fig. 1.1b shows U.S. and Japanese market shares in dynamic random-access memory chips (DRAMS)—which always have been the largest chip market. Here the United States had actually lost the world market, falling from 100 to 18 percent, while Japan rose from zero to market dominance at 78 percent. (Recall from the preceding illustration that it was the Japanese entry into the 16K memory chip market in 1975 that sent the MOS Technologies Company into Commodore's corporate arms in 1976 at "ten cents on the dollar.")

What is especially dramatic about this illustration is that the United States had invented all the *radical innovations* for the IC chip memory market—transistors, semiconductor chips, and DRAMS—while, in contrast, all the Japanese innovations in these had been *incremental improvements* in production processes.

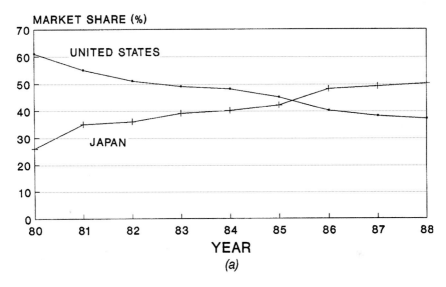

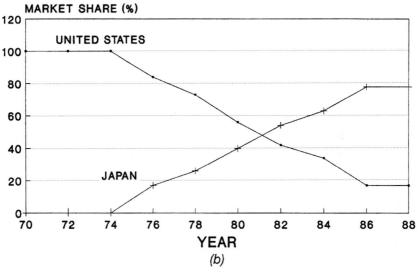

Figure 1.1 (a) Semiconductor production; (b) memory chip production (DRAMs).

The reasons the Japanese grabbed the mass memory chip market away from the United States involve principally the facts that they had (1) a competitive product design, (2) with superior production quality, (3) a strategic commitment to market share, and (4) a better management of the industrial value chain (funding upstream chip R&D with profits from downstream consumer electronics products).

The critical turning point occurred in 1981 when Japanese firms seized the mass memory market from American firms. This was a *tech-nology discontinuity*—the shift from one generation of technology to another generation of technology (in this case the shift from large-scale integration (LSI) technology to very-large-scale integration (VLSI) technology. The first next-generation technology product using VLSI was the 64K memory chip.

The Japanese firms were the first to market the 64K chip in 1981 and thereafter dominated the market. Also, they then had the ability to sustain lower profit margins in chips, because of Japanese vertical integration into consumer electronics, and survived a subsequent downturn in the chip market. (The consumer electronics market was another market the United States had earlier lost to the Japanese when consumer electronics technology was transistorized.)

In 1980, the Japanese and American firms had been neck-in-neck on a technical level and had dealt with the same technical risks. The difference in the success of the Japanese firms was organizational. Japanese firms in the electronics industry displayed a stronger ability to integrate business and technology strategies than did their American counterparts in the 1970s.

(This particular illustration is but one example of an unfortunate general pattern in the United States after World War II. In the period of 1950s through the 1980s, the United States led the world in radical innovation while losing its leadership in incremental innovation.)

Radical and Cyclic Innovation Processes

Early in the studies of innovation, Don Marquis (1969) identified three types of innovation, and I now distinguish four types:

1. Radical innovation
2. Incremental innovation
3. Systems innovation
4. Next-generation technology innovation

A *radical innovation* provides a brand-new functional capability, which is a discontinuity in then-current technological capabilities. New functionality provides opportunities for new business ventures and even for new industries. Examples of radical innovations include electron vacuum tubes, transistors, semiconductor integrated circuits, computers, lasers, and recombinant DNA techniques.

An *incremental innovation* improves the existing functional capability of an existing technology through improved performance, safety, and quality and lower cost. Examples of incremental innovations include additional grids in electron vacuum tubes, improved doping techniques in transistors, improved fabrication processes in semiconductor integrated circuits, improved memory devices in computers, and so on.

A *systems innovation* is a radical innovation that provides new functional capability based on reconfiguring existing technologies. An example is the systems innovation of the automobile, which put together existing carriage technology with bicycle technology and new gasoline engine technology.

Incremental innovations within a system also can sometimes create new technical generations of a system. Such an innovation is still a kind of systems innovation but not a radically new innovation. It is a systemic innovation that some have called a *next-generation technology (NGT) innovation* (Anderson and Tushman, 1990).

The preceding illustration about semiconductor chip production underscores the importance of being able to manage strategic technologies for both radical and incremental innovations. While they are both important, however, there are economic differences and organizational problems concerning them:

- While radical innovation creates new markets, incremental innovation captures markets in the long term.

- Radical and incremental innovation have to be managed differently.

- It is difficult for any given unit of a firm to manage both well at the same time.

- Because the locus for radical innovation is usually the corporate research laboratory and the loci for incremental innovation are usually the production divisions, it is difficult, but important, for diversified corporations to formally manage technology.

For example, in the 1970s while studying innovation in the automobile industry, William Abernathy (1978, p. 168) noted that the innovation processes which foster either radical or incremental innovations require different sets of circumstances and different kinds of management attention. Later on, other observers of innovation also emphasized the point that radical and incremental innovations require different management. For example, Ralph Gomory and Roland Schmitt (1988) pointed out that radical innovation followed from a kind of linear transfer of knowledge from science into technology, while product development required periodic improvements with incremental technological advances. Gomory and Schmitt also emphasized that the linear-process radical innovation is focused around a novel idea, an invention, while the cyclic and incremental product-development process is focused around product design. (I will discuss the natures of design and invention in later chapters.)

> *In the short term, competitiveness requires a well-managed cyclic incremental innovation process—the product-development cycle, but in the long term, competitiveness also requires a well-managed linear radical innovation process—next generations of technology.*

Logic of the Radical Innovation Process

Logically, radical innovation begins with the latest in science—new scientific knowledge. In radical innovation, there is a direct connection between how scientists experimented with a new scientific phenomenon and how inventors created new technology using the phenomenon. The instruments of science have often become the devices of new technology (the computer, the maser and laser, and recombinant DNA techniques are some examples). Thus, historically, in radical innovation there has often occurred a kind of linear transfer of knowledge and technique from science through engineering to technology. Figure 1.2 schematically represents this linear logic as one kind of innovation process for radical innovation.

The steps that can link the passing of information from one phase to another are listed across the middle of the figure as

Discovery and Understanding

Scientific Feasibility

Invention

Technical Feasibility Prototype

Functional Application Prototype

Engineering Prototype

Manufacturing Prototype

Pilot Production Process

Volume Product Production

The first step in moving from nature to product is the *discovery* in nature of an interesting object, process, property, or attribute. The next step is a sufficient *understanding* of the phenomenon to know how its variables influence other variables in order to systematically and consistently intervene and alter the course or state of the phenomenon.

With an awareness of the existence of nature and an understanding of interventions in such existence, one can *invent* a way to manipulate

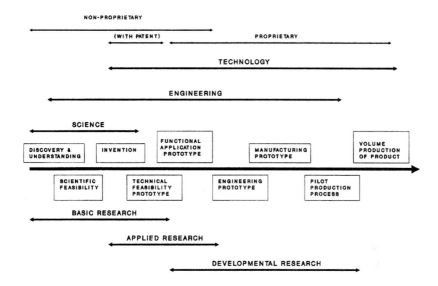

Figure 1.2   Linear process for radical innovation.

nature that may be useful to humanity. (For example, in the invention of the laser, a knowledge of the discrete quantum states of atoms and the transitional photon yielded between states provided a knowledge base for Townes and others to invent ways of stimulating the generation of coherent light—light amplification from stimulated electronic radiation, i.e., LASER.)

Invention is an idea that must next be reduced to practice, i.e., one must show that the technical idea is feasible and can be demonstrated. This is the purpose of a *technical feasibility prototype*. This is also sometimes called a *proof-of-principle technical demonstration*.

However, the mere technical feasibility of something that works is still not enough to proceed to a product. One must next establish that the technically feasible invention can work well enough for an application. Working *well enough* means that the performance of the invention is sufficient to be useful for a specific purpose. And this is the point of next creating a *functional prototype*.

Now, after a functional prototype shows sufficient performance, a product design can begin. The first goal of a novel product design is to establish completeness in product requirements. In addition to sufficient performance, a product idea must embody necessary features, have appropriate size and shape, be safe and convenient to operate, and be maintainable and repairable. The goal of the *engineering proto-*

type is to design and demonstrate that the product can be engineered as a complete product.

The next step on the way to product production is the creation of a *manufacturing prototype.* In this prototype, the engineering design is modified for manufacturability, considering tolerances, assembly, quality, and cost. If a new production process must be innovated to produce the product, then that process should be tested as a *pilot production process.*

After the manufacturing prototype is tested and pilot production processes are developed and tested, then *volume production* of the product begins.

There are several points to note about this spectrum of research that connects nature to product:

1. There is an overlap of research activities among science, engineering, and technology.
2. There is an overlap of research motivation among basic, applied, and developmental research.
3. There is an overlap of intellectual property interests between proprietary and nonproprietary research.

These overlaps make science, engineering, and technology dynamically interacting processes. Science is a general approach to understanding nature. Technology is a generic way of providing a functional capability of doing things. Engineering grounds technology in scientific understanding and general principles. Product development focuses on a specific embodiment of technology and engineering understanding and principles for a business purpose of providing value-added sales to customers. Accordingly, for purposes of forecasting and planning technological change, management must understand the differences between science, engineering, technology, and product—and how to gain cooperation among these activities. (These topics will be examined later in this book.)

And from a management perspective, the problems for technology strategy begin with the facts that product development is in the domain of strategic business units and the research basis for a next-generation technology is in the domain of the corporate research laboratory and that these are positioned differently in the logic of innovation.

Unless there is good and close working relationships between corporate research and product development in the strategic business units, product development may not be properly led by visions of next-generation technology.

Also, the logic of the process of radical innovation indicates that one cannot make a simple distinction between research for engineering and technology and for product development without also taking into account the intellectual property rights of the research. Information in product development about product design and production processes is very important proprietary material to the firm designing, producing, and selling the product. It is in this product-specific knowledge that the competitive differences between firms lie. Yet next-generation technologies occur at an industrial level. All firms in an industry share the generic knowledge about technology and engineering that underlies product and production design and control. The generic level of engineering/technology knowledge is generally nonproprietary. The overlap between proprietary and nonproprietary research usually occurs in the functional prototyping stage. It is here where the critical decisions about the next generation of technologies, their directions, and their pace occur.

Management of strategic technologies over the long term requires firms to participate in the national development of technology progress.

Logic of Cyclic Incremental Innovation Processes

In contrast to the logic of a linear radical innovation process proceeding from science to technology, the logic of cyclic incremental innovation processes differs substantially, cycling around technology and product.

As depicted in Fig. 1.3, the logic of cyclic innovation processes consists of five central concepts:

1. Technology anticipation
2. Technology acquisition
3. Technology implementation
4. Technology exploitation
5. Technology stimulation

It is necessary to first anticipate technological change so that one has time to prepare and take advantage of it. Next, it is necessary to acquire new technology for use by the firm. Then new technology can be embedded into new products, processes, or services. The next logical step is to commercially exploit the new technology through new product introductions and aggressive pricing/quality strategies. Customers' subsequent experiences with the new products suggest new needs and stimulate requirements for more new technology. The cyclic innovation

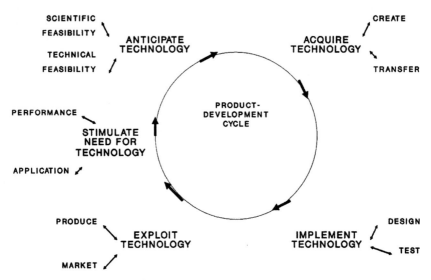

Figure 1.3 Cyclic incremental innovation process.

process is driven by incremental improvements to technology that periodically improve the *value* of products to customers.

For firms, these steps form a kind of circular pattern in the technical and business activities of the firm. I call this the *cyclic incremental innovation process,* and it forms the basis of what is usually called the *product-development cycle* (which will be reviewed in a later chapter).

Relating the Linear and Cyclic Innovation Processes: Next Generations of Technology

The linear radical innovation process provides dramatic, discontinuous impacts on the economy (enabling the creation of new markets or the entry of existing markets), whereas the cyclic incremental innovation process provides quiet, steady, continuous impacts on the economy (eventually enabling the capture and domination of markets). However, since technology is implemented in either innovation process, how can both processes be managed within a firm?

One way to do this is to use the concept of "generations of technology," for this requires both radical and incremental innovation. Figure 1.4 sketches how a linear innovation process for next-generation technology can provide technical information for a cyclic innovation process:

In the cyclic innovation process, the *stimulation of need for new tech-*

nology fosters the *vision of a next-generation technology* in the linear innovation process.

The *technical feasibility prototype* in the linear innovation process provides the grounds for *anticipation* of new technology in the cyclic innovation process.

The *functional prototype* in the linear innovation process provides the information for *acquiring* new technology in the cyclic innovation process.

The *implementation* phase in the cyclic innovation process designs the *engineering prototype* in the linear innovation process (which with concurrent engineering practice also fosters the *manufacturing prototypes* and *pilot production*).

Volume production in the linear innovation process of the next-generation product provides the opportunity for *exploitation* of the new technology in the cyclic innovation process.

By establishing "technical feasibility," the linear innovation process can provide *anticipation* to the cyclic innovation process for the new generation of a technology.

Once such demonstrations can be pointed out, then the next problem is estimating the timing of the introduction of new technology. And this is a critical problem, since the timing of technological innovation is one of its most important competitive assets. The step of functional prototyping in the linear innovation process establishes the timing for introduction of a next generation of a technology to be *acquired* by the product-development cycle. The engineering prototype of a new-generation technology product next allows the product-development cycle of a strategic business unit to implement the new technology.

Now in the cyclic innovation process, the critical step is *exploiting* a new technology. When one produces and markets the new product, the design, manufacture, timing of the introduction, pricing of the new product, and establishing distribution are all critical events. And the more radical the innovation in the new products, the more likely a correction to these events will be necessary. Accordingly, the accumulating market experience in the new product will likely require a fast recycle through the product-development cycle. The more the technology discontinuity in a next-generation technology, the more it becomes necessary for a redesign into a new product. (In the Commodore illustration, the Commodore 64 was a new product model developed in Commodore's product-development cycle and not a next-generation technology product. The MOS Technology game chips were next-generation technology products for that company.

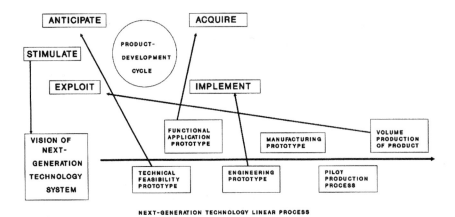

Figure 1.4 Integration cyclic innovation with linear innovation in a next-generation technology product.

The 64K memory chips which Tramiel used in the Commodore 64 were next-generation technology products in the semiconductor chip industry.)

> *For technology discontinuities, after the product-development cycle has produced a new product dependent on the new technology, (1) the cycle of technology anticipation, acquisition, implementation, and exploitation must start all over again soon after a new product is introduced, (2) for competitors will see the new product and its impact on the market and will quickly introduce "me too" products either with improved features or lowered price.*

In summary, formulating technology strategy requires methods for bringing the technical and other business-functional personnel together to manage both incremental and radical innovation:

1. A vision of a next-generation technological system and the planning of the directions and nature of research programs to generate the advanced ideas for that system.
2. An understanding of the business opportunities and competitive advantages of the new technology.
3. Management of the scientific and engineering activities that create new technology.
4. Transfer of the new technology into product design and production-process design and improvement.

Looking Ahead

Refining the logic of management to deal with the logic of technology requires that managers perceive and understand how the technologies in their businesses provide the productive capabilities of the businesses. We will be examining techniques for formulating and integrating technology strategy into business planning.

Logically, one should first forecast technological change, plan it, and then implement it. For pedagogical purposes, however, I will alter the order of presentation—beginning with planning technology and implementing technology change before studying technology forecasting and research planning. The reason for this is that knowing the applications for techniques can provide motivation for learning the techniques. The techniques for forecasting and planning technology can get quite esoteric, requiring one to delve deeply into the heart of research and invention. First knowing why and how to plan and implement new technology I hope may inspire patience in the reader for the later chapters aimed at understanding the methodological complexity in our modern scientific technology.

Practical Implications

The lessons to be learned from this chapter concern the basic principles for formulating technology strategy:

1. Strategic technologies are the rapidly changing corporate core-technical competencies that provide a competitive edge to the businesses of the corporation.
   - The selection of a strategic technology should be evaluated on either the economic goals of (1) creating a new market or (2) dominating an existing market.
   - Commercially successful innovation carefully matches technological superiority with market application, providing competitively priced, high-quality goods.
2. Four types of technological innovation play different economic roles in competitiveness:
   - A radical technological innovation may create a new industry.
   - A radical systems-type technological innovation also may create a new industry.
   - Incremental technological innovations maintain competitiveness in an industry.
   - A next-generation technology system innovation alters the competitive conditions of an existing industry.

3. The linear innovation process fosters radical innovation and is effective for the creation of or entry into new markets. Its logic links science to product, focused by an inventive idea:
   - Discovery and understanding
   - Scientific feasibility
   - Invention
   - Technical feasibility prototype
   - Functional application prototype
   - Engineering prototype
   - Manufacturing prototype
   - Pilot production process
   - Product production in volume

4. The cyclic innovation process fosters incremental innovation and is effective for capturing and dominating existing markets. Its logic is focused on existing products and markets, requiring periodic
   - Technology anticipation
   - Technology acquisition
   - Technology implementation
   - Technology exploitation

5. Implementing technology strategy for existing businesses requires managing both radical and incremental innovation processes through the concept of a next generation of a technology.

For Further Reflection

1. Identify a radical technology innovation (either a product or a process) that began a new industry.

2. List the scientific basis of the radical innovation (product and/or process).

3. List the subsequent incremental engineering improvements to the technology (both product and process).

4. Identify the innovating firm and country of the radical innovation.

5. Identify the firms and country that came to dominate the market(s).

6. Was the innovating firm/country the same as the later market-dominating firm(s)? If not, why not?

2

Corporate
Core Competencies

Product-development cycles (discussed in the next chapter) occur in the strategic business units of a firm, yet their technology roadmaps require a long-term technology strategy that is integrated with corporate research. Therefore, how does a diversified firm with a central research laboratory formulate a technology strategy that can contribute to innovation in the new products of the strategic business units? The essential thing is to create a *culture for technological change* throughout the whole corporation.

Let us remind ourselves that in the general concept of strategy there is an important distinction between a strategic plan and a strategic attitude. A *strategic plan* is preparatory to a specific action, but a *strategic attitude* is preparatory to all action.

> A *strategic plan* identifies specific goals over a specific time horizon for a specific set of actions.

> A *strategic attitude* prepares in general for any action by formulating a perception about the nature of action, by commitment to types of action, and by preparation and training for action.

For example, in military strategy, going into a war or a battle requires a strategic plan for that specific war or battle. The conditions of the battle provide a context for specific planning of deployment, attack, and defense. Yet before that battle or before any and all battles, a strategic attitude about the nature of military capability is required. A military strategic attitude requires preparation and training—including developing and acquiring weapons, organizing and training personnel, and devising and testing general military tactics.

A strategic attitude provides the capabilities for dealing with the risks and vagaries of action. Such an attitude is important for any action that is nonroutine and high in risk. Technology change is such a nonroutine and risky action. A strategic attitude about technology provides the corporate perceptions, commitments, and preparations for technological change.

> *For diversified corporations, the concept of* core technology competencies *implements a strategic technology attitude for the whole corporation.*

ILLUSTRATION
Strategy in NEC and GTE in the 1980s

As an illustration of the effect of strategic attitude on the growth and character of diversified firms, C. K. Prahalad and Gary Hamel (1990, p. 84) compared the performances of NEC and GTE:

> Consider the last ten years of GTE and NEC. In the early 1980s, GTE was well positioned to become a major player in the evolving information technology industry....In 1980, GTE's sales were $9.98 billion....NEC, in contrast, was much smaller, at $3.8 billion....Yet look at the positions of GTE and NEC in 1988. GTE's 1988 sales were $16.46 billion, and NEC's sales were considerably higher at $21.89 billion.

However, it was more than the relative levels of the two companies' sales that changed. NEC had emerged as a world leader in information technologies, whereas GTE had not (Prahalad and Hamel, 1990, p. 80):

> GTE has, in effect, become a telephone operating company with a position in defense and lighting products....NEC has emerged as the world leader in semiconductors and as a first-tier player in telecommunications products and computers. It has consolidated its position in mainframe computers....NEC is the only company in the world to be in the top five in revenue in telecommunications, semiconductors, and mainframes. Why did these two companies, starting with comparable business portfolios, perform so differently? Largely because NEC conceived of itself in terms of "core competencies" and GTE did not.

Prahalad and Hamel viewed the differences as arising from NEC's superior strategic capability. NEC created a corporate level committee to plan core corporate technical competencies and to oversee the development of core products for the businesses of the firm. This committee established groups across the individual businesses to coordinate the research and development (R&D) efforts for core products. This committee, which NEC called the "C&C Committee," identified three directions of technologies in computers, components, and communications: "Top management determined that computing would evolve from large mainframes to distributed processing, components from simple ICs to VLSI, and communications from mechanical cross-bar exchange to...digital systems" (Prahalad and Hamel, 1990, p. 80).

The strategic vision of the C&C Committee encompassed the convergence of technologies in computing, communications, and components businesses. The committee judged that there would be great opportunities for any company to serve all three markets. The committee anticipated the fact that two previously different industries, computing and communications, were going to come together into a single and more complex industry. The components for the two businesses were to become increasingly more complex, common to the two, and interrelated. With this technological vision, NEC positioned itself for a "restructuring" of the industries, and GTE did not. In this restructuring, NEC saw that the competitive factors lay not only in the systems-integrator sectors but also in the major-devices sectors and in the components and parts sectors. Thus NEC strategically positioned itself in all three sectors.

GTE, in contrast, had been managed only as a diversified set of businesses without core corporate competencies (Prahalad and Hamel, 1990, p. 81):

> No such clarity of strategic intent and strategic architecture [as at NEC] appeared to exist at GTE. Although senior executives discussed the implications of the evolving information technology industry, no commonly accepted view of which competencies would be required to compete in that industry were communicated widely. While significant staff work was done to identify key technologies, senior line managers continued to act as if they were managing independent business units. Decentralization made it difficult to focus on core competencies. Instead, individual businesses became increasingly dependent on outsiders for critical skills.

The senior management of GTE failed to develop the corporate strategic insight to fully exploit the evolving market and competitive opportunities in information technology. They failed to develop core technical competencies *common* to its several independently run businesses.

Corporate Core Competencies

Intuitive perceptions that underlie technology strategy attain realization when senior corporate management integrates the notion of core competencies into its corporate business strategies. Some corporations, even some with diversified products and businesses, have structured competencies that distinguish the firm as a whole. For example, Jo Didrichsen (1972) argued that some corporations have shown broad technological competence in a scientific area (such as Du Pont in chemistry), while others have shown a kind of branching technological competence (such as 3M in adhesives and surface coatings).

Later, Prahalad and Hamel (1990, p. 79) also argued that strategic competency was necessary for the success of diversified corporations under the new conditions of global competition: "During the 1980s, the top executives were judged on their ability to restructure, declutter, and delayer their corporations. In the 1990s, they'll be judged on their ability to identify, cultivate, and exploit the core competencies that make growth possible—indeed, they'll have to rethink the concept of the corporation itself."

Prahalad and Hamel defined a *core competency* as the collective knowledge in an organization about how to coordinate and integrate the multiple technologies required for product design and production. Such knowledge and skills are not located in one place in a firm but are spread among, developed, and maintained in the different research, engineering and production, and marketing units of a diversified firm.

In global competition, core corporate competencies are necessary for survival. For example, Prahalad and Hamel (1990, p. 86) argued that a purely financial management approach to diversified firms was no longer sufficient: "The new terms of competitive engagement cannot be understood using analytical tools devised to manage the diversified corporation of 20 years ago, when competition was primarily domestic....U.S. companies do not lack the technical resources to build competencies, but their top management often lacked the vision to build them and the administrative means for assembling resources spread across multiple businesses."

Historically, therefore, the development of core technological competencies has determined why some diversified firms have thrived as others have died. How does a corporation identify and develop core competencies? Before describing this procedure, it is necessary to ensure that we all understand a fundamental concept underlying the notion of core technological competency. This concept is that of a *strategic attitude*.

ILLUSTRATION
Sony's Strategic Technology Attitude

The origin of the Sony Corporation provides a clear illustration of the role of strategic attitudes and intuition in the development of a firm. Akio Morita was one of the two founders of Sony. In 1944, he was a university student studying applied physics under Professor Tsunesaburo Asada at Tokyo University. "At the university, Professor Asada was regarded as the expert in applied physics....when I entered the university, we were at war, and Professor Asada's laboratory had been pressed into service as a naval research facility" (Morita and Reingold, 1986, p. 21). It was in the last years of the war that Morita was finishing his education—and developing his interest in the new technology of electronics. "Professor Asada helped me more and more, and before long, I was able to help him in some small jobs for the navy, mainly electronics, which was closer to true physics than working with the old electrical circuits or electromechanical ones" (Morita and Reingold, 1986, p. 21).

It is important to note that Asada and Morita were both interested in applied physics. Applied physics is one of the interfaces between science and technology. It is in such interfaces that advanced technologies are created. Morita was being trained by Asada from a scientific orientation but motivated to create advanced technology. This provided a lasting influence on Morita as a basis for his strategic technology attitude. New and advanced technologies were then popular in Japan (Morita and Reingold, 1986, p. 21):

> Eventually he [Asada] began to write a short weekly column [for the Tokyo newspapers] elaborating on the latest developments in research and technology, at least those that were not secret....I often helped Professor Asada with his research, and occasionally when he was too busy I would write the column. I remembered that in one of these columns I discussed the theory of atomic energy..."if atomic energy were treated appropriately, an extremely powerful weapon could be made."...I did not know how far the scientific community in the United States and in Germany had come. And nobody in Japan knew about the Manhattan Project.

Morita was trained to be attentive to learning about the new advances in science and to imagining what technological use could be made of them. Morita viewed science as a source of ideas for new technologies.

Viewing science as a source for new technology is the first element of a strategic technology attitude.

As Morita neared graduation, he had not yet been drafted. The war was still going on, however, and Morita learned that he could enlist as a career technical officer in the navy by taking an examination and then be assigned to a research facility instead of a ship (which would allow him to finish his education). In early 1945, he was assigned to the optics laboratory in the Office of Aviation Technology at Yokosuka. There he worked on the problem of preventing static electric streaks on the films of aerial mapping cameras.

After graduating from the university, Morita next underwent a 4-month officer's indoctrination and training course and received the rank of lieutenant. He was assigned back to the optical division and began supervising a special unit working on thermal guidance weapons and on a night-vision gun sight. This was in a small town south of

Kamakura on Sagami Bay. "We were based in a big old country house in Zushi....The house was built on the Western style, faced with stucco, with a courtyard garden....The house was built at the foot of a cliff just above the beach, and I took a room at the nearby Nagisa Hotel....It seemed incongruous because sometimes it was as peaceful as any beach resort, yet we were right under the return path of the B-29's..." (Morita and Reingold, 1986, p. 29).

One of the members of the special project was an electronics engineer, Masaru Ibuka, who owned his own electronics company. Morita and Ibuka became colleagues and good friends. After the war, together they founded the Sony Corporation.

Morita was worried about the American bombers but lucky to be at Sagami Bay. "In July and August of 1945, there were raids over the Tokyo-Yokohama area almost every day and every night. We could watch the big silver B-29's passing overhead....The thing that concerned me a great deal then was that the military would not give up this war no matter how badly it was going and that the Miura Peninsula, where we were stationed, would become a bloody battleground, a last ditch battleground for the fanatical military..." (Morita and Reingold, 1986, p. 32).

Morita's luck held. He was on leave visiting his family in their ancestral home in Kosugaya when the two bombs fell on Nagasaki and Hiroshima. Grudgingly, the military went along with the Emperor's order to surrender. "I was shaken awake by my mother in the early morning....she said the Emperor Hirohito was going to make an announcement on the radio at noon. It was August 15. Even the announcement that the emperor would speak to the nation was stunning....The emperor's voice had never been heard by the Japanese people....Because I was, after all, a naval officer, I put on my full uniform, including my sword, and I stood at attention while we listened to the broadcast....the high, thin voice of His Majesty came through....The war was over" (Morita and Reingold, 1986, p. 34).

The postwar world was one of grim survival. Morita returned to his station on August 16. The American occupation forces arrived without incident. Morita waited at the station for days without orders. Finally, he received an order to send the staff home. He bartered equipment for railway tickets and sent the staff home. "The new period of peace was strange. The bombers did not come anymore, but many cities looked as though there was nothing more to bomb....The Morita family was fortunate because we had lost no one in the war and the company offices and factory in Nagoya, and even our home, survived with no serious bombing damage" (Morita and Reingold, 1986, p. 41).

The Morita family business was a brewery run by Morita's father. Although elder sons, such as Akio Morita, were expected to succeed their father in the business, Akio's father was still healthy and robust and did not yet need him. Besides, Akio's interests were in advanced electronics, so he accepted a teaching position in physics at the Tokyo Institute of Technology.

There he looked up his friend Ibuka. Ibuka was starting a new electronics company. Morita did not really want to teach. He wanted to create new technology. He decided to join Ibuka in founding a new company. He left the institute and worked with Ibuka. His father invested in the new firm, which first they called Tokyo Telecommunications Engineering Company and later renamed Sony.

Strategic Attitude and Planning

Let us pause in our illustration to review the roles of analysis and intuition in strategy. If you have read texts on strategy, you probably have seen a logical scheme for strategic planning that contains something like the following steps:

1. Define the mission or business of the organization.
2. Define the measures of performance for effectiveness of the organization.
3. Select and describe goals and desirable outcomes for the future of the organization.
4. Devise a strategic plan for attaining the goals, with progress toward those goals expressed in the performance measures.
5. Organize efforts and assign responsibilities to carry out the strategic plan.
6. Allocate resources of budgets and personnel to perform the efforts.
7. Monitor and control performance and, if necessary, revise the strategic plan.

This logic (or some other version of it) certainly captures the different steps in laying out strategy. It emphasizes an *analytical* approach to strategy. Analysis is an essential feature in strategy. Without analysis, a good plan cannot be created.

A plan is a systematic approach into the future—a kind of an intellectual roadmap indicating the routes one must take from the present situation to get to a future desired situation.

Roadmaps, however, presuppose an already explored territory. To the extent that one has visited a place before, reconstructing the steps to getting there is possible.

Analysis in strategy increases in importance in direct proportion to the repetitive nature of the action.

An example of this is a set of engineering construction projects. If one has designed the engineering system of a process previously and the new project is a customization of known designs, then the construction project can be carefully analyzed and planned using the prior experience of previous designs. In this case, project planning tools, such as PERT, are extremely useful in laying out the scheduled map for a project.

However, if a future action is unexplored, unique, and nonrepetitive, then analysis in strategy becomes less important and intuition becomes more important.

In the case of a unique future action, such as the introduction of radically innovative technology, strategic processes are more elusive in concept than the kind of analytically based planning process described above. In such cases, strategy becomes an adventure into unknown territory. Although one may start with a plan, much more important is preparation, supplies, flexibility and adaptability, and the ability to learn rapidly in completely new environments.

In exploration, the strategic attitude is more important to success than is the strategic plan.

Strategic attitudes are necessary to planning processes because (1) nothing ever really happens quite as planned, and (2) one often learns what are the real goals only after a strategic activity has begun. It is a strategic attitude that enables a planner to use the planning process effectively as a real guide to action:

- The strategic attitude allows the planner to exploit unplanned opportunities and to revise goals.
- The strategic attitude allows plans to accommodate unanticipated delays and problems.
- A delayed and altered plan can be rescheduled and redirected because the strategic attitude provides continuing vision.

ILLUSTRATION (*Continued*)
Starting Sony

In starting the new firm that was to become Sony, Morita and Ibuka were going into unexplored territory. Their new firm was a company with only a strategic technology attitude but no strategic technology plan.

Morita's partner, Masaru Ibuka, had been born in 1908 in Nikko City, north of Tokyo. His father was an engineer, but he had died when Ibuka was only 3 years old. Ibuka's mother was a graduate of Japan Women's College (Nihon Joshi Daigaku) and had taught kindergarten. Ibuka had always been fascinated with technology, and he had always been inventive: "Radio enthralled Ibuka, even as a boy. Dissatisfied with the simple galena-crystal receivers that most amateurs used, and eager to listen to...broadcasts from distant Osaka, he built a three-vacuum-tube set—but because vacuum tubes were scarce and expensive, the resourceful 17-year-old fabricated his own" (Macoby, 1991, p. 27). Ibuka studied engineering at Waseda University and patented the first of his many inventions while a student. By 1991, at the age of 83, he had acquired 104 patents: "Masaru Ibuka has invented and promoted a whole string of consumer electronics firsts: Japan's first all-transistor radio, a transistorized television, a videocassette recorder for home use, the Walkman personal stereo, and a compact disc player, among others....Also high on his list is his development of the Trinitron color television receiver and the Jumbotron large-screen television display" (Macoby, 1991, p. 27).

Back in 1946, when Morita and Ibuka formed their new firm with $500 dollars, they did not even have a product plan. They tried to think of a possible product: "During the search for a likely product to make, it was often suggested to Ibuka that we make a radio receiver...but Ibuka adamantly refused. His reasoning was that the major companies were likely to have a very fast recovery from the war and would make use of their own components in their own product first and sell parts to others later. Also, they would naturally keep their latest technology to themselves, trying to preserve their lead over their competitors as long as possible" (Morita and Reingold, 1986, p. 50).

Lead time is the name of the game for using technology in competitive strategy. Ibuka was anticipating the competitive conditions in the industrial value chain. Ibuka knew that the new company had to have a product in which it was the technology leader, not the competitor. A small firm has a chance against bigger competitors only with technology leadership.

But what technology edge? What product? Immediate products were necessary for cashflow survival. At first, Ibuka and Morita made shortwave radio adapters to enhance the medium-frequency radios that were then widely owned in Japan. Shortwave radios could receive broadcasts from other countries, an important source of news in occupied Japan. Shortwave radios were in short supply, and this product was a starting place for their business.

Next, Ibuka and Morita noted that many Japanese households had prewar phonographs that needed repair. They began making new motors and magnetic pickups. American products were arriving in Japan, and the American swing and jazz records were very popular. The parts business, however, did not offer much of a future. Ibuka and Morita still wanted to produce a completely new high-tech consumer product—this was their strategic technology attitude.

At this time, Ibuka and Morita knew about the wire recorder, which had been invented in Germany just before the war. Ibuka found a company, Sumitomo Metals Corporation, that could make the special kind of small, precise-diameter steel wire needed for such a recorder, so Ibuka and Morita decided to produce a wire recorder. However, there was a problem. Sumitomo was not interested in a small order from a new, untried company. Ibuka and Morita were discouraged, but they persevered. They continued to keep their company afloat in the parts business.

Meanwhile, the U.S. occupation forces had taken over the Japan Broadcasting Company, NHK, and needed new technical equipment. Ibuka was familiar with audio mixing units and submitted a bid to make one for the U.S. forces. He received the contract and made the mixing unit. When he delivered the unit to the NHK station, he saw there a new American high-tech product—a Wilcox-Gay tape recorder, which the U.S. Army had brought with it from the United States. It was the first tape recorder Ibuka had ever seen. As he looked it over, he could see immediately that it had technical advantages over the wire-based recorder.

In the wire recorder, the wire had to pass over the recording and playback heads at very high speeds to obtain decent fidelity in sound reproduction. The tape of the tape recorder, however, with its wide size, provided a much larger magnetic area for signal recording and therefore could be allowed to travel much more slowly and still provide good fidelity of reproduction. Also, to get a long enough playing time, wire reels had to be very big, whereas the larger magnetic area of the recording tape meant that tape reels could be much shorter for the same playing time. In addition, the wire could not be spliced, as could tape, in order to correct recording errors or to rearrange recorded sequences. For these reasons, Ibuka knew that tape would inevitably replace wire in audio recorders. This was the new high-tech product for which he had been searching! Moreover, Ibuka understood what kind of electronics, physics, and chemistry would be required to copy and develop the new technology. Ibuka decided that tape recorders would be his and Morita's new product.

One sees here a strategic technology attitude at work. Ibuka was committed to a high-tech consumer electronic product—no plan, just a commitment. He was looking for a new high-tech product with which to compete with the bigger but slower-moving established Japanese electronics firms. He saw an innovative foreign technology and decided to duplicate it. His decision was based on the distinct advantages of the new technology.

Strategic Technology Attitude—Perception, Competence, and Commitment

In any strategic attitude, three concepts are central—perception, competence, and commitment. Any strategic technology attitude revolves around perceiving the world as a series of relationships between technology, product, and applications. These three items form a kind of three-dimensional space in which to think about technology strategy.

In the perceptual space of applications, the market for a new technology does not begin until the products of that new technology provide an application for a customer that performs something of value at a price the customer can and is willing to pay for that value. Therefore, the first focus of technology strategy is to identify an initial application and the customer requirements for that application. In the perceptual space of technology, it is important to identify a phenomenal base and a system for a technology that provide adequate performance for the application. In the perceptual space of product, it is important to focus on which product of the technology system the firm will produce and sell. It is seldom possible or even desirable for a firm to produce all products that complete an application system.

Critical to a strategic attitude are the choices for the boundaries of the technology system and the product system. Also, the product system is incompletely determined by the choice of technology system (since one can chose wholes or parts of a technology system and one can add other technologies into the product system).

A technology system incompletely defines the product system, and the product system incompletely defines the application system.

These kinds of incompleteness are the sources of the greatest commercial risks in innovating new technology.

ILLUSTRATION (*Continued*)
Sony's First Product

There are many technical problems in copying a new technology, but Ibuka and Morita knew how to proceed—with systems analysis of the technology and with applied science. "We soon realized that the main trouble we had was that we didn't know anything about how to make the crucial part of the system, the recording tape. Tape was the heart of the new project and the tape was a mystery to us. From our early work on wire recorders, we were pretty confident about building the mechanical and electronic components for a tape recorder" (Morita and Reingold, 1986, p. 55).

The tape was a subsystem of the technology that involved physical and chemically based materials. The rest of the system was based on electronic circuitry and mechanical processes. Ibuka and Morita wanted to produce the tape as well as the machine in order to obtain a follow-on sales business. This commercial strategy to produce the tape as well as the tape recorders became apparent to them immediately after they had made the decision to produce tape recorders.

Ibuka and Morita's first technical problem had to do with the base material for the tape. Again, there also was a supply problem. The American firm 3M was then the source of the base tape material, but 3M also was the major tape producer. Ibuka and Morita thought that 3M might not want to supply tape to a small Japanese competitor.

There were severe materials shortages in postwar Japan. Ibuka and Morita found that they could not obtain cellophane in Japan, which was their first idea for a base. They tried it, even knowing beforehand that it might prove inadequate, and indeed, it turned out to be wholly inadequate. They cut it into long quarter-inch-wide strips and coated it with various ferromagnetic materials. The cellophane would stretch hopelessly, after only one or two passes on the machine. They even hired chemists to improve the cellophane, but without success.

They needed another material. Morita went to one of his cousins, Goro Kodera, who worked for the Honsu Paper Company. Morita asked Kodera if the company could produce a very strong, very thin, and very smooth kraft paper for their tape. Kodera said he would try. He did, and it worked. Ibuka and Morita had a base material, which they sliced into tape using razor blades at first.

The next problem was the magnetic coating to put on the paper tape. Nobutoshi Kihara, one of the young engineers Ibuka and Morita had hired, did empirical research by grinding up magnetic materials into a powder that was sprayed onto the tape. At first the magnetic materials were too powerful for the system. Then Kihara tried burning oxalic ferrite into ferric oxide. This worked! Morita and Kihara searched all the pharmaceutical wholesalers in their district of Tokyo, finding the only store that stocked oxalic ferrite. They bought two bottles of it, cooked it in a pan until it turned brown and black (the brown stuff being the ferric oxide they wanted and the black stuff being ferrous tetraoxide they did not want).

Next, they mixed the brown stuff with Japanese lacquer and tried to airbrush it onto the tape, but this did not work. Then they tried brushes, and they found that a brush made from the very fine bristles from the fur from a raccoon's belly did the best job.

Still it was terrible tape, of very poor fidelity. Ibuka and Morita really needed plastic tape. Finally, they did get a supply of plastic tape. With a supply of plastic tape and the ferric oxide coating, they had a recording-tape technology. This was the critical technology for their new product.

Tape became a major source of cashflow for Sony. Ibuka and Morita continued to put heavy development into it. Many years later, in 1965, IBM chose Sony as its tape supplier for IBM computer magnetic data storage.

Back to the year 1950, however. Ibuka and Morita's first tape machine, which used their new plastic-based magnetic tape, had turned out to be very bulky and heavy (about 75 lb) and expensive (170,000 yen). Next, they began learning something about the marketing of new and expensive high-tech products. Japanese consumers simply would not buy it. They had to look for another market.

At the time, there was an acute labor shortage of stenographers, because during the war so many students had been pushed from school into war material production. Ibuka and Morita demonstrated their new tape recorder to the Japan Supreme Court and immediately sold it 20 machines. This was the breakthrough sale for their new high-tech product.

We can see that from their technology strategic attitude to produce leading-edge high-tech consumer products, Ibuka and Morita were learning some product and marketing strategy as well. They redesigned their tape recorder into a medium-sized machine (a little larger than an attaché case). They also simplified it for a single speed and placed a much lower selling price on it. They then sold their modified product to schools, for English language instruction to Japanese students.

As their product and marketing strategy progressed, Ibuka and Morita next developed strategic attitudes about intellectual property. To obtain a high-quality recorded signal, they had purchased a license to an invention patented by Dr. Knszo Nagai—a high-frequency ac bias system for recording. This demagnetized the tape before it reached the recording head, reducing background and prior recording noise. At that time, the patent was owned by Anritsu Electric, and Ibuka and Morita bought half-rights in the patent from them in 1949. Eventually, when Americans imported American-made tape recorders into Japan using the ac bias system technology, the American firms had to pay royalties to Sony. This made Sony aggressive about intellectual property.

By 1950, then, the new company had products, tape recorders and magnetic recording tapes, and intellectual property. The company was using its strategic technology attitude, but it still did not have a strategic technology plan. This was to come next.

Strategy: Analysis versus Intuition

We have seen how Morita and Ibuka's intuition for high-tech electronic consumer products guided their search for new products. We will next see how that same intuition led to Sony's technology strategy. Let us pause, however, to further explore the roles of analysis and intuition in strategy.

In strategy, intuition arises from perception and commitment, and it creates a goal.

Analysis details the perception of and commitment to reach a strategic goal.

Strategic intuition is the synthesis of a committed perception of the world. *Strategic analysis* is a means of realizing a world shaped by a strategic intuition. Of the many Western students of strategy, Henry Mintzberg has been a particularly strong advocate of the importance of the role of intuition in the strategic process (Mintzberg, 1990). For example, in a published interview, Mintzberg categorized the strategic theorists who primarily emphasize the role of analysis in the strategic process as a kind of "design school of strategy." "The Design School represents an extreme end of a spectrum, where strategy is derived from deliberate thought. Grass-root emergent strategy is intuitive. It is just as extreme, but it is at the other end of the spectrum. To understand strategy, you have to accept both ends of the spectrum" (Campbell, 1991, p. 109).

Mintzberg argued that both analysis and intuition play critical roles in strategy: "I am not saying that analysis is bad or unnecessary. There is a danger, though, that you can preclude synthesis with too much analysis. The reason that strategic planning failed as a technique was because it was based on the assumption that you could get synthesis from analysis" (Campbell, 1990, p. 109).

Analysis and synthesis are complementary cognitive functions. *Analysis* takes apart a complex concept, reducing it to simpler component ideas. *Synthesis* takes discrete, apparently unrelated ideas and constructs them into a new coherent idea. Analysis divides, and synthesis unites. Thus they are not derivable from each other.

Analysis and intuition are complementary tools of cognition—of thinking and of recognition.

Intuition, which creates wholes, does not always create the correct totality, nor is intuitive strategy always successful. In business strategy, the market is ultimately the test of whether a strategic intuition is proper or improper, a success or a failure. The problem in business strategy is how to evaluate an intuition without prematurely destroying it. It is easier to use strategic analysis in evaluating intuitions to find what is wrong with them rather than what supports them.

Some firms have put elaborate procedures in place for developing a strategic plan and then have been disappointed with the outcome—erroneously concluding that strategic planning is useless. However, if they have not distinguished analytical strategy from intuitive strategy, they will not know whether the failure is one of strategic analysis or strategic intuition.

Although the market is ultimately the test, evaluation of a correct strategy prior to implementation is not easy. Strategy is always a top-management responsibility. One way that management fails in this responsibility is to blame every failure in strategy on the implementation efforts of lower-level staff. Actually, every failure in strategy is also a failure by top management, because good strategy should include strategy for implementation.

One symptom of failure in strategic attitude is seen when a strategic planning process results in a very voluminous planning report that is then neither read nor used. Another failure occurs when a large corporation creates an expensive corporate research laboratory that remains isolated from the firm's business units and is always being criticized by management as being unrelated to the firm's businesses. Another failure of strategic technology attitude in a diversified firm is seen when a strategic business unit's product development incorrectly implements research results from corporate research because the business divisions are looking sideways and backwards toward competitors' past technology and not forward toward technology that can leapfrog competitors' capabilities. Other breakdowns in strategic technology attitude occur when a firm's productivity lags and continues to lag behind competition, when new products do not get out in a timely manner, and when a firm is falling behind competitors in the race for technical competence.

An analytical strategic plan cannot substitute for an intuitive strategic attitude but must follow on it.

ILLUSTRATION (Concluded)
Sony's Strategic Technology
Competencies

In 1952, Ibuka and Morita decided to try exporting their tapes and recording machines to the United States. Ibuka visited the United States to study its markets, and since he had earlier read about the invention of the transistor at Bell Labs, he also visited Western Electric in New York (then the patent holder on the transistor). Ibuka was impressed with the new technology. He wanted it. In the following year, in 1953, Morita went to the United States to purchase a license to the transistor from Western Electric (for $25,000 dollars—a big sum to a new company in those days).

Ibuka had appreciated the great technical performance advantage that transistors had over vacuum tubes. A transistor was a fraction of the size and could operate at a fraction of the operating current of vacuum tubes. Ibuka and Morita knew that any business that made portable consumer electronics products eventually would have to change from vacuum tube circuits to transistorized circuits.

This was the beginning of Sony's technology strategy—transistorized circuitry and miniaturization.

Tubes to transistors—so obvious! Now we might say this in hindsight. Yet ponder this mystery: It is a historical fact that the U.S. consumer electronics industry, which in the 1950s was the greatest electronics industry in the world, had almost completely disappeared by 1980. And the reason is a failure of technology strategy. The U.S. firms then in the consumer electronics businesses generally failed to transform their products from tubes to transistors in a timely, committed manner. That little transistor—an American invention (and its follow-on key invention, the integrated circuit semiconductor chip)— was the technological key to the rise to world dominance of the Japanese consumer electronics industry and the corresponding demise of the American consumer electronics industry.

Ibuka and Morita then had a strategic technology attitude that focused on a core technology competency—transistorized electronics. "Soon after my [Morita's] return...the laborious work of creating a new type of transistor began in our research lab based on the Western Electric technology we had licensed. We had to raise the power of the transistor (and improve its frequency response)—otherwise it could not be used in a radio" (Morita and Reingold, 1986, p. 67).

So the radio—the consumer product that Ibuka would not produce a few years back (since at that time the new firm had no technologically competitive advantage)—would now become a second "flagship" product line, the pocket radio. "We managed to produce our first transistorized radio in 1955, and our first tiny 'pocketable' transistor radio in 1957. It was the world's smallest, but actually it was a bit bigger than a standard men's shirt pocket, and that gave us a problem....We liked the idea of a salesman being able to demonstrate how simple it would be to drop it into a shirt pocket....we had some shirts made for our salesmen with slightly larger than normal pockets, just big enough to slip the radio into" (Morita and Reingold, 1986, p. 71).

The technical problem with the original transistor invented at Bell Labs was its frequency response. The original transistors were constructed out of two kinds of semiconductors arranged like a sandwich, in which the middle slab controls the current flow between the outer two slabs. Since current in semiconductors can either be carried by electrons or by holes (holes are unfilled electronic orbits around atoms), one can design either hole-electron-hole carrier combinations (positive-negative-positive: *pnp*) or electron-hole-electron combinations (negative-positive-negative: *npn*).

The original Bell Labs transistor had a *pnp* sandwich of germanium-indium-germanium. Electrons (the negative carriers) inherently move faster through a semiconductor than holes (the positive carriers). The physical reason for this is that a hole "waits" for an electron to put into its empty orbit from a neighboring atom before that empty orbit "appears" to move from one atom to the next. This is an inherently slower process than a relatively freely moving electron passing by one atom after another.

Therefore, the first technical improvement the Sony researchers had to implement to make the new transistor technology useful was to speed up its signal-processing capability by using electrons rather than holes as carriers. The Sony researchers therefore reversed the order of the transistor sandwich: from a positive-negative-positive structure to a negative-positive-negative structure (indium-germanium-indium).

The development of the transistor by altering its phenomenal basis from "hole conduction" to "electron conduction" is an example of a science strategy in technological research.

Had Ibuka and Morita's new electronics firm been staffed with only electronics engineers and no scientists, it would not have had the staff to understand the new physics of semiconductors. The company would not have had the technical imagination to begin developing the transistor by reversing the materials combination to seek a higher-frequency response. Ibuka and Morita had established a firm that had not only an electronics technical capability but also an applied physics capability. That was how Morita was trained. Applied physics underlay Morita's strategic technology attitude.

The new firm had research physicists. In fact, the firm's physicists were so good that during the transistor research, one of them, Leo Esaki, discovered a new fundamental phenomenon of physics—quantum tunneling (in which electrons can sometimes tunnel through physical barriers that would bar them if they only obeyed classical physical laws and not quantum physics). In 1973, Esaki won the Nobel Prize in physics.

The next technical problem the researchers faced was the choice of materials for the bases of the transistor and their impurities. Without adding a small quantity of different atoms (called *doping*), neither germanium nor indium conducts electricity. The doped atoms (impurities) make these materials semiconducting, as opposed to nonconducting. Ibuka and Morita decided to discard the indium used in the Bell Labs' original version of the transistor. Indium had too low a melting point for use in a commercial transistor. They tried working with gallium, with antimony as its doping atom. This did not work well either. Next, they tried replacing the doping element antimony in the gallium with phosphorus. At first the results were not encouraging, but the researchers persisted. Eventually, they found just the right level of phosphorus doping. Then they had an *npn* transistor of gallium-germanium-gallium structure, with just the right amount of phosphorus atoms doping the gallium. Sony researchers had developed a higher-frequency germanium transistor that was commercially adequate for their pocket radio.

The development of the transistor for radio applications is an example of the Japanese acquisition of a foreign-invented technology and subsequent improvement of that technology for commercialization by applied research. This pattern of acquisition of foreign technology and subsequent improvement for commercialization was common in both early and later industrial development in Japan and led to its emergence first as a world military power and second as a world economic power.

This is also an example of the kind of development that all commercial technologies require in their early days to become usable. This kind of research constitutes the basis of the exponential portion of the technology S-curve of a new technology (which we will review in a later chapter).

With their new strategic technology of the transistor and the new pocket radio, Ibuka and Morita decided to change the name of their firm. They now had global aspirations, and they wanted a globally recognizable name, so they changed from the Tokyo Telecommunications Engineering Company to Sony.

When Sony introduced its transistorized pocket radio into the United States, it discovered that Texas Instruments also had independently innovated a transistorized pocket radio. But Texas Instruments had no strong commitment to the consumer market, and the company soon dropped the product. Sony, however, was committed to the consumer electronics market and soon began its climb to the position of world leader in consumer electronics.

This is the second element of a strategic attitude—commitment. Texas Instrument's commercial successes remained in the industrial and military markets, where their real heart was—their strategic attitude and commitment. Sony focused on the consumer electronics market and became an innovative, high-tech, top-quality consumer electronics firm and a giant, global company. Sony also introduced the first transistorized small black-and-white television set. In color television, Sony innovated a single-gun, three-color television tube. Sony also innovated the Walkman series of miniature audio players and the first home videocassette recorder (VCR), after the industrial version had been invented in America.

Ibuka and Morita had imbued their new company with a strategic technology attitude that searched for and focused new technologies on advanced consumer electronics products—with a corporate technology competency in transistorized and miniaturized products.

Strategic Attitude as a Preparation for Battle

What we have seen in the preceding illustration is a company based on and guided by the *perceptions* and *commitment* of its founders concerning the importance of advanced technology applied to consumer goods and the development of corporate *competency* in new technologies to carry out the founders' *goals*.

Perception, commitment, and competency are the fundamental components of a strategic attitude.

These components constitute the elements of "intuition" in strategy in the formulation of corporate goals. In Western culture, the term *intuition* is widely and frequently used and yet remains vague and ill-defined. How does one create totalities—conceive conceptual wholes—perceive and individuate objects? How does one synthesize?

Although in the Western management literature there are many books on creativity, there is no consensus on how synthesis occurs or how to facilitate synthesis. For example, despite Mintzberg's emphasis on the importance of intuition in strategy, he himself offered only a vague description of intuition: "Intuition is a deeply held sense that something is going to work. It is grounded in the context in which it is relevant and based on experience of that context. I cannot be intuitive about something I know nothing about" (Campbell, 1990, p. 109).

This negative judgment is probably the most certain thing one can say about intuition from the Western literature on intuition:

> *One cannot be intuitive if one lacks experience.*

Yet this is the nub of the problem of action. If one must be intuitive to perceive a correct strategy for action, how can one gain the experience that only action provides?

In contrast to this minimal discussion of intuitive strategy in Western literature, Eastern literature discusses intuitive strategy extensively. I think the reason for the differences in European/American and Japanese management thinking about strategy lie in their different philosophical roots. In medieval Europe, the literate caste was comprised of priests, and the warrior caste was generally illiterate. In feudal Japan after the Tokogawa shogunate, however, the warrior caste was generally literate, along with the priests. Of course, priests generally have little experience with action, whereas action is the business of warriors, actions of battle and war.

A famous example of the adoption of Zen Buddhism as a mode of action is Minyuomoto Musashi's *A Book of Five Rings*. It has been popular with Japanese management, and in the 1980s, it also became popular with American management. Musashi was a literate and highly cultivated warrior, but he was a warrior. Before he turned to writing, Musashi claimed he fought 60 sword duels without a single defeat.

Paraphrasing, this is Musashi's (1989) way for warriors to learn strategy:

1. Do not think dishonestly.
2. The "Way" is in training.
3. Become acquainted with every art.
4. Know the ways of all professions.
5. Distinguish between gain and loss in worldly matters.
6. Develop intuitive judgment and understanding for everything.
7. Perceive things that cannot be seen.
8. Pay attention even to trifles.
9. Do nothing that is of no use.

In what way do Musashi's principles constitute a kind of philosophy of action?

First, if one thinks dishonestly, one cannot be honest with oneself, and without self-honesty one cannot think clearly.

Second, the path to the attitude, the "Way," must be in training if the attitude is one of action, for without training, one cannot act with skill.

Third and fourth, action always occurs within a complicated, holistic context—a system in which many skills many be required to carry out an action. (This is why, for example, innovation teams should include personnel with technical, marketing, production, and financial skills.)

Fifth, action always requires some resources, tools, or supplies and results in outcomes that either deplete or replenish and/or improve previous resources. Thus all action must be strategically judged in terms of gain or loss of goals.

Sixth and seventh, intuitive facility in the heat of action is extremely important. Actions are complex, contradictory situations of flux and motion in which totalities must be instantly synthesized and comprehended. Hence a warrior's (or a manager's) intuitive ability is as important as his or her analytical ability. Musashi's precepts here argue that in the heat of battle, intuition and perception are critical cognitive functions and hence must be trained and exercised before battle.

Eighth, Musashi emphasizes that in action, a detail can sidetrack or defeat a whole project. (For example, "bugs" in software programs have often destroyed a whole company's product reputation. As another example, once a poorly mixed antifreeze solution caused thousands of new automobiles to be recalled, harming the company's reputation for quality.)

Finally, the ninth precept emphasizes that in action, economy is important, for action consumes resources. Action must be focused, disciplined, and economically executed.

In summary, we can see from this example of a strategic attitude for a Samurai warrior that the precepts are sensible ways to prepare for and behave in battle—they are efficient and effective guides to action, i.e., a philosophy of action.

Strategic Attitude and Corporate Culture

A corporate strategic attitude resides in the company's culture and management philosophy, which structures its basic perceptions and the commitments of its management. Any given company may or may not be conscious of its culture. If it is conscious and has procedures to inculcate culture, then it could be said to have a "Way."

Strategy-as-attitude, as "Way," is a philosophy of action that re-quires both an attitude and a perspective. The attitude is the continual awareness of the greater scheme of things during all moments of action. The perspective is the plan of action at the moment of action.

For example, Lowell Steele (1989) has emphasized that all compa-nies develop management conventions that include common assump-tions about the nature of the business. Managers in a firm share traditions about the way the company grew and assumptions about the way the firm gains competitive advantages. In addition, managers share conventions about guidance and operational control of the enter-prise and appropriate measures of performance. These kinds of as-sumptions and shared conventions constitute the culture of a given management team. To some extent, such shared conventions may be articulated by the company or remain inarticulate. When articulated as maxims or precepts, they form the "Way" of that company's culture.

ILLUSTRATION
The "Way" of General Electric

The preceding philosophical discussion may strike some as almost too ephemeral for a hard-headed technology strategy, yet we should re-mind ourselves that strategy is one of the principal responsibilities of corporate top management. An example of this is the management phi-losophy of Jack Welch when he was CEO of the General Electric Corporation in the 1980s. Stratford Sherman wrote an article on Welch's strategy: "His [Welch's] ideas are simple: Face reality. Communicate clearly. Control your own destiny. But put together, they could rewrite the book on how to run a big company" (Sherman, 1989, p. 39). Such precepts (face reality, communicate clearly, control your own destiny) are, you see, really elements of a philosophy of action—part of Welch's personal philosophy about the manager.

At the time, Welch's management philosophy attracted many ob-servers. For example, Noel Tichy and Ram Charan (1989, p. 112) also discussed Jack Welch's philosophy and its implementation:

> Jack E. Welch, Jr., chairman and CEO of General Electric [in 1989], leads one of the world's largest corporations. It is a very different corporation from the one he inherited in 1981. GE is now built around 14 distinct busi-nesses—including aircraft engines, medical systems, engineering plastics, major appliances, NBC television, and financial services. They reflect the aggressive strategic redirection Welch unveiled soon after he became CEO.

Sherman describes Welch's reputation after that strategic redirec-tion as follows (Sherman, 1989, p. 39):

Neutron Jack, as he is sometimes called, is widely regarded as one of the world's most ruthless managers. The truth is more complex. Some of his actions are indeed harsh, and he antagonized people inside the company and out by fixing something they didn't think was broke. What is becoming clear only now is how those moves fit into a larger plan to strengthen the enterprise and to make its remaining employees more secure.

Jack Welch graduated with a Ph.D. in chemistry from the University of Illinois. He went to work for GE as an engineer in the plastics business at Pittsfield, Pennsylvania. GE had invented a new thermoplastic called Lexan with an exceptional structural strength. It was a technical innovation that could develop into a large business for GE, but at the time Lexan had no markets and few sales. Welch saw its potential and pushed its commercialization. At age 27, he gained managerial responsibility for a GE's plastics business (as a profit-and-loss center). He remained in Pittsfield for 17 years, increasing the plastics business at an average earnings growth of 33 percent a year.

In 1977, Welch was promoted to GE's corporate headquarters as a senior vice president for consumer products. The CEO of GE then was Reginold H. Jones. Jones had become GE's president in 1970 and had moved GE from a state of chronic cashflow shortage to financial strength by 1977.

In the 1960s, GE had taken on, at the same time, three major new technological areas: mainframe computers, nuclear energy, and commercial jet engines. All three had gobbled up money for R&D, but GE had succeeded commercially only in jet engines. IBM beat GE in the computer market, and the nuclear industry turned environmentally and politically sour (Banks, 1984). Jones closed down the computer and nuclear energy businesses to get GE's financial health back in shape.

Jones had a financial background, and in 1980 he chose as his successor the technically trained Welsh. "...when Jones asked his staff to list 20 candidates to succeed him, Welch's name didn't come up. Too young. Too unusual. But the maverick's performance had attracted the CEO's attention....In choosing Welch as his successor, he went for an engineer who was comfortable with technology, had a record of sizzling financial performance, and was a proven master at managing change" (Sherman, 1989, p. 42).

Jones had viewed the strategic problems of GE as alternately requiring strategic focuses upon financial and technical attitudes, with an appreciation that both must be strong. As the new CEO, Welch continued change at GE (Tichy and Charan, 1989, p. 112.):

> In 1981, Welch declared that the company would focus its operations on three "strategic circles"—core manufacturing units such as lighting and

locomotives, technology-intensive businesses, and services—and that each of its businesses would rank first or second in its global market. GE has achieved world market–share leadership in nearly all of its 14 businesses.

While Jones had put GE back on its financial feet, the GE that Welch was to lead then was typical of the large financial conglomerates that many managers assembled in the 1960s and 1980s—diversified but without any integration or synergy between businesses. Welch rationalized its businesses into 14 areas. Welch believed that a strategic attitude for a conglomerated company must have not only a financial strategy but also a market and technology attitude: "He [Welch] sees global markets inevitably coming to be dominated by fewer, ever more formidable players—steamrollers like Philips and Siemens and Toshiba. To prosper in this world, Welch believes, GE must achieve competitive advantages that allow it to rank first or second in every market it serves. So often is this simple concept repeated around GE, people express it as a single, seven-syllable world: 'number-one-an-number-two'" (Sherman, 1989, p. 40).

In 1989, in both the U.S. and world markets, GE was first in aircraft engines, circuit breakers, electric motors, engineering plastics, industrial and power systems, locomotives, and medical diagnostic imaging. "Welch loves big, complex businesses with only a few competitors" (Sherman, 1989, p. 41).

When Welch was asked what he thought made a good manager, he replied: "I prefer the term business leader. Good business leaders create a vision, articulate the vision, passionately own the vision, relentlessly drive it to completion" (Tichy and Charan, 1989, p. 113).

This is the expression of a manager with a strategic attitude—an intuitive vision. The traditional functions of management are to plan, acquire resources, organize, implement, control, and supervise. Welch emphasizes that even before planning there is vision. Vision derives from strategic attitudes.

Sherman (1989, p. 50) summarized the six rules of Welch's strategic attitude:

- Face reality as it is, not as it was or as you wish it were.
- Be candid with everyone.
- Don't manage, lead.
- Change before you have to.
- If you don't have a competitive advantage, don't compete.
- Control your own destiny, or someone else will.

Now compare these with Musashi's classic principles listed on p. 50. One sees the similarity in attitude—even though Welch was a manager of the twentieth century and Musashi a warrior of the eighteenth century. What they have in common is a philosophy of action.

The "Dao," the "Way," the "Strategic Attitude" is the commitment, preparation, and perception before any battle and before all battles—the philosophy of action.

The Business as an Enterprise

The basic strategic attitude that managers hold toward their business defines the firm's culture through management's perception of and commitment to action. This is the idea behind the concept of the business as an enterprise. For example, Lowell Steele emphasized the importance of looking at a business as an enterprise: "Every business is based ultimately on a few simple ideas, principles, or even assumptions. They address the fundamentals of the business: What products or services do we provide? Who are our customers? How do we compete? How do we define success? How do we behave toward each other? In the aggregate these fundamental features could be termed *the concept of the enterprise*" (Steele, 1988, p. 69).

Steele points out that to a great extent the answers to these fundamental questions of the enterprise are implicit in the shared beliefs and conventions that constitute the culture of the firm. From experience in operations, managers develop a culture of shared beliefs and conventions about how the firm should operate. These conventions include assumptions about the nature of the business, the way competitive advantages are gained, a sense of how and why the company became what it is, and procedures for guidance and operational control of the enterprise.

Strategic Attitudes and Competitiveness

The environment for competition continues to change as technology grows and the world industrializes. Many students of modern economic change have noted the rapid pace of technological and economic change. For example, Kim Clark (1989) emphasized some of the changing features of competition, among which are

- A continuing and growing worldwide dissemination of scientific and technical knowledge.

- An increasing number of global competitors competing in different national markets.

- At the same time, the fragmentation of mass markets into market niches and rapidly changing customer preferences as a wider variety of products are offered.

- A continuing revolution in computer and communications technologies that provide corporate capabilities of more rapid responsiveness and greater flexibility.

- The proliferation of the number of technologies that may be relevant to any given product, including mechanical, electronic, and software technologies and choices of materials.

In light of these changes, Clark proposed five precepts for corporate strategy:

1. Managers should understand the technological core of their business and envision that as a strategic advantage.

2. Managers should take a broad, worldwide view of technical competence, seeking out the best technology wherever it can be found.

3. Managers should focus on time as the critical factor in using innovation for competitive advantage.

4. Managers should discipline their business function around the function of production (in production the technical knowledge of the company is focused on a value-adding activity for the customer).

5. Managers should integrate all business functions through the information system of the firm.

We can see that Clark's precepts for a strategic technology attitude require deepening managers' concerns about technology, widening their horizons on technical change, focusing their attention on timeliness, and centering technical activities around the science of manufacturing. Clark emphasized that management's fundamental responsibility with regard to the technical competence of an organization is to deliberately build such a competence.

As another example, T. G. Eshenbach and G. A. Geistauts offered precepts for engineers. They argued that the technological perspective of engineers should be broadened to view their companies as kinds of sociotechnical systems (1987, p. 63):

- Think of the firm as a total system.

- Focus on the interaction between the firm and its environment.

- Concentrate on the firm's most fundamental questions and issues, including the basic mission and a definition of the business and goals.

- Be explicit about value judgments in technology assessments and R&D cost/benefit analyses.
- Emphasize anticipatory adaptive control for the firm to optimize long-run performance in the face of inherent uncertainty.
- Articulate a philosophy of management that represents a permanent commitment to integrative, systematic long-range planning.
- Develop an ongoing planning process, wherein strategy is continuously reexamined.

Note the emphasis on perceiving the business from a systems point of view, envisioning the firm as an economic value-adding transformation. In addition, Eshenbach and Geistauts emphasized that engineers should focus on the interaction between technology and business goals. They also advocated an emphasis on anticipatory attitudes, formalized in a planning process. Technology planning processes in firms should be designed to foster a business strategic attitude in the firm's research scientists and engineers to enable them to collaborate strategically with business managers.

As a third example of strategic precepts for managing technology, Lowell Steele (1989, p. 345) offered what he thought a "technologically effective" enterprise should be capable of:

- Taking a systems view of technology.
- Being aware of the dynamics of maturation of technologies and industries.
- Being explicit about how the firm uses technology for a competitive advantage.
- Articulating a clear sense of what businesses the firm is in.
- Knowing who are the firm's competitors.
- Being aware of the changing nature of competition.
- Being relentless in the pursuit of excellence.
- Effectively dealing with uncertainty and ambiguity.

For all these commentators on corporate strategic attitude, their precepts constitute elements of their philosophy of action, their "Way," their "Dao." Clark's "Way" drew from the perspective of the manager and emphasized the need for a manager to be aware of technology and to have a commitment to globalization, time, and production. Eshenbach and Geistauts' "Way" drew from the perspective of the engineer and emphasized the need for an engineer to be aware of the business system and to have a commitment to adaptive control and long-range planning. Steele's "Way" drew from the interface of research and business and emphasized the need for a balance between business and engineering views on an enterprise, with commitments to competitiveness and excellence.

Which "Way" is the best? They are all best depending on one's experience of the world, position, and commitment to action. Each sees the world from the perspective of a particular experiential base of action. For technology strategy, management personnel require a "Way" so that they may be more aware of and attendant to technology as a competitive factor. Technical personnel require a "Way" so that they may be more aware of and attendant to the business implications of technology as part of the enterprise. Formulating strategic attitudes in the corporation requires bringing each group into a "Way" in order for all groups to cooperate in the total business enterprise.

Management's strategic precepts about the nature of the world and the value of action determine the focus of corporate perception, commitment, and preparation.

Developing Core Competencies

Concepts such as core competencies and core products enable the management of a diversified firm to systematically search for technological synergy throughout the firm. Formulation of a strategic attitude about core corporate competencies and the identification of core corporate products requires the leadership of senior management.

A committee of top managers needs to be constituted in order to identify corporate core competencies and products (a CCC&P committee). The committee should be composed of the senior executives in each business group of the diversified firm and senior scientists from each business group and the corporate laboratory. The committee should retreat for a 2-day session to review the industrial value chains of the different business groups and to forecast technical directions in these value chains (topics that will be reviewed in later chapters). From this review, the committee should identify technological trends that would lead to restructuring of any or all of these industrial value chains.

The committee should next organize task forces of first-line researchers, sales managers, and production management from the different business groups to further detail and evaluate the trends. The task forces should identify core products for the corporation that would provide competitive edges both vertically and horizontally for many businesses of the firm.

The CCC&P committee should use the material from the task forces to identify a small set of corporate competencies and core products and then formulate a strategic corporate business plan to acquire and implement these competencies. The CCC&P committee next should identify any restructuring of industrial sectors in which these competencies may be of advantage and should provide this vision to strategic business units to be used in their planning processes. In the annual corporate planning process, the allocation of resources should be reexamined in light of both the identified core competencies and products and the envisioned restructuring of industrial value chains.

Core products are key components (or materials or subsystems or services) produced in some of the businesses of the firm that provide competitive advantages in the products of other businesses of the firm. *Core competencies* are the knowledge, skills, and facilities necessary to design and produce core products.

Core technological competencies are corporate assets, and as assets, they facilitate corporate access to a variety of markets and businesses. For competitive advantage, a core technological competence should be difficult for a competitor to imitate.

In addition, the identification of core technological competencies requires management to look upstream in the economic value chain and to decide where and what to make or buy for one's business products. Management also must look downstream in the economic value chain toward the final customer applications to determine which technologies most visible to the customer affect application performance. (We will explore in detail the interaction of technological change and industrial structure in a later chapter.)

Failing to identify core competencies is a kind of opportunity loss for a company. Such a failure is due to the inability of management to conceive of the company as other than a mere collection of discrete businesses. If management cannot conceive of strategic totalities—other than only financial control—the concept of the corporation becomes merely that of a holding company. Management then will fail to utilize other competitive factors, such as technology, as corporate competitive weapons.

A deliberate corporate search for core products in a diversified corporation is one way to determine how the diversified corporation's businesses can create synergy to provide competitive advantages in their respective industrial-sector value chains. The concept of a corporation's core products can be used to judge how much vertical integration to institute and where that vertical integration can provide a competitive advantage. In the earlier illustration on NEC, the diversified corporation's strategic attitude was to be a major player in the fusion of computers and communications. Core products to both these industries were semiconductor chips.

Vertical integration in an industrial value chain is always a strategic option for any large corporation, but vertical integration has both advantages and disadvantages. The advantages are that vertical integration can improve control over components and resources; but the disadvantages are that it also may increase the cost of components and resources and reduce innovativeness. Vertical integration is therefore a two-edge sword. The idea of core products requires a strategic focus on where vertical integration will provide a competitive advantage in innovation, price, and quality for its businesses.

Technology planning in the corporation requires both a top-down procedure (core competencies) at the corporate level and bottom-up procedures (technology roadmaps) at the strategic business unit level—in order to identify corporate-wide technology competencies that can provide a competitive edge to the businesses of the firm.

Practical Implications

The lessons learned in this chapter are (1) about how to create a strategic technology attitude in a firm and (2) how to implement a corporate strategic attitude as core competencies in a diversified firm.

1. Technology planning in a firm should be underpinned by a corporate culture that facilitates a strategic technology attitude.

    a. A strategic attitude prepares in general for any action by formulating perception, commitment, and preparation for action.
    b. Strategic attitudes are necessary to planning processes because (1) nothing ever really happens quite as planned, and (2) one often learns what the real goals are only after a strategic activity has begun.
    c. A strategic attitude enables a planner to use a planning process effectively as a real guide to action:

        (1) By allowing the planner to exploit unplanned opportunities and to revise goals.
        (2) By allowing plans to cope with unanticipated delays and problems.
        (3) By enabling a delayed and altered plan to be rescheduled and redirected vision.

2. To the extent the common practices and shared conventions that a company's culture develops are articulated by senior management,

they become the "Way" of that company's culture. Examples of strategic precepts include

a.   *On intuition:*

    (1) Do not think dishonestly.
    (2) Face reality as it is—not as it was or as you wish it were.
    (3) Perceive things that cannot be seen.
    (4) Pay attention even to trifles.
    (5) Deal with uncertainty and ambiguity.
    (6) Develop intuitive judgment and understanding for everything.
    (7) The "Way" is in training.

b.   *On action*

    (1) Time is the essence.
    (2) Develop an ongoing planning process wherein strategy is continuously reexamined.
    (3) Emphasize anticipatory and adaptive control in the face of inherent uncertainty.
    (4) Control your own destiny, or someone else will.

c.   *On business*

    (1) Distinguish between gain and loss in worldly matters.
    (2) Think of the firm as a total system.
    (3) Focus on the interaction between the firm and its environment.
    (4) Articulate a clear sense of what businesses a firm is in.
    (5) Be relentless in the pursuit of excellence.
    (6) Discipline technical functions around the science of production.
    (3) Integrate operations around the information system.

d.   *On competition*

    (1) Know who your competitors are.
    (2) Be aware of the changing nature of competition.
    (3) If you do not have a competitive advantage, do not compete.

e.   *On technology*

    (1) Become acquainted with every art.
    (2) Know the "Ways" of all professions.
    (3) Take a systems view of technology.
    (4) Take a global view on the distribution of technical competence.

(5) Be aware of the dynamics of maturation of technologies and industries.
(6) Be explicit about how one uses technology for a competitive advantage.
(7) Know the technological core of a company, and link it to strategic intent.

3. The concept of core technology competencies in a firm articulates the strategic attitude for implementing strategic technology planning into corporate planning.

For Further Reflection

1. Identify two industrial value chains that historically originated independent of each other but later fused due to technological advances. (Include industries whether they were originally in manufacturing or in services.)

2. Identify the nature of the technology advances that drove the fusion. What were the core competencies and core products in this process?

3. Did the new technologies arise in one or the other of the industrial value chains or in neither?

4. What happened to the dominant firms in the separate industrial value chains after the chains merged? Did any survive as dominant in the new industrial value chain, or did new firms in the new technology rise to dominance?

Managing Innovation Within the Product-Development Cycle

As we saw in the Commodore 64 illustration in Chapter 1, new technology is introduced into products during the product-development process. This process of recurrent introduction of new products has been called the *product-development cycle*. Technological innovation must be managed within the product-development cycle.

In a firm, technology forecasting, planning, and implementation are not a single event, but rather a set of repetitious activities taking place within the ongoing activities of the firm. Current products are being produced, new product models are being designed, production improvement is being planned and implemented, and new generations of product lines and/or new product lines are being researched, planned, and designed. Innovation in the product-development cycle is a cyclic innovation process focused on incremental innovations.

Recall, however, that radical innovation is critical to long-term competitiveness and that the integration of the two processes can be accomplished by planning the next generation of technologies jointly between corporate research and the divisional development laboratories of the strategic business units. The procedure for doing this has been called a *technology road map*, which will be the subject of this chapter.

What makes the integration of linear and cyclic innovation processes in a firm so critical is the element of *time*. There is always only a timely window of opportunity when innovating new technology in a product can gain a positive competitive advantage. A speedy product-development process is necessary to use technological innovation as a competitive advantage. Moreover, once in that window, the product must have been correctly designed for the customer in terms of proper application focus, adequate performance and features, and reasonable pricing, safety, and dependability.

Speed and correctness are the criteria for a good product-development process and are essential when innovating new technology.

Why are speed and correctness not an easy thing to achieve in the product-development cycle? The reasons are the uncertainties, variations, and changes that must be managed in the cycle:

- There are uncertainties about timing in the anticipation of new technology—When will it be ready for embodiment in new products?

- There are uncertainties about the performance of new technologies—How much actual achievement will the technology provide in a new product?

- There are uncertainties in the customer requirements in the new product—How will the new technology alter the customer's use of the product?

- There are uncertainties in the tradeoffs between performance and cost that the product designer can provide.

- There are uncertainties in the production of the product from the inherent variability in the physical production processes.

Jim Solberg (1992) has emphasized that these kinds of uncertainties taken together are the source of delays and errors in the overall product design and production system. He suggested that each of these provides feedbacks of uncertainty in the prior decisions affecting parts of the overall system, as depicted in Fig. 3.1. These uncertainties result in changes and in demands for new work by preceding units that introduce delays into new product innovation.

For example, production variation can create feedback loops in the product-development process that create delays. A new product designed for production may turn out not to be producible to quality and cost without improving the production process, and therefore, production may be delayed until research creates the needed production improvement. Cost/quality problems may require changes in product plans or redesigns of a product.

The performance/cost tradeoffs in product design also may create delays by requiring further product development from research or alterations in product plans and schedules for introduction. Changes in customers or in customer requirements also can produce delays in planning new products by demanding more research for the creation for new products. Finally, the technical risks in research may result in research projects being delayed or in not achieving their desired per-

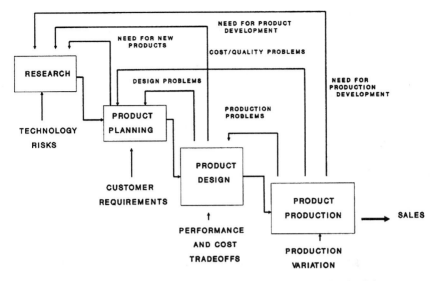

Figure 3.1 Sources of delays in the product-development process (feedback loops introduce delays).

formance level, which will delay the whole downstream product-development process.

Solberg and others have suggested that the development of procedures and software tools to promote concurrency and virtual and rapid prototyping in the product-development cycle will speed up and aim toward correctness of the activities in the product-development cycle.

1. *Fast cycle company*—Manage the firm using time as a competitive factor.
2. *Design in parallel*—Use concurrent engineering procedures that encourage the performance of product design activities in parallel.
3. *Produce it right the first time*—Design the product for quality, manufacturability, and life cycle.
4. *Design in virtual time*—Develop and use computer aides that model and simulate the product's configuration and functioning and also rapidly create physical prototypes.
5. *Manage design bugs*—Manage the design process to proactively seek out design bugs.
6. *Technology road maps*—Use strategic technology planning to develop products that embody next generations of technology.

ILLUSTRATION
Toyota as a Fast-Product-Cycle
Company

As an illustration of a company using time as a competitive advantage, Bower and Hout (1988, p. 111) described Toyota:

> To illustrate, let's look at Toyota, a classic fast-cycle company....[The] heart of the auto business consists of four interrelated cycles: product development, ordering, plant scheduling, and production. Over the years, Toyota has designed its organization to speed information, decisions, and materials through each of these critical operating cycles, individually and as parts of the whole. The result is better organizational performance on the dimensions that matter to customers—cost, quality, responsiveness, innovation.

For Toyota's product development, self-organizing and multifunctional teams focused on a particular model series. They accepted full responsibility for the whole cycle of product development—making the style, performance, and cost decisions and establishing schedules and reviews. In addition, the product-development teams selected and managed the supplier input, bringing suppliers into the design process early. As a result, in 1988 Toyota was then capable of a 3-year product-development cycle. (At that time, the average car-development cycle of U.S. automobile manufacturers was 5 years.)

A fast response time was used not only for product development but also for production control in Toyota. Toyota dealers in Japan were connected on-line to Toyota's factory scheduling system. As soon as an order was taken, the information on the selected model and options were entered immediately into the scheduling information. Toyota's purpose in integrating real-time sales information into production scheduling was to minimize sharp fluctuations in the daily volume of production while also minimizing inventories. Toyota could produce on each production line a full mix of models in the same assembly system utilizing flexible manufacturing cells.

Bower and Hout saw that Toyota's attention to responsiveness pervaded its organization. Toyota aimed at introducing new products faster than competitors. And through continually trying out new models, Toyota could observe what the customers bought. This enabled the company to keep current with changing market trends.

Competitiveness and Timing of Innovation

Bower and Hout called companies who use time as a sustainable competitive advantage "fast-cycle companies." Speeding up the response time of companies to changes in customer needs and the economic environment requires more than simply working faster. It requires working differently—thinking about why it takes time to respond, whether responses are correct, and how to respond more quickly and correctly. The sustainable competitive advantage from attending to time is through better and more quickly satisfying customers. Fast-cycle companies develop new products sooner than competitors, process customer orders into deliveries faster than competitors, are more sensitive to customer needs than competitors, and make decisions faster than competitors on how to add value in their products/services to the customer.

This new emphasis on time as a competitive factor has occurred because of the computer and communications service technologies. These technologies enable firms to communicate and operate faster than previously and to deal with more complex issues. Many other observers of the effect of service technologies on competitiveness also have seen the use of time as an important competitive advantage.

For example, Roger W. Schmenner (1988) argued that a management focus on the time of throughput in the whole organization makes managers reduce inventories, setup times, and lot sizes and encourages improved quality through reducing rework and redesign.

In another example, Roy Merrills (1989, pp. 108–109) reported on how Northern Telecom, Inc., focused on time in competitiveness:

> In 1986, we discovered something important about our company. We found that all the things that were vital to our long-term competitiveness had one thing in common: time....Today [1989] we manufacture products in about half the time it took just two or three years ago. Inventory and overhead have dropped, and quality has improved. Overall customer satisfaction has steadily risen....

A third example of an observer emphasizing timeliness and speed in competition is Joseph Vesey (1991, p. 33): "The new competitors are time-to-market 'accelerators.' Their focus is on SPEED: speed in engineering, in production, in sales response, and in customer service."

Vesey cited studies presenting examples in which getting a product to market even 6 months late (but on budget) reduced expected gross profit by a 33 percent, whereas getting to market early (even by a month) improved expected gross profit by 12 percent. Vesey suggested that getting an organization to be speedy required management to create an organizational environment in which (1) change and innovation were natural, everyday occurrences and (2) appropriate technologies were introduced to enable personnel to respond rapidly and correctly to changing market needs.

The consensus of all the observers concerning the use of time as a competitive factor was based on the fact that to use innovation in an *aggressive* competitive posture, innovative products and processes must be introduced *before* competitors act. Even for a defensive competitive posture, innovative products and processes must be introduced as soon as possible after competitors innovate and before significant market share is lost. Timing is important to any competitive posture.

> *The competitive importance of a fast product-development cycle lies in a firm's ability to respond to market changes faster than competitors: correcting product mistakes, refining product successes, and emulating competitors' product successes.*

In innovating new products, the product is never perfect the first time (because of uncertainties in design balances and customer applications of new technology—which will be discussed in a later chapter). Therefore, it is important to be able to alter a product quickly as sales begin and after one learns what should have been the design balance and what is the proper application focus. Even when one has been relatively correct in innovating a new product, one should still react quickly to refine the product to maintain a lead against competitors who will be entering with a "me too" product. Also, competitors will have successes that require emulation before they seize too much of the market and outpace the original innovation. A fast product-development process is necessary to keep up with competitor's leads.

One of the major impacts that the service technologies (computers, information, and communications) are having on society in the twentieth century has been to make time an increasingly important factor in competitiveness. Addressing this competitive climate requires management commitment to the fact that responsiveness and timeliness are important and should be rewarded. A management team that facilitates fast response operates in a leadership style that encourages the formation of self-organizing, multifunctional teams to take charge of entire corporate projects. Decentralizing the product-development process into self-organizing, multifunctional teams responsible for specific aspects of the development of a new product requires a change in management philosophy. (This has been called a *concurrent engineering approach,* which we will next review.)

Technological innovation creates competitiveness only when used in a timely fashion. Market research methodology has been developed to study the nature of existing markets, but not for brand new markets. Such methods are inappropriate for the study of changing or new markets because they measure only what was and not what will be. Analysis of consumer patterns and preferences in existing markets only describes the nature of the current markets and cannot forecast the nature of future markets if the markets are changing. At the same time, however, product innovation creates market change. In fact, the more radical the product innovation, the more dramatic is the market change and the less its direction of change can be anticipated by market research. Traditional methods of market analysis therefore cannot anticipate the impact on markets of product innovation.

Experimenting with a market requires a company to be able to develop and change products in a short product-development cycle. For innovating new-technology products, it is necessary for a company to be able to quickly try out new product ideas on customers and then revise them based on customer experience with the new technology.

The service technologies that provide capabilities for fast-response companies focus on (1) rapid product-development cycles, (2) manufacturing science bases in unit production processes, (3) intelligent control in operating systems, and (4) teamwork and communications that integrate the business enterprise.

The technologies important to rapid product-development procedures must get the product development right quickly. These include computer-aided design and engineering technologies and organizational procedures such as concurrent engineering practices. The technologies important to improving quality and throughput in unit production processes are experimentation and modeling of the physics and chemistry of production processes that can be used in the control of production. The technologies important to intelligent control use manufacturing science–based knowledge in the real-time control of production and include "smart" sensors, "smart" actuators, and algorithmic techniques capable of bringing together expert knowledge, scientific models, and rapid learning from control experience. The technologies important to communications for integrating the business enterprises are communications standards for transmission of design, manufacturing, and service information; data base organization and query technologies; demand-pull techniques in scheduling; and real-time and virtual-reality techniques in person and group communications.

Quickly developing a new product and redesigning an existing product are important responses to competitor's initiatives and market opportunities. A manufacturing science base allows the new product to be produced to high quality and at low cost. Intelligent control in operating systems allows the exploitation of observation, analysis, and information in the operation of manufacturing processes and organizational procedures. Communications that integrate the whole business enterprise from sales through production to research facilitate an organizational response as a whole to opportunities and threats in a rapidly paced competitive environment.

ILLUSTRATION
Ford Taurus Project in the 1980s

In 1979–1980, the U.S. automobile industry was in crisis, with cars that were too big and fuel inefficient. In addition, the industry was plagued with low productivity, high costs, and long product-development times. In 1980, Ford Motor Company began trying the formation of self-organizing, multifunctional teams for the development of the new Taurus/Sable models. At the time, the company called the approach a simultaneous approach to product development, or concurrent engineering.

Lew Veraldi (1988, p. 1), who was the leader of the Ford Taurus product-development team, described that project:

> We're very honored by all the attention about Team Taurus and by what it says about Ford Motor Company. But you must stop and ask yourself, why should a large company like Ford, which has been developing new cars for over eighty years, decide to change the way it does business?...the need to change was "survival." That gets everyone's attention.

In the economic turbulence of the 1970s resulting from the rapid rise in oil prices, Ford raised product design to the attention of senior management. However, this only added another 6 months to an already 5- to 6-year product-development cycle, and the Japanese firms could design and introduce a new car in less than 4 years. Veraldi (1988, p. 1) described Ford's condition at that time: "Remember when Taurus began, it was 1979–1980. Ford's image for quality was not very good. In addition, we were in the process of losing over $1 billion for two years in a row. That's a record that I believe still stands."

Ford's product-development process was essentially linear, as was then usual in the U.S. automobile industry, beginning with (1) concept generation, (2) product planning, (3) product engineering, (4) process engineering, and finally (5) full-scale production (Clark and Fujimoto, 1988). Veraldi (1988, p. 1) summarized the linear product-development process then in place at Ford as follows: "...the old organizational structure [was] where designers do their thing—Design then turns that over to the Engineers. After the engineers do their job, Manufacturing is told to go mass produce the product, and Marketing is then told to go sell it....What you have is each group of specialists operating in isolation of one another."

The problem with this linear style was the number of design changes (and redesigns) it encouraged and consequently the long time it took to complete the development with all those changes. "Moreover, what someone designs and styles may be quite another matter to Engineering. And by the time it reaches Manufacturing, there may be some practical problems inherent in the design that make manufacturing a nightmare....Marketing may well discover two or three reasons why the consumer doesn't like the product, and it is too late to make any changes" (Veraldi, 1988, p. 1).

Changes in Ford's management philosophy began only after Henry Ford II retired in 1980 (Quinn and Paquette, 1988, pp. 1–2):

> Ford, founded in 1903 by Henry Ford, was one of the very few large U.S. corporations where the top management position was traditionally held by a descendant of the founder....At Ford, the bottom-line was important, and the company culture was not people-oriented....Ford's style, as in many other companies, remained authoritarian....As one former engineer put it, "When I was with Ford Manufacturing building the 1970s Mustangs, we didn't see the car we were going to make until eight or nine months before production was to start. And designers didn't want our ideas, either!"

After Henry Ford II went through three presidents, Donald Peterson became president and Phillip Caldwell became chairman of the board. Earlier in the 1970s, Caldwell had been head of international operations (1973) and was assigned the task of developing the Ford Fiesta in Germany. The cost of development of the Fiesta came to $840 million and was then the most expensive car-development project at Ford. Caldwell had chosen a design engineer to lead the project, and he was Lew Veraldi.

Veraldi had worked 25 years in Ford's design systems. He had experienced the frustration of taking a design to manufacturing to be told that it was not designed right to be manufactured. In the Fiesta project, he called in the manufacturing people at the beginning of the design and asked them: "Before we put this design on paper, how do you, the manufacturing and assembly people, want us to proceed to make your job easier?" (Quinn and Paquette, 1988, p. 3).

Veraldi had tried concurrency between design and manufacturing in the Fiesta project and had liked the benefits of the experience. He had understood the importance of early and close cooperation between product design and manufacturing and saw that he could make that cooperation happen. Veraldi was then convinced of the real advantages to designing a car simultaneously between product engineering and process engineering and marketing—rather than in sequential isolation. Such an approach avoided many late changes or desirable changes that were identified too late to be made—and it saved money in the product development of the Fiesta car.

Later, in 1979, the new senior management under Caldwell and Petersen decided to replace Ford's midsized cars with innovative new products. The Taurus project was to be the first of these models, originally conceived as a five-passenger car with a four-cylinder engine. This was a high-risk development project, and essential to its success would be a change in management philosophy (Veraldi, 1988, p. 1): "...the second ingredient on what it took to change. Commitment of upper management. The courage, I believe, of Mr. Philip Caldwell then Chairman and Mr. Don Petersen then President to take a risk—in many ways, to bet the Company—to spend $3 billion when we were losing $1 billion per year."

In 1979, Caldwell asked Veraldi to develop the initial designs for the Taurus/Sable cars. In the summer of 1980, Veraldi and his group presented their concepts for the new car to top management. After Veraldi's previous experience with concurrent engineering in the Fiesta project, senior management agreed to create a management team (later to be called "Team Taurus") to try out concurrent engineering ideas on the Taurus development project.

Concurrent Engineering

Let us take a break from this illustration to review the concept of *concurrent engineering* (or *simultaneous engineering,* as it was called at Ford). A concurrent engineering product-development approach requires the formation of multifunctional teams for managing complete development of a new product. The product-development project management team should consist of representatives from the different relevant activities, such as engineering product design, engineering process design, manufacturing, marketing and sales, finance, and research. The job of this product-development program management group is to formulate the design requirements and specifications while taking into account manufacturing and marketing considerations as well as finance and research considerations as early in the process as possible.

The product-development program management group first conducts a competitive benchmarking analysis of competing products in all price categories and establishes a list of "best of breed" performance criteria and features for the product. They next establish a product-development schedule and early prototyping goals and means. Then they identify early sources of supplies and draw on their expertise and suggestions about part and subassembly design. The group also solicits suggestions from dealers, service firms that repair and maintain the product, and insurers for product improvement and feature desirability. And the group also assembles representative customers and solicits and analyzes their reactions to current and competing products and possible prototypes of the new product.

After this, the product-development program management group compiles a "want" list from all this input, both internal and external, of desirable product performance, features, and configuration and sub-classifies this list into priorities of (1) must have, (2) very desirable, and (3) desirable if possible within costs. Finally, the group continues to manage the product-development process by encouraging team work and cooperation in developing a product rapidly and with highest attainable quality and lowest cost.

Concurrent engineering practices can facilitate the implementation of new technology in products by improving communication and cooperation among research, product design, manufacturing, marketing / sales, and finance.

In 1991, Boston University had a conference with manufacturers to summarize lessons that were being learned in speeding products to market (based on a project conducted on the subject by Stephen R. Rosenthal and Artemis March, 1991). The conferees particularly emphasized how important the early phases of product development were to the ultimate cost and quality of the product.

The conferees also discussed lessons about collaboration in concurrent engineering:

- Selection of the team is a complex problem that is critical to eventual project success.

- Senior management should concentrate its involvement in the early stages on defining the product concept, but senior management should resist calling all the plays, eliminating choices for the team.

- Attention must be paid to overcoming the barriers to communication in the "separate thought worlds and meanings" that the different cultures of specialists bring with them (as discussed in Chap. 1 in terms of the intellectual differences between technical and business-functional personnel).

- Special attention also must be paid to the difficulties in creating cross-functional team collaboration in organizational settings that are multisite or organized by functional specialization and that have older, long-established traditions of sequential responsibility in product development.

- After learning to collaborate effectively with each other, the team members must still build credibility, cooperation, and support from their home departments.

The conferees also discussed lessons learned about balancing performance, quality, and costs in concurrent engineering projects:

- The targets for product success should not be confused with means, so the product specifications should be established early, but the design team should be allowed to design the product.

- Initially, the importance of using concurrent program management teams is to allow all product targets to be considered simultaneously for proper tradeoffs and balance in setting product specifications.

- It is important for the product technology requirements to be explicit and clear very early so that representatives of the downstream activities (such as manufacturing and marketing) can decide whether the requirements are feasible and achievable within target cost constraints and the product can be produced in existing facilities.

- Because of the risks inherent in new technology, contingency plans are necessary in case the new technology proves to be still immature, with early product prototyping and early testing to determine whether it is immature.

- Attention should be paid to thoroughly testing new production processes and training workers before volume production begins.

- Cost-reduction targets should be set for a mature product initially (rather than merely an untargeted cost-reduction type of program).

- It is important to integrate vendors into the concurrent engineering teams.

ILLUSTRATION (Continued)
Ford's "Team Taurus"

The Team Taurus group consisted of a program management group, headed by Lew Veraldi and consisting of key players, including John Risk (car product development planing director), A. L. Guthrie (chief engineer), John Telnack (chief designer), and Philip Benton (later president of Ford automotive group). Veraldi (1988, p. 2) stated its purpose:

> Team Taurus's organization...[was] designed to promote continuous interaction between Design, Engineering, Manufacturing and Marketing along with top management, the legal, purchasing and service organizations. Overall coordination or direction is by a Car Product Development group whose primary function is getting all of the team to work together.

The importance of organizing this kind of initially inclusive participation was to bring downstream judgments into the design process early. The car program management group set the initial management goals for the project (Quinn and Paquette, 1988):

- "Best-in-class" performance for the car

- First prototype to contain 100 percent of prototype parts to completely test the manufacturing prototype

- Improving the focus and timing of key decisions for rapid product development

- Reducing the complexity of the product to improve quality

- Eliminating late and avoidable design changes by bringing in manufacturing considerations early in the design

- Controlling design changes to as few as possible or late in the program to avoid extending the product-development time

- Creating a small business–like working situation for the project to improve commitment

To target the goal of "best-in-class," the team began with competitive benchmarking of competing products (Veraldi, 1988, p. 5):

> ...the team sought out the best vehicles in the world and evaluated more than 400 characteristics on each of them to identity those vehicles that were the best in the world for particular items. These items ranged from door closing efforts, to the feel of the heater control, to the underhood appearance. The cars identified included BMW, Mercedes, Toyota Cressida and Audi 5000. Once completed, the task of the Taurus Team was to implement design and/or processes that met or exceeded those "Best Objectives."

The features on this group of cars provided standards for the features of the Taurus, and Veraldi saw success in meeting many of these standards. Another of the first things the group did was to create a "want" list from all relevant constituencies, including (internally) Ford designers, body and assembly engineers, line workers, marketing managers and dealers, and legal and safety experts and (externally) parts suppliers, insurance companies, independent service people, ergonomics experts, and consumers.

As an example of this process, Veraldi (1988, p. 2) described the Taurus Team's visiting manufacturing personnel:

> To ensure that the best ideas of the people who would be designing and building the car would be considered, our engineers developed a comprehensive "WANT" system. We visited the Atlanta Assembly Plant and spoke to the hourly and salaried personnel who would eventually build the car....For example, the assembly workers told us that to achieve consistent door openings and tight door fits, a one-piece bodyside was necessary. So we reduced the number of components on the bodyside from six components to two. The doors are also one piece for improved quality and consistent build.

The number of components within which a designer can design a part is often variable. Here information from manufacturing experience made the designers aware of the importance of reducing the number of components (a topic that will be reviewed in a later chapter).

The "want" list built up to 1400 items, of which the team concentrated on 500. Veraldi included Ford's dealers and suppliers to the simultaneous process. The Taurus simultaneous engineering project also included early market research (Veraldi, 1988, pp. 3–4):

> But what about the customer? These would be the most consumer researched vehicles in Ford's history....The market research provided guidance but never a definitive answer because we were asking people about something that would have to be new, fresh and desirable in five years. We had to interpret the feedback and attempt to eliminate this bias. In the end, we decided to take the risk of being too advanced....The station wagon was a real dilemma [for example]....Our original styling theme was determined to be too drastic...so we decided to back-off a bit and settled on a new aero-wagon that was a little more tempered in styling.

Best-of-Breed Design Features

Let us pause again in the illustration to review the concepts of competitive benchmarking and best-of-breed features in product development. A product is a system embedding not only the principal transformational function of a generic technology system but also other technology systems as product subsystems and features—the product system. Product designs are always compromises between performance and cost. From a competitive perspective, however, it is foolish to design a product system that has any feature that is distinctly inferior to any feature of a competing product in the same price class. Moreover, there is a competitive advantage to providing, in a lower-price-class product, the quality or features of product systems in higher price classes.

The systematic identification and listing of product-system features and their highest-quality expression in a product system of any price class is called competitive benchmarking.

Competitive benchmarking allows a designer to see the best features in a current product line but does not prevent surprises that competitors may have in store in their next model. In performing a competitive benchmarking effort, the design team also should use *technology forecasting* to guess what competitors might be able to add in the next model.

For competitive purposes, those features whose quality are most noticeable by the customer are those which should have priority in designing best-of-breed product features.

*The design into a product system (of a given price class) the high-
est quality of features of all product-system price classes is called
a* best-of-breed product design.

Of course, given cost constraints in a lower-cost product system, one
can seldom attain all the best-of-breed features. However, here is a
good place to use technological innovation in product design.

*Technological innovation which allows best-of-breed designs in
lower price classes adds product differentiation competitiveness
to price competitiveness in a product.*

ILLUSTRATION (*Concluded*)
Ford's "Team Taurus"

One of the key changes in design specifications that came from Team
Taurus's concurrent engineering approach was a decision in April of
1981 to refocus the Taurus as a six-passenger, six-cylinder family car
(after gasoline prices then began to stabilize). Quinn and Paquette
(1988, p. 14) describe this decision: "Much of Ford's future was riding
on the success of the $3 billion Taurus/Sable program, carrying Ford's
aerodynamic styling into the crucial family car market. Ford aimed the
new cars at families in their late 30s, whose income was rising, and
who wanted to drive something more sophisticated than a traditional
American car."

Another goal of the Taurus project was very important—creating a
complete prototype very early: "The final consumer research before introduction was a ride-and-drive research conducted in Florida. This,
perhaps, was the most important and successful research of the entire
program, and it was only possible because we had prototype cars
closely resembling our intended production vehicles seven months earlier than in past programs. These prototypes were taken to Florida,
where a group of potential customers were pre-selected based on demographics and were asked to evaluate these prototype vehicles....the
results of this research were invaluable" (Veraldi, 1988, p. 4).

There had been uncertainty about the style and design balances
adopted in the design. Testing the prototype provided both confidence
and some chance to make changes before mass production began: "It
confirmed major decisions like the acceptance of the European favored
ride and handling characteristics to things like appreciation of the gas-
cylinder hood-assist and the easy-to-service engine compartment. It
wasn't until these prototype vehicles were shown alongside competitive
production cars that we began to get the positive reactions from people

that told us we were on our way to offering people a car with major potential for success" (Veraldi, 1988, p. 4).

These product-development techniques brought the Taurus project in on time and below budget. Introduction of the Taurus car by the Ford Division (and its variation Sable by the Mercury Division) cost $250 million less than the design budgets. Both products were successful in the marketplace. In 1988, Lew Veraldi was promoted to a vice president of Ford, in charge of luxury and large car engineering and planning.

In this illustration we examined three important concepts to improve the speed of the product-development cycle: concurrent engineering, competitive benchmarking and best-of-breed designs, and rapid product prototyping.

Management of the Product-Development Cycle

Management of the product-development cycle is critical to being a fast-cycle company. For example, in 1991, a committee of the U.S. National Research Council reported on the importance of managing the product-development process, particularly with new computer-aided design capabilities. The Committee on Engineering Design Theory and Methodology reviewed modern computer-aided design capabilities for product development and argued that attention also should be paid to how these aides were used in engineering management of product development: "Members of the committee visited several U.S. firms that use engineering design as a way to achieve competitive advantage. Information obtained from these visits, together with the collective experience of the committee, suggests that designing for competitive advantage requires much more than the adoption and use of new [computerized] design practices" (NRC, 1991a, p. 1).

The committee emphasized that in addition to the modern technologies of computer-aided design and manufacturing, management of the production introduction process also required (1) continuous and incremental improvements in manufacturing and (2) appropriate project management techniques and design practices, as well as a design setting tailored to the technical nature of product lines.

In addition to reports such as the preceding, other studies have emphasized the importance of proper management of the product-development process. Robert Cooper studied a sample of 203 new-product projects in 125 firms. He interviewed senior managers, who identified a typical success and failure product in each firm: "Success and failure were defined from a financial standpoint. For each project, the project leader and some team members were personally interviewed....The final sample consisted of 125 successes and 80 failures" (Cooper, 1990, p. 28).

Cooper identified several factors that financially successful product-development projects had in common:

1. Superior products that delivered unique benefits to the user were more often commercially successful than "me too" products.
2. Well-defined product specifications allowed a clearly focused product-development effort.
3. The quality of execution of the technical activities in development, testing, and pilot production affected the quality of the product.
4. Technological synergy between the firm's technical and production capabilities contributed to successful projects.
5. Marketing synergy between technical personnel and the firm's sales force facilitated the development of successful products.
6. The quality of execution of marketing activities also was important to product success.
7. Products that were targeted for attractive markets in terms of inherent profitability added to success.

Note that the first and last of these factors have to do with the relationship between product design and the market—a superior product aimed at a financially attractive market provides both competitiveness and profitability. Factors 5 and 6 have to do with marketing activity. Sales efforts are helped by products that fit well into an existing distribution system and by properly managed marketing activities that test and adapt a product to a market. Factors 2, 3, and 4 have to do with the quality of the technical development process. Good management of the technical activities in product development help produce good products. And good project management includes having a clearly focused product definition, managing well the different phases of the product-development process, and having the proper technical skills to execute the project.

The success of a new product requires not only a good product design but also good management of both the product-development and marketing processes. The product-development cycle should be managed to provide innovation in product performance, innovation in productivity, responsiveness to market changes, and competitive advantages.

ILLUSTRATION
Information Flow in a Product-Development Cycle in a Computer Firm

As an example of the type of information flows in a product-development process, Dundar Kocaoglu, M. Guven Iyigun, and Chuck Valceschini described the product-development process of a computer firm in 1990 (Kocaoglu et al., 1990). They summarized the flow of information within the firm when they began their study, and the process is illustrated in Fig. 3.2.

In the figure, we see that product ideas should begin with marketing and research interacting with development engineering—together jointly conceiving of an idea for a new product with a customer and defining system requirements for that new product. Next, a core team is assembled and interacts with development engineering to refine the product-system requirements into a set of product specifications. Then the core team is expanded to a product team that performs product development and any pilot production process changes needed to produce the product. After that, high-volume production can begin.

During their research on the product-development process within the computer firm, Kocaoglu and associates suggested improving the information flow with computerized design aides. First, they suggested that computerized project management tools could facilitate the planning, scheduling, and interactions between research, development engineering, and the core and product teams. Next, they suggested that computer-aided engineering tools could assist development engineering and the core and product teams in delineating the product specifications and designs. Then they suggested that the analysis of previous product designs could be codified to provide some company standards for best design and development practices, i.e., product-development standards. These would be useful in order to continue to improve management of the design and development process. Finally, Kocaoglu and

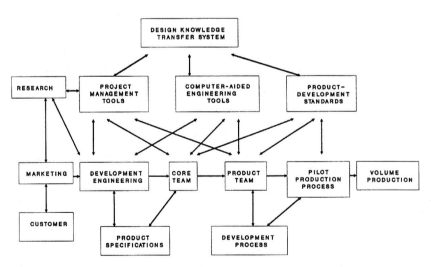

Figure 3.2 Computer aides in the information flow in a product-development cycle.

associates recommended that the whole set of computer-aided tools for design and development be formulated and maintained within a design knowledge transfer system that would coordinate the use of these tools and provide access to information from one tool to the next.

What we see in this illustration is that improvement of the service technology in a product-development process (through the use of computer-aided engineering and design software) also provides an opportunity for expanding the information flows and cooperation between research and product development and manufacturing and marketing—to improve the overall coordination and effectiveness of product development.

Innovation and Project Management

In implementing innovative new technology into products, project management techniques should address the risks of innovation. In preceding illustrations we saw the importance of team composition in the product design phase (which should include representation from all the functions). We also saw the importance of a clear definition of the product's technical requirements at the beginning of the project. However, the more innovative the technology to be designed into the product, the more design tradeoffs are available and the less certain are the initial technical specifications. How can one incorporate such considerations into traditional project management techniques?

Let us first review the elements of project management. Project management requires four phases, each of which involves specific steps:

1. Project planning:
   a. Specific objectives of the project
   b. Specific customers for the project
   c. Definite start and stop dates
   d. Staffing requirements
   e. Reporting lines
   f. Budget
2. Project staffing:
   a. Recruiting
   b. Organizing
   c. Developing the culture of the project team
   d. Coordinating liaisons and developing shared supervision of team members between the project and the functional homes of the team members
3. Project monitoring:
   a. Developing measures of technical progress in the project

   b. Developing a modular task structure for organizing the work
   c. Developing techniques to track technical progress (such as
       PERT charts, Gantt charts, etc.)
   d. Developing modes for problem solving and debugging during
       the project
   e. Developing a program for modular testing of the system
   f. Developing proper documentation for the project as the project
       progresses
4. Project completion:
   a. Prototyping the product of the project
   b. Demonstrating the prototype's performance
   c. Transferring the completed project to the project's customer
   d. Assisting the project's customer with the implementation of the
       next phase of the project or placing the product into use

   In addition, project success requires

1. Achieving technical performance

2. Achieving results within the window of opportunity

3. Achieving results within a budget that can produce economic benefit

4. Achieving results that satisfy the project customer

5. Granting effective rewards for successful performance to project
    personnel

   Within this general framework of project management, implementing new technology introduces the need for special recognition of and
management of technical risk. For example, Aaron Shenhar (1988,
p. 2) has emphasized identifying the degree of technical risk in the
planning phase of the project: "The technology used in projects is one of
the main aspects which should receive special attention, since there are
great differences among projects. Some projects use well-established
technologies, while others employ new and sometimes even immature
ones, and involve enormous uncertainties and risks."
   Shenhar suggested classifying a development project by the degree
of risk in the technology being implemented:

1. Low-tech projects

2. Medium-tech projects

3. High-tech projects

4. Super high-tech projects

A *low-tech project* is one in which no new technology is used. A *medium-
tech project* is one in which some new technology is used within an ex-

isting system but involves a relatively minor change in the whole technology of the system. A *high-tech project* is one that uses only key technologies as components but in which the integration of these technologies is a first-time achievement. A *super high-tech project* is one in which new key technologies must be developed and proved along with their integration into a new first-time system.

> *For projects that differ according to type of technology risk, project planning should differ in form to take into account the technology risk in project schedule and goals.*

For low-tech projects, planning should use the experience of prior projects of the same type for estimating scheduling, budgeting, and staffing and for arranging for completion and transfer. Here a PERT chart is very useful, wherein past experience can be used to estimate probabilities of risk within the chart.

For a medium-tech project, it is first important to determine how much competitive advantage the new technology provides in the project: nonessential or essential. If the new technology is nonessential, then a substituting technology should be within the scope of the design and ready if substantial problems develop with the new technology. If the new technology is essential, then it is important to plan and schedule the development and testing of the new technology before system integration is far advanced. System development should not be started until confidence is demonstrated that the new technology will perform properly and on time for the whole project schedule. Here probabilities in a PERT chart should only be used for established technologies, not for the new technology.

For a high-tech project, it is important to plan and schedule system integration in a modular, subsystem fashion so that subsystems can be tested thoroughly and debugged before final system integration and assembly occur.

For a super high-tech project, planning should be in the form of a series of "go/no-go" decisions. Each stage of the project development should consist first of developing and testing a key technology before the full project system integration program is launched. Whenever a key technology fails, a no-go decision should be made to keep the full-integration program on hold until either the key technology can be made to work or an adequate substituting technology can be implemented.

> *For low-, medium-, and high-tech projects, development can be scheduled for fixed dates (with appropriate regard to project design for testing in a modular fashion a new-technology component or subassembly), but for a super high-tech project, full*

*system development should not be scheduled until all key tech-
nologies have been successfully proven or substituted.*

New Technology and
Product-Development Plans

Long-term product planning is necessary if technology strategy is to be
included in the product-development cycle. For example, Steven
Wheelwright and W. Earl Sasser (1989) have emphasized the importance of long-term product planning—mapping out the evolution of
products in a product line. They pointed out that products often evolve
from a core product, e.g., when a model is enhanced for a higher-priced
line or stripped of some features for a price-reduced line. They also
argue that one should plan the evolution of a core product in generations of products. The core-product system will express the generic
technology system, and higher- or lower-priced versions will differ in
the subsidiary technologies of features. Next-generation products differ in dramatically improved performance or new features, improved
technologies, or reduced cost.

Dramatically improved performance or lower cost will technically obsolete previous generations of the product. If an added or new-technology feature opens new applications for the product, then the product
also can make obsolete a previous generation (as the electric starter did
in automobiles).

Wheelwright and Sasser called the diagramatic planning of product
lines as branched products and generations of products a *new product-development map.* Such a map is particularly useful for the long-term product-development process, when the anticipated technical goals for product
performance and features are listed in the product boxes. It is also helpful
to summarize needed technological developments against new product
ideas to identify where concurrent engineering cooperation between research and product design and manufacturing should be planned.

ILLUSTRATION
Motorola's Product/Technology
Roadmaps

In the 1980s, Motorola used a combination of techniques to combine
product planning with technology planning for formulating a technology strategy. Motorola called the outcome of the process a *technology
roadmap* for their long-term product development (Willyard and
McClees, 1987, p. 13):

> Motorola is a technology-based company and has, for each of its many
> businesses, a strategic plan which is often based on the anticipated ad-

vancements that will be made in certain technologies. Our products involve the application of scientific principles, directed by our people, to solve the problems of our customers, to improve their productivity, or to let them do things they could not do otherwise.

Motorola's systematic approach to technology and product planning arose from the growing technological complexity of their products and processes: "Because our products and processes were becoming much more complex over the years, we realized there was the danger that we could neglect some important element of technology. This potential danger gave rise to corporate-wide processes we call the 'Technology Roadmaps'" (Willyard and McClees, 1987, p. 13).

The formal outputs of the process were documents that recorded the information generated in the process: "The purpose of these documents is to encourage our business managers to give proper attention to their technological future, as well as to provide them a vehicle with which to organize their forecasting process" (Willyard and McClees, 1987, p. 13).

Motorola's technology roadmaps facilitated the orientation of and internal communication among managers in engineering, manufacturing, and marketing. Together they could track the evolution and direction of their technology/product/market planning. Their purposes in the roadmap process were to (1) forecast the progress of technology, (2) obtain an objective evaluation of Motorola's technology capabilities, and (3) compare Motorola's capabilities with those of competitors, currently and in the future.

Each technology roadmap document was composed of several sections (Willyard and McClees, 1987):

1. Description of Business
2. Technology Forecast
3. Technology Roadmap Matrix
4. Quality
5. Allocation of Resources
6. Patent Portfolio
7. Product Descriptions, Status Reports, and Summary Charts
8. Minority Report

The "Description of Business" section included the following:

- *Business mission.* A statement providing an answer to the question: "What is expected for the business to achieve greater success?"
- *Strategies.* This part summarized the timing and application of re-

sources to a product line to generate a competence distinctive from that of the competition.

- *Market share.* This part summarized the present shares of world-wide markets by Motorola and major competitors as well as trends of change.

- *Sales history and forecast.* This part displayed forecasts based on strategic assumptions.

- *Product life-cycle curves.* These expressed the anticipated length of time from conception of a product through development to market introduction, growth, decline, and withdrawal from the market. (Motorola used the product life-cycle histories from past products upon which to base educated guesses about current or planned products.)

- *Product plan.* This part summarized the milestones for product development, introduction, and redesign.

- *Production-experience curve.* This projected the planned rate of decline of production costs over the volumes and lifetime of the product.

- *Competition.* This part emphasized understanding competitors' current strengths and weaknesses and the directions of their technological development. (Motorola summarized this information in a competitor/technology matrix.)

For technology forecasts, Motorola used the technology S-curve format for summarizing projected technical characteristics both of products and key product components (such as IC chip technologies). Motorola summarized their product plan as sets of milestones and product characteristics.

In a 1989 interview, George Fisher, then CEO of Motorola, discussed Motorola's technology planning process (Avishai and Taylor, 1989, p. 110):

> We are such a major force in the radio communications business that we virtually define the standard. The basics of the business are not all that complex, although the technical details are quite demanding. In portable equipment....there are four critical forces: size, weight, current drain (driving function of the size and weight of the battery), and cost. So in our formal technology planning, for particular products, we develop 5-year and 10-year technology roadmaps of those forces.

What Motorola was projecting were the key technical performance parameters underlying their product line of portable radio communications. Because the technologies were being planned, the opportunities for new products to embody the technical changes also could be planned: "Our ability to introduce a portable telephone that's the size of a wallet and weighs less than 11 ounces—that was driven by the technologists in our cellular and semiconductor groups and the people in corporate research" (Avishai and Taylor, 1989, p. 110).

The point of Motorola's formal product-development process was to ensure that technology planning and product planning were closely coupled (Avishai and Taylor, 1989, p. 110):

> Most of us who have been brought up in the portable equipment business—serving the needs of people and machines on the move—have developed pretty refined instincts as to what the next steps have to be technologically. When we go to customers with a new product, we don't expect many surprises....This is a business we know very, very well.

This last comment underlies the central point of innovating technology in new products—one does not want to be badly surprised when the new product is introduced to a customer. Bad surprises are of two kinds: (1) a product that does not focus on a customer need or (2) a product that focuses on a customer need too early.

For the first problem, Fisher emphasized the importance of having Motorola's technologists visit customers frequently (Avishai and Taylor, 1989, p. 10):

> ...we've established a massive program of increasing customer visits at all levels of the organization. We want everyone in Motorola, from top to bottom, to get out and see customers—to talk with them directly and understand their business better....At our last policy committee meeting, we had an updated report on customer visits. That report showed that our technologists were still not visiting as many customers as we like....So we've been trying to figure out how to promote more effective interaction between our technologists and our customers.

However, even talking to customers still requires an interpretation of what the customer really would want if new technology made it possible: "...customers will always pull you in directions of interest to them. As a technology leader, sometimes we have to show people what we can do. The portable telephone is a good example. Motorola jumped way out in front on the car phone....We didn't have thousands of people saying, 'We want portable telephones that look a certain way and have particular kinds of features.' We presented it to the world" (Avishai and Taylor, 1989, p. 109).

Yet the problem is also not to get too far out in front of the customer (Avishai and Taylor, 1989, p. 109):

> Many technologists have this problem. They look at a technology, get very excited and say it's not a question of "whether," it's a question of "when." But

"when" turns out to be the big question. In the early 1970s, when I was still at AT&T Bell Labs, we had a grand vision of picture-phones....some of the technologists at Bell Labs really got caught up with it. They thought, "This is where the world is going!" Well, the world is going that way, no doubt about it. But we were much too early, even technically. The picture-phone needed a high-performance switching and transmission system that just didn't exist.

As the 1990s began, Motorola was one of the American firms that was performing well in global competition. For example, one popular business journal reported at that time on Motorola's product successes (Therrien, 1989, p. 108):

> The wins are beginning to stack up. In just the past six months, Motorola knocked the wind out of...telecommunication rivals with two new products, both marvels of miniaturization. First came the MicroTac cellular phone, a Star Trekky unit that slips into a coat pocket and flips open for use. It is far sleeker and a third lighter than the next closest portable phone....This summer, the company followed that act with the first wristwatch pager. This ultrasmall product, which resembles the wrist-radio familiar to Dick Tracy fans, has also garnered accolades.

The new capability of Motorola to manage all three aspects of innovation in the product-development cycle—technology leadership implemented in advanced products, product quality designed into the products, and rapid and correct responsiveness to market needs—was seen as important and new (Therrien, 1989, p. 108):

> Compared with Motorola's fortunes of just a few years ago, this is a radical reversal....In a few short years, Motorola [had] slid from being No. 2 in worldwide chip sales to tagging along behind NEC, Toshiba and archrival Texas Instruments....The payoff [now] has been startling. In semiconductors, Motorola is now king of the U.S. hill, No. 4 in the world....And in Southeast Asia, the world's fastest-growing chip market, Motorola has become the No. 3 supplier. Whether judged by U.S. or by Japanese standards, Motorola builds an usually broad line of chips.

Procedure: Product/Technology Roadmap

A technology roadmap for a division is a set of documents summarizing the product-development and manufacturing-development plans of the division. Such documents include

1. Technology and market forecasts
2. Product and competitor benchmarks and manufacturing competitive benchmarks
3. Product life-cycle estimates
4. Divisional profitability gap analysis

5. Product-development maps
6. New product project plans
7. Manufacturing improvement project plans
8. Divisional marketing plans
9. Divisional business plans

The technology roadmap expresses the technology planning at the divisional level. However, it also should fit within the larger planning process of the diversified corporation, and it is to technology planning in a diversified corporation that we attend in the next chapter.

ILLUSTRATION
R&D Management Practices
in Japanese Firms

William Finan and Jeffrey Frey (1989, p. 1) studied the research and development (R&D) practices of several Japanese electronics firms at the end of the 1980s: "In the microelectronics engineering community in the United States, there is a consensus that Japan's semiconductor industry has moved to the forefront in a number of key semiconductor-related technologies. These technical advancements have, in turn, been effectively translated into commercial advantage by the leading Japanese microelectronics companies."

In the 1980s the Japanese firms had taken the world market away from American firms in the semiconductor chip memory market of DRAMs. This had occurred not so much from superior product design but from superior production innovations—getting new products into production sooner, with higher quality and lower price, than had the competing American firms: "Spurring their advances in technology seems to be a facility for integrating the activities of research, development, and commercialization...also of transferring technology to production" (Finan and Frey, 1989, p. 1).

Finan and Frey focused on comparing Japanese R&D practices with American practices. One of the significant differences was that the Japanese firms used their corporate research laboratories for training new engineers. In Japan, corporate research laboratories are often called *advanced development labs* (*ADL*): "Some firms use the ADL as the entry point for new engineers joining the firm....a new recruit might spend up to three years...at the ADL learning about the company's technologies and general business before...[moving] on into an operating division" (Finan and Frey, 1989, p. 54).

There are several advantages to using the corporate research laboratory for training. With this approach, a new engineer becomes familiar not only with the current technologies of a company but also with the future technologies. Also, a new engineer becomes familiar with the research function, understanding how to use it and to call on it for help in carrying out his or her subsequent duties in the operating divisions. Personal contacts made at the research laboratory provide the young engineer with access to research people for later cooperation.

Few American firms at that time used corporate research for a training function. Consequently, one of the most common problems in American R&D practice was the poor (or even almost total lack of) relationship between corporate research and the operating divisions. The isolation of U.S. firms' corporate laboratories was a frequently cited problem in technology-transfer difficulties in American industry in the 1980s (see, for example, Rubenstein, 1989). Japanese firms did not have a problem with the isolation of research from operating divisions. For example, one respondent in the study said: "In Japan there is no preference between R&D and the operating divisions. Roughly 40 people transfer [each year] from R&D to the operating division and 29 people will transfer back—not necessarily [ones who worked] in the labs before" (Finan and Frey, 1989, p. 88).

Another significant difference in R&D management practices was the relatively longer patience of Japanese management for research projects: "Where one begins to see a real difference in management practice is in the apparent willingness on the part of Japanese management to allow research projects to have sufficiently long lives, with greater stability in resources supplied over the duration of the project to allow time for all of the technical parameters to be thoroughly evaluated" (Finan and Frey, 1989, p. 74).

One of the sources of this patience was a better integration of research into business planning: "The ADL may also project a broad range of technologies that might be of importance to the firm 10 to 15 years ahead. One Japanese semiconductor firm uses what they call the 'core' technology approach" (Finan and Frey, 1989, p. 55).

The concept of core technology was discussed in the previous chapter. From the operating divisions, these core technologies are incorporated into product planning through the division's technology roadmap: "Technology Roadmap Preparation....firms build this type of strategic view of technology into their annual laboratory budget/planning cycle. The result is that the ADL becomes almost like a central library—a central repository—for information cutting across all the basic technologies the firm expects will be important to supporting its overall business objectives" (Finan and Frey, 1989, p. 55).

This strategic view on technologies provides the operating divisional managers with the ability to anticipate problems: "Japanese managers, at all levels, recognize the need to start projects designed to address technical bottlenecks before they become near-term problems" (Finan and Frey, 1989, p. 74).

A third significant difference was the relatively greater Japanese emphasis on manufacturing innovation as part of the product-development cycle: "There is a general, continuing net flow of personnel—especially younger engineers—out of the labs into manufacturing" (Finan and Frey, 1989, p. 87).

In the United States, manufacturing personnel had been relatively separate from research and product-design personnel. In Japan, greater value and prestige were placed on manufacturing and on providing manufacturing with engineering personnel who had experience in research and product development.

The Japanese also used teams for transferring process research from the laboratory into manufacturing: "Transfer of major process advances is accomplished by establishing teams within the labs and moving the team through the stages of development into manufacturing operations" (Finan and Frey, 1989, p. 88).

Research and the
Product-Development Process

In managing innovation in the product-development process, operating personnel should be familiar with and know how to use the research function:

- Product development should plan long term for its products and production processes, with a view to using research for competitive advantage.

- Process research should be valued as highly as product research and should be a part of both corporate research and divisional long-range plans.

- Divisional technology roadmaps are useful tools for integrating long-term research and product development with business planning.

- The corporate research laboratory should be the first assignment for new engineers in the firm—to begin becoming familiar with the company's future technologies and with the research function in the product-development process.

- Movement of personnel between research and operating divisions (both as permanent and temporary moves) is a useful means of facilitating communication and cooperation in the product-development process between research and operating divisions.

- The use of teams to move product and process research from corporate research into operating divisions is a useful procedure for technology transfer in the product-development process.

Practical Implications

The lessons to be learned from this chapter are about the desired features of the product-development process that can properly integrate innovation into new products.

1. To economically exploit innovation within the design of new products and services, both a fast-response organization and a proper product-development process are required.

   - Fast-response organizations treat time as a competitive advantage.
   - The ways to compress time in product development are to perform activities in parallel, do it right the first time, do it in virtual time, and implement process innovations in an incremental and continuous improvement mode.

2. The early phases of the product-development process are critical in terms of cost and quality of the product design.

   - A concurrent engineering product-development team can speed the product-development process and improve the quality of design.
   - Senior management should concentrate on and limit their involvement in the early stages to defining the product concept.
   - Service technologies used in product design, such as computer-aided design and engineering, can assist concurrent engineering.
   - Rapid prototyping and user experimentation with prototypes are another important feature in a fast and correct product-development process.
   - Competitive benchmarking should include technology forecasting in the product-development process.

3. Managing innovation within the product-design process requires distinguishing the type of technology risk and planning the project schedule differently to deal with the different types of technical risks.

   - In managing innovation in the product-development process, engineering personnel should be familiar with and know how to use the research function.

4. Creating technology roadmaps for product planning is a useful procedure for incorporating technology strategy into product planning.

- The proper use of the corporate research laboratory and its integration in the product development process is critical to long-term product success.

For Further Reflection

1. Identify two industries that have existed for at least 70 years, and list their product types over time.

2. What has been the major source of competitive change in each industry, product types or production processes?

3. In reviewing the histories of the two industries, could one have anticipated at a given period of time the major technological changes in each industry? If so, how would that have been possible?

4. Identify several companies in each industry, long ago and currently existing. Which long-ago companies still exist, and which do not. How did the successful companies that made the changes plan their change?

Planning Technology

Once the core competencies of a corporation have been identified at the corporate level, corporate research can then plan technological progress for maintaining and advancing a specific core competency. This chapter addresses technology planning in a diversified firm.

Planning procedures are important in technology because the key to the commercial success of technological innovation depends on planning technology within the larger context of a good business plan. The competitiveness of a business enterprise arises from balanced and coordinated strength in all aspects of the enterprise. In radically new technologies, the firms that first planned and put together a properly balanced enterprise have had such a powerful advantage that they became leaders in their new industries.

ILLUSTRATION
Air Products Corporation

The origin of Air Products Corporation provides an illustration of technology planning within the context of a good business plan. Before World War II, oxygen had become an important gas for industrial use—principally for steel production and for welding. At that time, oxygen was produced at a central plant, bottled, and shipped by rail or truck to industrial customers. An employee of a firm supplying oxygen, Leonard Pool, thought of a better way. He conceived of designing a smaller and more reliable oxygen plant that could be located at the plant sites of major customers and operated by semiskilled labor.

Pool wanted to redesign the architecture of the technology system of oxygen supply to eliminate shipment of the product. Although the oxygen itself was light, the steel bottles required to contain the pressurized oxygen were heavy. The shipment cost of bottled oxygen was a major cost. "He [Pool] reasoned that if you eliminated the transportation costs associated with high pressure cylinders, or the cost of converting oxygen gas to liquid and then delivering it by road tanker, you could reduce the price of oxygen by more than 50 percent. Lower prices would then lead to new applications..." (Baker, 1986, p. 18).

Fifty percent reduction! That is a competitive advantage. Pool's employer for some reason was not impressed with the idea. Pool, however, was an entrepreneur, and he left to start his own company—Air Products. His business strategy for the new company was to center on his innovative technology strategy. At that time, oxygen plants were complicated and hazardous to operate. Three European firms held the technology. Pool recruited several bright engineers recently graduated from the university.

The company began just before World War II, and when the war started, the company won a design competition sponsored by the U.S. government to build packaged plants to supply oxygen for high-altitude flying. Dexter Baker of Air Products described the benefits of this contract: "Air Products didn't make much money during the war—but we did learn how to design and build reliable plants that could be operated by semi-skilled labor at remote locations all over the world" (Baker, 1986, p. 18).

After the war, with solid experience and a debugged technology, the small company devised a strategy to finance its commercial expansion. The company structured a supply contract in which its customers agreed to purchase all products produced by the on-site plant. Since the company's customers were large and financially strong, the company was allowed to use the signed contracts as collateral for bank loans to finance the construction of the on-site plants. Thus the company had a self-financing scheme for business expansion. (In contrast, many entrepreneurs lose control of their high-tech firms to outside investors during the business-expansion phase.) Air Products' technology/business strategy worked (Baker, 1986, p. 18):

> The advantages of on-site supply were so superior to hauled-in gas products that the steel, automotive, chemical and glass companies were prepared to sign our supply contracts....The lower-cost supply method of the on-site plants did open many new markets. The steel industry adopted oxygen blown processes on a large scale. So did the chemical and glass industries. Our plants became larger....

Air Products continued a coordinated business/technology strategy based on the technology system of on-site gas production. The company added more capacity to the plants to sell off-site cylinder or liquid oxygen products. This provided a product-expansion strategy. "...[This] allowed us to enter the merchant market on much more favorable terms than existing regional producers, whose plants were a fraction of the size of our tonnage/piggyback plants and whose energy costs were several times greater" (Baker, 1986, p. 18).

Elementary gas products (such as oxygen, nitrogen, hydrogen, helium, etc.) are not differentiable from different producers (except as to purity) and are thus commodity-type products. Hence a cost/quality leadership competitive strategy is essential. First, Air Products used a reconfiguration of the technology system (eliminating distribution by on-site plants) to gain a decisive cost advantage. Next, Air Products used a piggyback addition to the technology system to produce liquefied gas products—again with a competitive cost advantage. The lower-cost strategy was decisive and allowed Air Products to begin acquiring the competition. "We used this supply method to create a cost advantage which, in turn, enabled us to persuade some 14 regional gas producers to shut down their smaller plants and purchase their requirements from Air Products. Many of these regional companies also elected to merge with Air Products and thereby helped create a new nationwide industrial gas company" (Baker, 1986, p. 19).

Continuing its technology strategy of focusing on the technology system and products of the system, Air Products continued to expand business (Baker, 1986, p. 19):

> Our fourth major innovation was to create new markets for our industrial gas products. When Air Products entered the industrial gas business,...nitrogen gas—co-produced with oxygen by the liquefaction of air process but at some four times the volume—was traditionally vented back to the atmosphere....We reasoned that if we could develop a market for nitrogen which was equal in volume to oxygen, we could average down production costs by selling twice the volume of gas products from each cubic foot of processed atmospheric air.

Nitrogen was essentially free as a by-product, so why not sell it? Air Products still focused on the technology system of atmospheric gas production for the basis of their product diversification. The company established an R&D organization with the mission to develop new applications for nitrogen. This was so successful that currently the company sells three times as much liquid nitrogen as liquid oxygen.

Air Products first used a technology strategy for a commodity-type business, lowering production cost. Next, they used a technology strategy for expanding markets for commodity products through performing research to develop customer applications of their products.

First Movers in a New-Technology Industry

The Air Products illustration is an example of a firm that moved first in a new industry to exploit the new technology with the economic advantages of scale and scope. The term *first movers* in an industry has been used for those firms which aggressively plan technological advancement to exploit economies of scale and scope. For example, Alfred Chandler used the term in his study of the role of management in business history: "Those who first made these large investments—the companies I call 'first movers'—quickly dominated their industries and continued to do so for decades. Those who failed to make these investments rarely became competitive at home or in international markets; nor did the industries in which they operated" (Chandler, 1990, p. 132).

In moving first toward dominance in a new industry, it is necessary (1) to devise a business strategy that exploits a technology plan to gain economies of scale and scope and (2) to make the appropriate investments in the business system to gain these advantages before competitors. The determining factors of success for first movers are *completeness* in strategy and *appropriate size* in investment. Completeness in strategy requires implementing not only technical change but also appropriate development of production capacity, marketing capability, and managerial skills. The size of the business investments necessary is that which creates economies of scale and scope as competitive advantages.

The completeness in investments necessary for a firm to be a first mover are

- Investments in advancing technology
- Investments in large-scale production capacity
- Investments in national (and international) distribution capability
- Investments in developing management talent to grow the new industry

Competitive advantages arise from such a balance of skilled management improving and producing a quality product at a low cost that captures a dominant market share. In this way, a first mover benefits from lower costs while competitors are struggling with building plants of comparable size. The first movers have already begun to debug the new scaled-up production processes and are moving down their learning curve just as competitors are beginning to design and construct their plants. First movers also gain brand recognition and market share and establish distribution systems before their competitors. And the managers of first movers gain experience and refine their procedures while competitors are only hiring their staff and building their organizations.

Economies of scale can lower production costs when the technologies of production provide an improvement in efficiency with size. Not all technologies provide increased efficiency as a function of size. Generally, there will be an optimal scale of production for a particular class of products and production processes—dependent on several factors, such as the nature of the product, the nature of the transformation processes, the availability and costs of resources, the possibilities and benefits of automation, the costs of labor, the costs of transportation, etc.

Economies of scope can lower costs when production capacity can be expanded to more markets by producing a variety of products. When a variety of products expands the markets available to the production facility, the fixed costs of production and the investments in gaining market access can be spread over several products, thereby lowering the fixed and invested cost per unit of a given type of product.

In a new-technology industry, when a firm is the first to complete its business system to gain economic advantages of scale and scope, a first-mover competitive advantage is gained.

ILLUSTRATION
First Movers in the New European Chemical Industry of the Nineteenth Century

The innovation of chemically produced dyes provided the technical foundations for the origin of the modern chemical industry. Alfred Chandler (1990, p. 133) described how the German firms became first movers in the new chemical industry at the turn of the nineteenth century (even though originally the British firms had several advantages):

> An Englishman, William Perkin, invented the first man-made dyes in 1856, and in the 1860s and 1870s Britain had almost every comparative advantage in the new industry. Dyes are made from coal, and Britain had the largest supplies of high-quality coal in Europe. Its huge domestic textile industry constituted the world's largest market for the new dyes. All it lacked was experienced chemists, and British entrepreneurs had little trouble hiring trained German chemists for their factories. By an economic criterion, British entrepreneurs should soon have dominated the world in this new industry.

But this did not happen. "Instead German companies—Bayer, BASF, Hoechst, and three small companies—took the lead. Why? Because they made the essential investments in production, distribution and management that the British industrialists failed to make" (Chandler, 1990, p. 133).

After World War I, a British firm, Imperial Chemical Industries (ICI), was formed and did eventually become the dominant chemical firm in Great Britain. But ICI was not a first mover. It was formed as a government-sponsored cartel from a government-established firm, British Dyestuffs, which had been set up during World War I. "By 1900, the German chemical companies had driven nearly all the pioneering companies in Britain out of business....Then came World War I. To meet acute shortages the [sole] British producer expanded its output. The government set up another enterprise, British Dyestuffs Corporation, and then, in 1918, acquired the private company....Once the war was over, however, the German companies came back quickly" (Chandler, 1990, p. 139).

The British Dyestuffs Corporation (set up by the government and not a first mover, as were the German firms) could not compete with the Germans. Even though it had the production capability and the market, it lacked the technical and managerial capability. "By 1921, the newly appointed head of British Dyestuffs told the [English] Board of Trade that he saw little hope of making the company competitive unless it acquired technical skills from the Germans. The latter were willing enough to strike a deal, but their proposal would give them [the Germans] de facto control" (Chandler, 1990, p. 139).

The British had just fought a 4-year war with Germany, and the last thing the British government wanted to do was to turn over control of an essential industry to German firms, so the German offers were not accepted. Yet British Dyestuffs still desperately needed private control and good management. The company was privatized in 1925 but performed poorly. In 1926 it was merged with two of the then largest British chemical companies, Nobel Industries and Brunner Mond, to form Imperial Chemical Industries (ICI).

Putting the three largest British chemical companies together, with government blessing, set the stage for a European cartel. "By then, both industry and government saw the merger as an essential response to the [German] creation of I.G. Farben, which consolidated Bayer, BASF, Hoechst, and five smaller German chemical companies" (Chandler, 1990, p. 139).

After the German defeat and the desperate economic situation in Germany, with France and England requiring war reparations, the first-mover firms of Germany formed a German cartel/firm and the British firms formed the British cartel/firm, ICI.

Technology Strategy: Corporate-wide and in Strategic Business Units

In planning within a diversified firm, there may be very different and even opposite goals at the two levels of the firm. Lowell Steele (1989) is one author who has emphasized the opposition of goals in a diversified firm between management and its strategic business units. He even argued that in a modern diversified corporation, the overall strategic goals of the firm have to be considered as two distinct subgoals: (1) survival of the firm and (2) competitiveness of the businesses of the firm.

The goal for survival of the firm as a whole may or may not include survival of any given strategic business unit. (For example, a previous chapter discussed Welch's restructuring GE for survival, which reduced the number of businesses from 100 to 14 and eliminated 100,000 jobs.) The conditions of competition will be different for the different businesses in the firm.

Steele also emphasized that the two requirements for survival of a firm are *prosperity* and *change*. A firm obviously cannot survive if it is not profitable, with a positive cashflow over the long term. In addition, however, a firm must be able to change, since technology, products, competitors, and economic conditions all change over time. Therefore, the strategic goals at the corporate level of the conglomerate should focus on both prosperity and change.

Steele also argued that competitiveness cannot be dealt with *directly* at the conglomerate level but only at the business level of the firm— where value to the customer is produced. At the business level, the firm is in direct contact with its customers. The two requirements for a business to be competitive are to provide economic value to its customers and to attain profitability. But then Steele also did not advocate corporate core competencies. Still, it is important to note these four basic goals in corporate strategy: prosperity, change, adding value to customers, and profitability. The way to integrate these different goals in a diversified corporation lies in the notions of core competencies and core products, as discussed in Chap. 2.

ILLUSTRATION (*Concluded*)
Change at Imperial Chemical Industries

A diversified firm that did not develop core competencies during diversification was the previously described British firm of Imperial Chemical Industries (ICI). Recall that ICI had never been a first mover in the chemical industry. As a cartel, however, ICI did grow to be the major chemical firm of the United Kingdom. In 1989, for example, its sales were 13.2 billion pounds sterling ($21 billion), with pretax profits of 1.5 billion pounds. Also, ICI was the fourth largest company in the world's $750 billion chemical industry, composed of 70 business units. However, even despite its size, ICI's future was in trouble (Economist, 1990).

The post-World War II European Economic Community had finally put a stop to European cartels in 1970. Having to operate under the new conditions of competition, ICI was a large chemical conglomerate that had great difficulty formulating a successful strategy.

By 1990, ICI's continuing problem with strategy had resulted in a disparate set of businesses—that analysts estimated if taken apart would be worth twice as much as the current stock value of 7 billion pounds. Moreover, its price-earnings ratio was 25 percent below the London industrial average. The stock market's judgment was that despite ICI's enormous size, it was underperforming.

ICI produced the commodity chemicals and plastics that supplied the rest of Britain's expanding manufacturing industry, and most of its assets were in Britain. "Although IG Farben was split into three after 1945, and the prewar 'understanding' between it, ICI and Du Pont supposedly had come to an end, the chemical giants continued right up until the 1980s to run their markets much as they always had. ICI sold to Britain and the ex-empire, the Germans (Bayer, BASF and Hoechst) to continental Europe, and Du Pont in the Americas. And nobody worried much about competition" (Economist, 1990, pp. 21–22).

The change in competitive conditions came in the 1970s from two sources: the sharp rise in energy prices forced by the OPEC oil cartel of the 1970s and the maturing of the technology life cycle in commodity chemicals. In 1980, after the second oil shock, the chemical industry of the world fell into recession as a result of increased prices for feedstocks, falling demand and prices, and excess capacity. Moreover, new materials inventions in plastics were not forthcoming, with the last major plastic to be discovered being polypropylene in 1954.

The reaction of ICI management to the recession lay in its previous strategic attitude—it tried to continue acting as a member of a cartel. "In 1980, ICI reported losses in petrochemicals and plastics....The industry's response was the old one: even more market rigging. But this time it was stopped. The European Commission uncovered illegal price- and/or market-fixing cartels in...PVC (polypropylene polyvinyl chloride) and low-density polyethylene, to name only the largest cases. The fines that it imposed, totaling some $140 million, hit most of the names that matter in European chemicals" (Economist, 1990, p. 22).

After 1980, ICI knew it had to change its strategic attitude, and it began a new plan. It invested more in specialty chemicals (such as drugs and farm chemicals) and less in the traditional commodity chemicals (for industrial supply and plastics). The business was reorganized as nine international business groups. Later, in 1986, ICI merged its bulk industrial chemical, petrochemical, and plastics operations into a single group and set it up as a separate company, ICI Chemicals and Polymers. The commodity chemicals business was consolidated, just in case ICI might want to sell it off. ICI did sell its polyethylene business and moved to sell off its PVC business—retreating from the highly competitive, low-margin, but large-volume mature-technology businesses. Thus ICI's strategic focus was to be in specialty chemicals, where technological innovation continued and high-profit-margin products could be found.

However, the problem with ICI's strategic decision in 1986 to focus on specialty chemicals was that each specialty market is small for a giant like ICI. ICI needed very large markets to produce large sales (as could only be found in the chemical commodity markets).

In this illustration, we do not see any attention to building core competency across the firm. This left ICI with a management paradox. Could a large firm like ICI really have its future as a large collection of small markets?

In 1980, only 30 percent of ICI's profits had come from specialty chemicals, and by 1989, this had expanded to 50 percent. Moreover, the British sales, which in 1980 had made up 42 percent of total sales, had declined to 22 percent. ICI had changed into half a specialty chemicals company and half a global commodities company. So did ICI's story of the 1980s end happily? Unfortunately, no. A British financier considered it as a conglomerate ripe to be taken apart: "Long before Sir James Godsmith was said to be eyeing ICI last September, with unbundling on his mind, Sir Denys Henderson, chairman since April 1987, had set up a task forces to discover if there were better ways of running the giant British chemicals conglomerate" (Economist, 1990, p. 21).

The strategy from 1987 to 1990, however, had not accomplished what was intended, a strong ICI: "...after a period of consolidation, the board [in 1990] has decided ICI is again ripe for another shakeup....The board has a lot on its plate. Can ICI justify retaining all its operations? Can it continue to pursue the market without forsaking traditional businesses...? Is there a better way of managing what it does keep? ICI has no obvious answers, only difficult choices....It may, however, divest. But what? ICI has first to decide where its focus should lie. With its old commodity chemicals, which though slow-growing, remain indispensable to many industrial customers? Or with its eight consumer and specialty businesses which are mostly faster growing and should...bring higher profit margins?" (Economist, 1990, p. 22).

The strategic problem the ICI board faced was that commodity and specialty businesses required different strategies. Their planning process was obviously not enlightening this problem nor facilitating any new synthesis of strategy. ICI's board was merely flipflopping from one strategy to another, reorganizing and reorganizing. And through it all, the overall corporate performance suffered.

Technology Planning for Businesses: Commodity versus Specialty Products

Recall that, generally, the two common competitive strategies are (1) product differentiation and (2) cost/quality leadership. Running businesses in which the products are relatively undifferentiated (commodity products) and businesses in which the products are technically differentiated (specialty products) requires different strategies for both technology and business.

A way that has been used to summarize the state of the competitive goals of different businesses in a diversified firm is to present the businesses plotted on a three-dimensional profitability and market-share and market-size chart, as depicted in Fig. 4.1. Of course, every diversified firm would like to have all its businesses in the quadrant of high-profit, large market share, and large market size. For large markets, however, this is only possible for a strong, basic, patented product (and only during the 17 years of the patent life) or for a cartel or monopoly.

Business strategy for a commodity-type business should focus on maintaining both market share and profitability. Technology planning should focus on improving productivity, lowering costs, and improving product

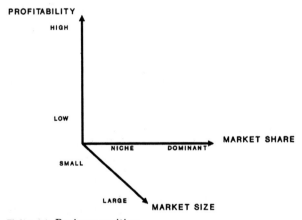

Figure 4.1 Business position.

quality. Technology strategy also may focus on developing new product applications to increase markets and to broaden the customer base.

Since a commodity product–type business produces undifferentiated products, technology strategy for such a business should be focused primarily on the cost and quality of production.

Business strategy for a specialty product–type business should focus on increasing market share by both improving the performance of the product and lowering its cost. Technology strategy should focus predominantly on improving and differentiating the product. Research also should focus on improving productivity in anticipation of the time when prices must be cut to fend off competition.

Since a specialty product–type business initially has a small but focused market, technology strategy should be focused on expanding the market by improving performance while preparing for competition by lowering cost of production.

Financial evaluation of commodity-type businesses should focus on maintaining cashflow and market share while keeping a positive profitability. In contrast, in a specialty business, one can try to optimize the profit margin and return on investment while increasing market share.

When products transform from specialty performance to commodity, a company loses control over the profitability as a result of competitors increasing supply over demand with look-alike products.

Therefore, return on investment and gross margins are always lower in commodity businesses than in specialty businesses. To try to seek the same level of profitability in the two types of business is fruitless. (This was the problem with the ICI board in the preceding illustration.) However, the total cashflow in commodity businesses is likely to be much higher than in specialty businesses because of the size of the commodities markets. (This led to the dilemma of the ICI board wanting high profitability and large markets in all businesses, whether commodity or specialty.)

In the long term in a strictly specialty business, only those who dominate a market niche survive. In the commodity business, however, only the big and few survive. The strategic dilemma is that most high-profit-margin products will occur in small markets and that in large markets most products will be low-profit-margin products (due to the intense competition in large markets).

What is wrong with expecting any business to be forever high tech *and* high profit is a strategic misunderstanding—large markets will always become low-margin businesses, after the basic technologies in the primary products mature and become commodity-type products. And even when a basic technology is still changing, a product may become even temporarily commodity-like. Even commodity-like businesses are very worthwhile, however, since in large markets (even though profit margins may be small) cashflow can still be very large.

In business, cashflow is always more fundamental than profit, and a shortage of cash is always the immediate cause of business failure.

In addition to cashflow, businesses can be compared on the basis of their rates of return on investment (ROI). Since all businesses require capital to begin and for maintenance, ROI shows which businesses provide most return on invested capital or on the future need to invest capital. Earlier we noted that the ROI of a firm is constrained by the ROI of the industry. If a diversified firm manages only by comparing ROI rates from each business and sets a standard ROI for all businesses, it must eventually eliminate all its present businesses and be seeking new businesses eternally.

Over time, the attainable rate of ROI for new investment in any industry will decline as its core technologies mature, and its markets saturate, and competition creates excess capacity in the industry. This pattern will be described in more detail in a later chapter, where life cycles in technology are discussed. At different times in the technology life cycle and under different competitive conditions, different industries can provide different rates of return on investment. A strategic commitment to stay in a business must be based on several criteria, of which rate of ROI is only one. Technology strategy for large-volume businesses should be defensive in focus. Products are defended by R&D, which continually improves the quality of the product, lowers its cost, and broadens its applications. It is difficult for a conglomerated corporation to run both commodity and specialty businesses, unless their differences in technology and financial strategies are properly taken into account.

Over the long run, a diversified firm must plan its technology to provide all its businesses it intends to keep *with the competitive means of being both low-cost, high-quality leaders as well as product differentiators.*

Business size can be an asset or a liability. It depends on why the size occurred and how it is used. Business largeness in general does not necessarily carry the size advantages of a first mover. For example, Alfred Chandler pointed out: "If size makes sense...why have so many large

U.S. companies done so poorly in the last few decades? Why is size so often a disadvantage rather than an asset? One answer is that like any human institution, managerial enterprises can stagnate. In the 1920s, Henry Ford destroyed his company's first-mover advantages with a series of wrongheaded decisions, including the firing of his most effective executives" (Chandler, 1990, p. 138).

In addition to institutional stagnation through poor leadership, another reason that size can be a disadvantage in competitiveness is that largeness is sometimes created for the wrong economic reasons: "More serious to the long-term health of American companies and industries was the diversification movement of the 1960s—and the chain of events it helped to set off. When senior management choose to grow through diversification—to acquire businesses in which they had few if any organizational capabilities to give them a competitive edge..." (Chandler, 1990, p. 138).

Technology planning should provide the businesses of a corporation with competitive edges. To the extent that a firm diversifies by building from its technical and marketing strengths, its diversification can strengthen the whole firm and provide paths to new opportunities.

ILLUSTRATION
Research and Diversification at Ethyl

An illustration of a firm that used corporate technology planning to strategically diversify is the Ethyl Corporation. Originally, Ethyl began as a joint-venture firm between General Motors and Exxon to produce the lead antiknock gasoline additive (which had been invented at General Motors for high-compression gasoline engines but which General Motors did not wish itself to produce). Ethyl first sold antiknock gasoline in 1923, and for nearly 40 years it was a one-product company. In early 1960s, Ethyl was acquired by the Albemarle Paper Manufacturing Company, and pressure was there to diversity to survive (Gottwald, 1987).

At the time of the merger of Ethyl with Albemarle Paper, Floyd D. Gottwald, Jr. was elected executive vice president of the newly merged Ethyl Corporation. He had originally joined Albemarle Paper Manufacturing Company in 1943 as a chemist and had advanced to president. Later, he commented on the merger that set Ethyl to diversifying (Gottwald, 1987, p. 27):

> When Albemarle purchased Ethyl in 1962, Ethyl was clearly a one-product company with a wealth of pent-up talent restless to exert itself. Under the previous joint owners, GM and Exxon, there had been virtually no op-

portunity for commercialization of the many possibilities that had emerged from 40 years of research on improving or finding a better anti-knock. For good reasons of their own, the previous owners had preferred to keep Ethyl a one-product company. Our change of perspective in 1962, as we sought to diversify, could not have been more dramatic.

The problem was to choose what businesses to acquire. Ethyl chose to diversify toward areas in which it had a strong underlying research base. The Ethyl research program had grown out of the company's original focus on lead antiknock compounds, whose original rights had been acquired from General Motors (Moser, 1987).

Figure 4.2 shows the two families of technologies that Ethyl developed—branching from the lead antiknock chemistry and from the aluminum alkyl chemicals. This kind of research branching focused the company's original business acquisitions. And these acquisitions further stimulated research branching.

As Ethyl began acquiring businesses, the original focus was the company's existing research base in chemicals and paper. In 1963, for plastics, Ethyl acquired Visqueen Film. In 1967, for paper, the company acquired Oxford Paper, and in 1968, it acquired IMCO Container.

The company had an interesting research program in aluminum that eventually did not work out; in the short term, however, it motivated the company to acquire the William L. Bonnel Company in 1966 and Capital Product Corporation in 1970. And after these initial acquisitions of the 1960s, Ethyl continued acquiring businesses in the 1970s. "As we reached the mid-1970s and Ethyl's success seemed as-

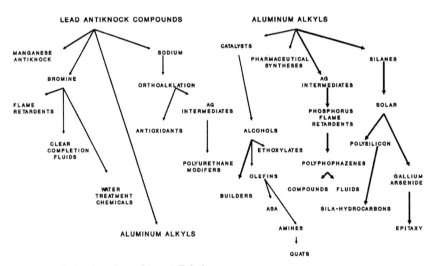

Figure 4.2 Technology branching at Ethyl.

sured, our acquisition program entered phase two—broadening the base. In 1975, we acquired the Edwin Cooper Division of Burmah Oil to give added strength to our existing lube additive lines. Harwicke Chemical, purchased in 1978, expanded our insecticide business to include synthetic pyrethroids. In 1980, we acquired Saytech, Inc., to extend Ethyl's basic bromine position into flame retardants" (Gottwald, 1987, p. 27).

The result of this diversification program over 20 years was a dramatic alteration of the structure of Ethyl's businesses. Moreover, one may note that the diversification was fortunate in timing, since later in the 1970s the U.S. government began regulating lead additives out of gasoline as a health hazard.

The diversification program at Ethyl also looked at many other businesses that the company chose not to acquire. In these cases, the businesses had no relation to Ethyl's research strengths. An important factor in Ethyl's acquisition program was that the company had not only a financial strategy but also a technology plan.

In this illustration, we saw a company, Ethyl, originally founded on a core product (antiknock lead compounds), which in its technology strategy built a technical skill to maintain its product. Later, as its technical skills built the basis for developing new business, its owners (GM and Du Pont) had no interest in diversifying the company. After its sale to Albemarle Paper, business diversification was guided by the technical strengths in each of the two merged companies, Ethyl and Albemarle.

Goals of Technology Planning

Over the long term, all firms must change in order to survive. Moreover, when business diversification is based on technological and research capabilities, the likelihood that such diversification will succeed commercially is enhanced. One strategy for diversification is to build businesses on the technical strengths of a corporation. This produces a corporation with an integrated business and technology strategy—integrated around core technical competencies.

In general, the goals of any technology planning include

1. Maintaining technical competence in existing businesses through
   a. Incremental improvement of existing products
   b. Incremental improvement of existing processes
2. Expansion of markets in existing businesses or launching of new businesses through

    a. Innovation of new products

    b. Innovation of new processes

3. Provision of distinctive competitive advantages to strategic business units with technologies from outside their respective industrial sectors through

    a. Vertical integration from core products and skills

    b. Horizontal transfer of core products and skills

To accomplish these ends, research activities need to

1. Deepen the understanding of the science beneath the existing corporate technologies

2. Invent or acquire improvements in the existing technologies

3. Maintain a window on science for new or substituting technologies

4. Invent or acquire new technologies for new corporate products and businesses

Technology planning in a firm occurs as a process within the annual business planning cycle of the firm. Setting up such a process requires special attention because the nature of research and the nature of business activities are considerably different. The technology planning process must integrate the two very different contexts of business planning and research planning.

ILLUSTRATION
Technology Planning
at Eastman Chemicals

As an illustration of a technology planning process at a divisional level in a large firm, let us review the process initiated by Harry W. Coover in the Eastman Chemicals Division of Eastman Kodak (the business-group level of a large diversified corporation) (Coover, 1986).

    In 1965, Harry W. Coover was named director for research at Eastman Chemicals Division, which at that time was successful, with 1965 sales of $368 million (an increase of 15 percent over the prior year) and after-tax earnings that were 16 percent of sales. Coover (1986, p. 12) thought that a more systematic approach to research planning could improve operations:

> Our approach to research at that time was much like everybody else's. Research was somewhat of an island unto itself with very little interaction with other company functions. And the director of research was not con-

sidered an integral part of the top level management team....when I looked forward into the late 1960s and early 1970s I began to wonder: *How can research be made central to determining the company's future?*

This question expressed Coover's strategic technology attitude—a determination to make research central to the company's future. His problem was how to create a management process for planning innovation systematically and routinely. The technology planning process Coover instituted in the research laboratory of Eastman Chemicals Division focused on three goals: (1) to generate the new products, (2) to ensure that all products were technologically advanced, and (3) to maintain a knowledge base in the sciences underlying company technologies.

Note that these goals cover several of the preceding general goals for technology planning:

1. The maintenance of technical competence in core businesses
2. Incremental improvement of existing products of businesses
3. Incremental improvement of existing processes of businesses
4. Innovation of new products
5. Innovation of new processes
6. Innovation of new businesses

Note also that the business emphasis of the goals also satisfies the two goals of providing economic value to customers and obtaining a competitive rate of return.

Expression of these goals as the *mission* of the R&D laboratory was important to changing the culture of the laboratory: "...we defined our responsibility as both applied and basic [technology and science]. But we maintained that all research must be consistent with current and future corporate directions....We were coming out of our ivory tower and getting involved in the 'business of our business.' That meant a greater reliance on *planning*" (Coover, 1986, p. 13).

This is a fundamental point. The usefulness of a strategic attitude and a formal set of planning procedures for technology is that they improve the integration of technology with business strategy.

The purpose of the technology planning process is to integrate technical change with business development.

However, this does not mean that creativity or serendipity in research efforts need be stifled by a ponderous, stupid, heavy-handed bureaucratic style of centralized command planning. "In research, we usually aren't presented with well-charted territory. No one has clearly defined

the new products that need to be invented, or the problems that need
to be solved. Research management has the responsibility of developing ways of perceiving and conveying to the research scientists those
things that the corporation needs from research" (Coover, 1986, p. 13).
Coover is saying that a main goal of a technology planning process is to
inculcate a strategic technology attitude *throughout the research organization.*

Although technology invention cannot be planned the way one plans
a battle, the creative attention of invention can be focused—much as
one focuses training and perception before any battle. Recall that this is
the purpose of a strategic technology attitude, as discussed in Chap. 2.

How did Coover focus the attention of the researchers on corporate
business goals? First, he organized projects so as to focus on businesses. "To tackle the overall planning problem, we found that we had
to subdivide the total job into separate business segments or strategic
business units. Most of these segments correspond to current lines of
business, and a few of them are new areas of potential business interest to our company. These segments are broad enough to be challenging, but narrow enough to be manageable" (Coover, 1986, p. 13).

The business context of research projects need not follow exactly the
organization of the businesses in the corporation. For technology planning, the conception of business groupings should be to challenge and
stimulate research imagination as to the potential crosscutting impact
of technology change in the corporation as a whole.

The next step in Coover's technology planning process was to have
his staff make technology and market projections for each business
group—forecasting. The kinds of factors they considered in their projections were technical changes, economic changes, market changes,
and political and social trends.

They considered all factors that might have an impact on the value
to the customer of a product. Next, they identified what would be required technically to maintain the competitiveness of a product and
what new products might be required to maintain or enlarge a market
share. They also did a critical analysis of current and potential competitors in the business segment, estimating their strengths and weaknesses. To make these forecasts, they formed teams of five or six people
with members from R&D, marketing, manufacturing, and finance and
a staff person skilled in forecasting techniques.

Management responsibility for the studies resided in R&D. In addition, in potentially new business areas with which they were unfamil-
iar, the staff prepared a business area analysis, attempting to learn
and understand a possible target business area. The next step was then
to organize a panel to consider the studies. "The results of the study are

presented to an opportunities panel composed of our most knowledge-able people from marketing, manufacturing, general management and R&D. The panel decides which market segments, if any, offer the greatest potential" (Coover, 1986, p. 14).

After one creates forecasts, one must organize a special panel to specifically review and identify opportunities, and that panel must be cross-functional and forward-looking. The market segments that were identified for greatest potential then provided market-oriented opportunities on which research could concentrate. The advantage of this procedure is that research areas are systematically focused by adding information from other functional areas of the firm—marketing, manufacturing, and finance. Collectively, their identification of market potential is likely to be much better than if research personnel do it on their own.

The usefulness of these projections was to establish areas of markets in which needs could be identified. "Needs; projected products and processes; and innovation projects....A need is simply a description of a valid new product or process for our company" (Coover, 1986, p. 15).

Coover's planning process was clearly intended to facilitate the synthesis of market need and technical opportunity. Coover understood that the planning process should foster focused creativity on the identification of needs. "People throughout the company...play a role in describing these needs....we are seeking ideas for *value-added products and next-generation products* that represent the major opportunities of the future"(Coover, 1986, p. 15).

In addition to the ability of this planning approach to make research scientists and engineers sensitive to and focused on future market needs, Coover also instituted a major programmatic effort to get them out into the marketplace. R&D personnel were assigned to each market segment of major interest, which required them to be expert not only in their disciplines but also in their assigned markets.

Coover intended to improve the researchers' ability to communicate with senior management. (To facilitate this communication, Coover used a technique called *profit gap analysis,* which will be described next.)

Technology Planning at the Level of the Strategic Business Unit (Divisional Level)

The heart of a technology planning process is to create and foster a strategic technology attitude. Recall that technological innovation is known to have two sources—technology push and market pull—and also recall that most successful innovations coordinate technological opportunities with market needs.

Christopher Freeman (1974, p. 193) summarized the debate over technology push versus market pull: "Perhaps the highest level generalization that it is safe to make about technological innovation is that it must involve synthesis of some kind of [market] need with some kind of technical possibility." Since research personnel are technically responsible for technology push, being informed by this kind of a planning process (which included teams of people from marketing, manufacturing, and finance) about market needs provided a way for the research people to be informed and motivated by market pull.

For a new product, the "need" describes the kinds of features and performance required for the product to find utility in the marketplace. For a new process, the "need" describes the desirable process characteristics and/or changing economic conditions that necessitate new process techniques. These "need" descriptions then provide market-oriented and market-targeted opportunities toward which the R&D personnel can direct their research and inventive creativity.

The technology planning process *in the corporate research organization does not "plan" research projects so much as it creates a business-focused* strategic attitude *in the research organization and personnel.*

A business-focused strategic attitude provides a culture within which the individual researchers are motivated to think of and to plan their own R&D projects that are both technologically creative and business relevant.

In dealing with the uncertainty, creativity, and inventiveness that really characterize technological invention and change, relevant formal strategic planning processes should facilitate creativity, not constrain it. Creativity and invention fundamentally involve surprise, serendipity, and the "bright idea never before thought of."

Although research creativity and invention cannot be planned, the areas of research focus and the conditions for fostering creativity *can be planned.*

The major function of a technology planning process is setting the appropriate context for research planning that emphasizes the economic impact of technical progress.

One way of presenting an analysis of a firm's strategic goal for profitability is to delineate the present and expected sources of profits by product. (A later chapter reviews the concept that most products have a finite product life.) The expected profits from the products of a firm can be projected in the form of a gap analysis, as depicted in Fig. 4.3. The total profits of a company are plotted over time, adding in the contribution to profits from each product A, B, C, etc. Since most products have a finite life time in the market (for reasons discussed in a later chapter), eventually, present product A will be withdrawn from the market, as will other current products such as B. Meanwhile, new products such as C will be introduced into the market. Looking out in time, the difference between expected total profits and desired total profits is the so-called *profit gap.*

Product gap analysis is a procedure to chart, over time, the past and projected relative contributions to the profits of a business of each existing product and to compare this total with the desired total profit.

At the strategic business unit level, the technology planning process has three purposes:

1. To encourage a market-oriented strategic attitude for research creativity that focuses on market needs
2. To encourage interaction and integration of research personnel with other business functional personal in the planning process
3. To present the expected contribution of the research efforts to the firm's business to high-level management through a profit gap analysis

ILLUSTRATION
Technology Planning at NEC

Divisional-level technology planning in a diversified corporation should fit into a larger technology planning process for the whole corporation. As an illustration, let us review technology planning at NEC Corporation.

Michiyuki Uenohara described the technology planning process in place at NEC in 1991. At that time, Uenohara was an executive advisor to NEC Corporation in Tokyo and chairman of the board of trustees of the NEC Research Institute for Advanced Management Systems and the NEC Research Institute in Princeton, New Jersey. He had received a B.E. degree from Nihon University and an M.S. and Ph.D. from Ohio State University. At first he went to work for Bell Labs, and then he joined NEC in 1967, subsequently managing NEC's Central Research Laboratories. He was elected to NEC's board, with responsibility for corporate R&D.

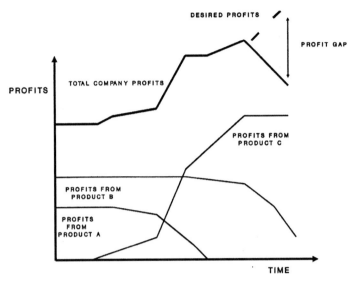

Figure 4.3 Profit gap analysis showing total company profits from profit contributions of each product.

NEC focused on technology for its core businesses, attending to the links from research to product and market (Uenohara, 1991, p. 18):

> The key to competing successfully in industry is overall strength—each link in the chain from basic research to development to production to marketing must be as strong as the next. I believe that overall strength is the secret of the success of Japanese industry in general and of the NEC Corporation in particular....our core business areas are communications equipment and systems, computers and industrial systems, electron devices and home electronics. These businesses contributed respectively 25, 44, 18, and 13 percent of our sales in 1989.

This selection of businesses was integrated by a strategic vision of an information-oriented society. "NEC is contributing to the betterment of the highly information-oriented society that is to come. We strongly believe that information, including software, will play an important role in this future society, but hardware innovation is basic to the improvement of information productivity" (Uenohara, 1991, p. 18).

NEC had over 190 companies that formed the NEC group. Total employment was around 160,000, with 35,000 engineers. NEC had distributed the R&D activities throughout the company and affiliated companies.

The corporate R&D laboratory was comprised of the research and development group, the computer and communications software development group, and the production engineering group. The corporate laboratory was primarily engaged in long-term R&D for future products—the products after the next products. The divisional laboratories focused on the next products. Overall, NEC spent 10 percent of sales on R&D. The corporate laboratories received 65 percent of funds from corporate headquarters, 33 percent from manufacturing groups, and 2 percent from government. Also, it is interesting to note that both software and production were singled out as both corporate research and divisional responsibilities.

NEC's businesses were clustered into 10 manufacturing groups, with 4 groups for communications, 2 for electronic devices, 1 for home electronics, and 1 for special projects. Together these constituted 60 product divisions and 25 divisional laboratories. Thus the management of technology strategy in this large diversified firm required sophisticated planning and implementation procedures.

The cooperation between the corporate research laboratories and the divisional laboratories of the strategic business units occurred in joint projects (Uenohara, 1991, p. 20):

> Generic technologies are developed far ahead of manufacturing division's needs in each professional laboratory. As soon as basic technologies are reasonably well developed, new core product models that effectively utilize such basic technologies are proposed to various manufacturing divisions. After hard negotiation and evaluation, joint development projects are initiated and development resources concentrated from both laboratories and manufacturing divisions. Laboratories provide key technologies and brains, and manufacturing divisions provide product development skills and established peripheral technologies.

However, to make such joint projects possible, long-range planning was required so that the generic technologies could be focused on next-generation products. "Educating R&D managers to the point where they are able to identify the correlations between the market and the generic technologies and then getting them to develop these technologies requires a time-consuming effort and long-range planning....Formulating the strategy for R&D programs involves high-level interaction among marketing, operating and R&D groups. Considerable information must be gathered about future market trends, potential product ideas for business growth, and science and technology trends" (Uenohara, 1991, p. 21).

The key concept in this long-range planning is the notion of core technologies to the corporation as a whole. "The heart of the strategy is the core technology program initiated by me [Uenohara] 17 years ago. The core technology program is developed from extensive analyses of both the market and the technology. A major analysis is done every 10 years for a two-year period, and minor modifications are made every year if necessary" (Uenohara, 1991, p. 21).

Formulation of the core technologies strategy is the responsibility of corporate research management, assisted by a planning office. "The corporate laboratory top management is responsible for establishing the core technology program....However, since it requires extensive analysis and evaluation, the R&D planning office handles most of the process. The office members consist of a few managers, who are laboratory general manager candidates, a small number of analytical experts, and most senior researchers from every laboratory who hold two jobs concurrently—research leader and planner" (Uenohara, 1991, p. 20).

This illustration again emphasizes that technology strategy is a management responsibility housed in corporate research laboratory management, assisted by an analytical staff, but primarily composed of *research leaders from every divisional laboratory*. The planning of core technology competencies requires involving both corporate research staff and strategic business unit product-development staff in the formulation of technology strategy. This is an example of the general principle that planning works well in any organization and on any activity only when it is both a top-down and a bottom-up process.

In the NEC planning office, the research leaders from the divisions worked together with senior researchers from corporate research to forecast both technology and market opportunities. Half the researchers/planners were laboratory manager candidates, and the remainder were specialist candidates. They analyzed the strengths of all the strategic business units. They then analyzed science and technology trends and listed the most important technology areas. Finally, they correlated the technologies with the business units. These correlations were used to identity NEC's key strategic technologies.

One sees in this illustration a focus on forecasting both market and technology trends and on identifying matrices (correlation tables) between key strategic technologies and market/product needs. (The remaining chapters of this book review useful techniques for forecasting market and technology trends and correlating the two.)

In a large diversified firm such as NEC, one may end with a large number of key strategic technologies from such a planning exercise, and these must be rationalized. "Strategic key technologies that should be pursued are grouped into a limited number of core technologies, each of which can be created and nurtured by a professional group. The strategic importance, minimum number of committed key basic technologies, important core products, and interdisciplinary relationships among other core technologies are defined for each core technology. The core technology program is presented to laboratory managers and divisional managers before the five-year plan is developed at each division" (Uenohara, 1991, p. 21).

This core technology program at NEC has been important for providing the strategic vision of NEC as an information company. The number of core technologies for a diversified firm will likely be large, but it is finite. In 1975, NEC defined 27 core technologies. In 1983, the company included software technologies and increased the number to 30 core technologies. In 1990, the number again increased to 34 core technologies. Thus, because of the number of core technologies and businesses in a diversified firm, the activity of managing strategic technologies is a complex activity that requires formal attention and procedures by management.

A way to deal with complexity is to create conceptual hierarchies. Just as key strategic technologies were grouped into sets of core technologies, core technologies can be further grouped into strategic technology domains. "A group of core technologies demands a common strategy for better communication and collaboration. Hence, we have defined six strategic technology domains (STDs)" (Uenohara, 1991, p. 22).

A strategic technology domain *is an area of rapidly changing research that underlies a group of strategic technologies.*

Technology Planning Process
in a Diversified Firm

As we saw in the preceding illustration concerning NEC, strategic technology planning within a diversified corporation is complicated. The core technologies organized as strategy technology domains (or as I have called them, the *core corporate technical competencies*) provide the basis for (1) long-range planning of the divisional laboratories in the strategic business units and (2) long-range planning of corporate research.

The problem in corporate strategic technology planning requires both identification of corporate core technology competencies and planning of long-range research programs for technical progress in those competencies. These projects result in core products (or core processes or core services) for the strategic business units to use competitively in their products and services to their customers.

Figure 4.4 depicts a scheme for a corporate-wide technology planning process that integrates both core technology competencies created by corporate research and the technology roadmaps of strategic business

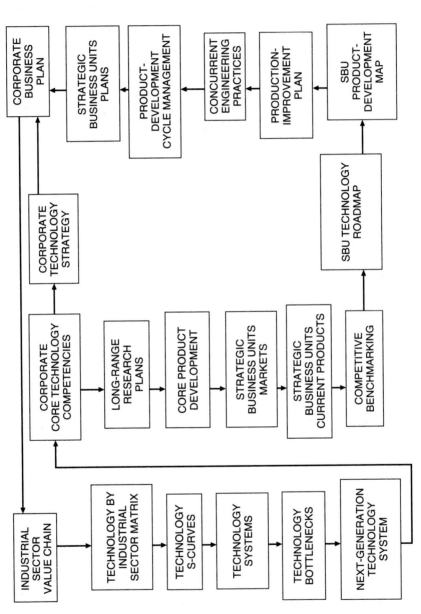

Figure 4.4 Technology planning process in a diversified firm.

units. Corporate business planning must be both the beginning and ending points in a cyclic planning process as indicated in the top right of the figure—the *corporate business plans.* The economic context of the businesses of the firm lie on the left side of the figure in the *economic structures* of the industrial value chains in which the businesses of the corporation participate.

The center of a strategic technology process for a diversified corporation turns on

- Core technology competencies of the whole firm
- Corporate core products (or core processes or core services) that give strategic business units competitive advantages
- Market focus of the strategic business units
- Competitive benchmarking for both the corporation and its business units

Indicated down the left side of the scheme in Fig. 4.4 are different techniques that can assist the technology planning required to strategically identify core competencies for the corporation (and these will all be reviewed in later chapters):

- Technology by industrial-sector matrix
- Technology *S*-curves
- Technology systems
- Technology bottlenecks
- Visions of next generations of technology systems

To the right side of the scheme in Fig. 4.4 are the steps in the product-development cycles of the strategic business units that utilize technology and core products/processes/services of the corporation in their own business plans:

- Product-development maps
- Production-improvement plans
- Concurrent engineering practices
- Product-development cycle management
- Strategic business units plans

And in the middle are the procedures that connect technology forecasting to technology implementation:

- Formulation of corporate core technology competencies
- Long-range research plans

- Corporate core product development
- Review of markets of strategic business units
- Review of current products of strategic business units
- Competitive benchmarking
- Formulation of a corporate technology strategy
- Formulation of technology roadmaps for each strategic business unit

The kinds of issues indicated in each of the boxes of the technology planning scheme include

1. *Corporate business plans*
   a. The businesses of a firm should be planned within groupings of coherent industrial sectors, indicating the generic product lines of the sector and connections between customers and suppliers.
   b. The dominant theme of corporate-level business strategy should be reconciling long-term survival with short-term profitability.
   c. Survival requires both corporate prosperity and corporate change; short-term profitability requires competitiveness and productivity.
   d. Technology strategy at the corporate level should be focused on the need and opportunity for technical change and the contributions of such change to long-term corporate prosperity.
2. *Industrial value-adding chains*
   a. The concept of industrial system value chains helps a diversified firm to understand how to include technology within its corporate-level strategies through a regrouping of its businesses into relations within industrial value chains.
   b. For the current businesses of the firm, each industrial sector relation (from resources to final customer) should be described in detail.
3. *Technology/industrial sector matrices*
   a. Generic-level technological change (either as core technologies or as substituting technologies) that affects any of the industrial sectors should be described.
   b. Lists should be made of the generic core and substituting technologies against the industrial sectors to identify all the potential industrial relevancies of technical changes.
4. *Technology S-curves*
   a. For the relevant technologies identified in the technology/sector matrix, it is important to anticipate rates of change in core and substituting technologies.

   b. In particular, one should identify new and/or rapidly changing technologies as distinct from slowly changing and mature technologies.

   c. Appropriate and generic technology performance measures should be constructed to express the potential impact of technological progress on functional applications.

   d. These should be projected against time, identifying natural limits and the reasons for such limits.

5. *Technology systems*
   a. Although the rates of change of technologies are summarized in generic and key performance parameters, the actual changes in technologies occur in system aspects of the technologies.
   b. It is necessary to describe the relevant technologies as systems in order to connect technology forecasting with technology planning.

6. *Technology bottlenecks*
   a. The current technical bottlenecks to progress in a technology system can be identified, and if they are overcome, the overall performance of the technology system will be improved.
   b. Such bottlenecks are the points for planning research to improve technologies.

7. *Next-generation technology systems*
   a. Most progress will be incremental within current technology systems, but it is also important to look further into the future to envision the next generation of a technology system.
   b. Plan for incremental improvement toward a next generation of technology so that products do not reach an eventual technical dead end (and ultimately a failed business).

8. *Corporate core technology competencies*
   a. Core technology competencies express the corporation's ability to integrate technologies basic to the critical features of products and production generic to several businesses of the corporation.
   b. They identify the technological distinctiveness of the corporation (with respect to competitors) in the diverse businesses and markets of the firm's strategic business units.
   c. The attitude of senior management about technology is expressed in the core competencies of the firm and is realized in the culture for technological change created and reinforced by the CEO team.
   d. If simply ignored or treated superficially by the CEO team, the diversified firm may miss opportunities for technological synergism between its businesses.

9. *Corporate technology strategy*
   a. Corporate technology strategies express the forecasted directions of technological change in the strategic technologies of the corporation.
   b. They also identify the forecasted impacts of strategic technology changes on the businesses and competitive position of the firm.
   c. They also identify the long-term capital investments required to exploit the technological changes as a corporate first mover.
10. *Long-term research plans*
    a. Within the commitment to corporate core technology competencies, a firm should plan its long-term research to acquire, maintain, and advance core competencies.
    b. A corporate research laboratory is required to organized research for core competencies that cut across business units.
    c. The long-term research program plans should include
       (1) Areas of research thrusts
       (2) Methodological approaches in these thrusts
       (3) Technical goals and targets
       (4) Technical risks and tests of technical feasibility
       (5) Plans for incorporating successful results into core products/processes/services
       (6) Profit gap analyses of the impact of research on products
       (7) Required resources and schedules for progress
       (8) Organization and management responsibility for implementing progress
11. *Core corporate products/processes/services*
    a. Core products/processes/services are the tangible embodiment of core competencies as components, materials, production processes, software systems, and design capabilities used by strategic business units in their products or production or services.
    b. Core products/processes/services enable business units to compete with technological factors beyond their particular industrial sector.
12. *Strategic business unit markets*
    a. Strategic business unit markets implement the direct competitiveness aspect of corporate strategy.
    b. Strategic business unit competitive strategy attends to the options chosen to offer attractive value to the customers of the strategic business unit market focus.
    c. Factors such as product attributes, cost, service, and segmented customer focus target strategic business unit markets.

13. *Strategic business units current products:* The current product lines of strategic business units are reviewed to the extent they may exploit the corporate core products/processes/services.

14. *Competitive Benchmarking:* In addition to forecasting technology change, a firm also needs to forecast
    a. The commercial impact of such change on the competitive environment of the firm.
    b. Anticipated changes in current or new markets due to technical progress (in core and substituting technologies by industrial sector).
    c. Strengths and weaknesses of competitors and any changes in competitive factors of industrial organization that technical progress may facilitate.
    d. Scenarios for possibilities of restructuring competition (through market niching or new business ventures or product substitution or vertical or horizontal integration within and between industrial value chains).

15. *Strategic business unit technology roadmap:* Each strategic business unit creates a technology roadmap for the unit indicating the long-range product/process development of the unit.

16. *Strategic business unit product-development map*
    a. The implementation of research results will be in products, which together with customer needs, provide the basis for the development and evolution of product lines.
    b. Each strategic business unit should plan its long-term product development as a map depicting core products, product branching and next-generation models, and new product lines.
    c. The product-development map facilitates research cooperation with the corporate research laboratory and the use of their core product/process/services developments.

17. *Production-improvement plans:* The implementation of research results will also be in production processes and systems, which provide the capabilities of mass producing the products at low cost, high quality, and quick responsiveness.

18. *Concurrent engineering practices:* The computerization of engineering design and manufacturing are increasingly providing the enabling technologies for coordinating product design and manufacturing—a necessity for rapidly market-responsive companies.

19. *Product-development cycle management:* All the tools and procedures for anticipating, planning, and implementing technological innovation will be of no use unless the firm's product-development cycles are well managed.

20. *Strategic business unit plans*
    a. Business plans of the strategic business units should integrate business planning with technology planning as technology roadmaps.
    b. Technology roadmaps should include
        (1) Technology and market forecasts
        (2) Product and competitor benchmarks and manufacturing competitive benchmarks
        (3) Product life-cycle estimates
        (4) Divisional profitability gap analysis
        (5) Product-development maps
        (6) New product project plans
        (7) Manufacturing-improvement project plans
        (8) Divisional marketing plans
        (9) Divisional business plans

A formal procedure for formulating technology strategy and integrating it into product and business planning can provide a systematic method to utilize technology as a competitive edge. Managing strategic technologies can develop core technical competencies that provide competitive advantages to the businesses of the firm—beyond the ordinary boundaries of the firm's respective industries.

The remainder of this book will review the concepts and techniques noted in this scheme that are useful for formulating a technology strategy and integrating it into business planning. First, however, the next two chapters turn to techniques for implementing the results from technology plans into new products, services, and production improvements.

Practical Implications

The lessons to be learned in this chapter focus on the logical technology planning processes at both the divisional and corporate levels of the firm.

1. In a new industry, the first firms to make the appropriate investments in a complete business system, including technology, have gained dominant positions through exploiting the advantages of economies of scale and scope. Technology plans should have goals of
    a. Maintaining technical competence in core businesses
    b. Innovating new products and businesses

2. Running businesses in which the products are either commodity or specialty products requires different technology strategies.

   a. For a commodity product type of business, technology strategy should be focused primarily on the cost and quality of production.

   b. In a specialty product type of business, technology strategy should first be focused on expanding the market by improving performance and second on preparing for competition by lowering the cost of production.

3. A strategic technology plan consists of
   a. The representation of a core technology as a system
   b. The identification of research efforts on the technical bottlenecks
   c. The setting of technical goals for system improvement to be attained by the research efforts.

4. As sketched in Fig. 4.4, a technology-integrated formal planning procedure for a diversified firm can assist formulation of technology strategy at both the corporate level and the strategic business unit level.

For Further Reflection

1. Identify a large conglomerated firm, and obtain a copy of its annual report. List the businesses in the firm.

2. Can you then identify some core technology competencies that would provide distinctive competitive advantages to several businesses of the firm?

3. Which businesses of the firm could use these core products/processes/services to provide them with a competitive advantage?

4. Does the conglomerated firm actually have a business in the core products/processes/services of this core technology and so could utilize a core competency?

Foundation of Technology: Invention

Before we proceed to the topics of forecasting and implementing technological change, we must be sure that we all understand how technology begins—invention. Understanding *invention* requires us to appreciate the key concepts of *morphology* and *function*. These concepts are central to all techniques for forecasting and planning technological change.

> Invention *is the creative process in which new logical ways are imagined to manipulate nature for human purposes.*

ILLUSTRATION
Ghosts of the Machine

The potentials and limitations of all technologies ultimately reside in both nature and logic. For example, using "motorcycle maintenance" and "Zen Buddhism" as kinds of metaphors in a popular book about technology in the 1970s (*Zen and the Art of Motorcycle Maintenance*), Robert Pirsig (1974) tried to communicate the relationship between nature and logic in technology: "Not everyone understands what a completely rational process this is, this maintenance of a motorcycle....A motorcycle functions entirely in accordance with the laws of reason, and a study of the art of motorcycle maintenance is really a miniature study of the art of rationality itself."

By the "art of rationality," Pirsig emphasized that the design and operation of any technological device (such as a motorcycle) is an implementation into matter of a system of ideas:

> That's all the motorcycle is, a system of concepts, worked out in steel. There's no part in it, no shape in it, that is not out of someone's

mind....Shapes, like this tappet, are what you *arrive* at, what you give to the steel. Steel has no more shape than this old pile of dirt on the engine here. These shapes are all out of someone's mind. That's important to see. The steel? Hell, even the steel is out of someone's mind. There's no steel in nature. Anyone from the Bronze Age could have told you that. All nature has is a *potential* for steel! There's nothing else there. But what's "potential"? That's also in someone's mind!...Ghosts.

A system of concepts worked out in matter is a kind of functional logic. Technology is about the potentials in nature, the potential of logic to order nature—the "ghosts" of nature.

Technology and the Potentials of Nature

Technological invention arranges to bring the potentials of nature into existence. The fascination with the power of technology to shape nature has long been a part of human culture. For example, one can recall that the ancient Greeks believed that the invention of fire was a gift to humans from the Titan Prometheus. Prometheus was then eternally punished by the gods, by being chained to a mountain and exposed to the elements of nature.

Metaphorically speaking, all technological invention has been a kind of Promethean activity: taking power over nature from the domain of the gods and giving that power to human control. Now what any of these metaphors about technology (such as Pirsig's "ghosts of nature's potentials" or the Greeks' "Promethean fire") really means is—in every technology one can look for both a kind of logic and a kind of natural form.

The human act of first creating a new mapping of the logic of function to physical form is invention.

ILLUSTRATION
The Promethean Fates of the Inventors
of the Computer

The invention of the computer was an example of a Promethean kind of achievement—putting mind into the machine. It profoundly altered the capability of humanity to acquire and process information and control organizations and devices. Central to its invention were the persons of Von Neumann, Goedel, Turing, Mauchly, and Eckert (Heppenheimer, 1990).

John Von Neumann created the world's first stored-program electronic computer. He was influenced by the earlier ideas of Kurt Goedel and Allen Turing. He also drew on the ideas and work of John Mauchly and J. Presper Eckert.

Mauchly and Eckert created one of the world's first all electron vacuum tube special-purpose computers (at about the same time, other special-purpose computers also were being created independently in England by Turing and in Germany by Konrad Zuse). Mauchly also borrowed earlier ideas from John Atanasoff. Mauchly and Eckert, in turn, were influenced by Von Neumann's ideas. Moreover, Mauchly and Eckert created the world's first *commercially successful* stored-program electronic computer, the UNIVAC.

The drama behind all these events did indeed show Promethean dimensions. One even might be tempted to some mythic speculation. Perhaps, as displeased as the gods were with Prometheus, the gods also were displeased with the inventors of the computer. The ends of Goedel and Turning and Von Neumann and Mauchly were all horrible.

The historical facts are these. In 1943, the logical scheme of the general-purpose computer was first described in a famous report entitled "First Draft of a Report on the EDVAC," written by John Von Neumann. It was about a machine being planned by John Mauchly and J. Presper Eckert. EDVAC is an acronym for electronic discrete variable automatic computer. Although the EDVAC had been planned to be the world's first general-purpose programmable electron vacuum tube computer, it was never built. The ideas for EDVAC eventually turned up in the Mauchly/Eckert UNIVAC machine. And just before the UNIVAC, Von Neumann supervised the building of the first programmable electronic computer at the Institute of Advanced Studies at Princeton. At first, Von Neumann, Mauchly, and Eckert had been intellectual collaborators, but they later split over their own claims to fame.

Mauchly was very bitter that Von Neumann never acknowledged him as a coinventor of the stored-program concept. Adding injury to insult, Von Neumann even testified against Mauchly in an important trial that challenged Mauchly and Eckert's patent claim from the ENIAC. The judge ruled that Mauchly and Eckert were not the true inventors of the electronic computer. Because of this, there were never any basic patents on the computer. This ruling also contributed to Mauchly's final penury. "'Lawyers keep making money,' said Mauchly toward the end. 'We've got down to the point where maybe we can buy some hot dogs with our Social Security'" (Heppenheimer, 1990, p. 16). And even the gods seemed to have added to Mauchly's woes. In 1980 he died from a genetic disease that had terribly disfigured him.

Terrible also was the gods' fate for Von Neumann. "For Von Neumann, it was even worse. In the summer of 1955, he was diagnosed with bone cancer, which soon brought on excruciating pain. In the words of his friend Edward Teller, 'I think that Von Neumann suffered more when his mind would no longer function than I have ever seen any human being suffer.' Toward the end there was panic and uncontrolled terror" (Heppenheimer, 1990, p. 16).

The gods also were hard on Allen Turing, the person who had first conceptualized general computation as programmable operations. "Convicted in England of soliciting sexual favors from a teen-age boy, he was given a choice of prison or hormone treatments. He chose the hormones and soon found his breasts growing. Driven to despair, he made up a batch of cyanide in a home laboratory and died an apparent suicide" (Heppenheimer, 1990, p. 16).

Even the famous logician Kurt Goedel, whose work influenced the computational ideas of both Turing and Von Neumann, was personally tormented. "...it was his own personal demons that would drive him to death....After his wife underwent surgery and was placed in a nursing home, in 1977, Goedel refused to take any food. He starved himself to death" (Heppenheimer, 1990, p. 16).

Invention is a wonderful thing. Apparently, however, one must watch out for the wrath of the gods!

Intellectual Property

Let us pause in our illustration of the invention of the computer to review patent law. In law, the right to use an invention is protected and limited only to the holder of the patent to the invention or to those whom the patent holder may license use. In the United States, only the inventor may file for a patent. In many other countries, anyone can file for a patent, whether or not he or she is the inventor. In Europe and Japan, the first to file is awarded the patent.

Establishing priority to invention has often resulted in patent conflicts in the United States, because technological invention has often involved the cross-fertilization of ideas by several individuals. Often there have been cases of independent simultaneous invention of the same device by different individuals.

The reason behind multiple invention lie both in the patterns of scientific and technical progress and in societal problems and market needs. The progress in science or technology is usually widely shared and can often stimulate new ideas for inventions in several individuals who are familiar with them. Thus ideas in *technology push* are often shared by researchers who can imagine the potential applications of the research.

In addition, the problems and market demands of a situation also can simultaneously motivate different people to invent solutions. Thus ideas in *market pull* are also often shared by several inventors.

In the United States, four classes of intellectual property are recognized in law:

1. Patents
2. Copyrights
3. Trademarks
4. Trade secrets

Patents protect inventions of new and useful *ideas. Copyrights* protect the *expression* of ideas but not the ideas expressed. *Trademarks* are *registered identifications* of products or corporate identities. *Trade secrets* are commercially important information that is gained about one company by another *without permission and by wrongful means,* such as spying, etc.

The government provides the legal right of exclusivity to use in a patent to the inventor for a finite term in return for the inventor disclosing full details of the invention. The patent concept evolved in England in the eighteenth century in order for details of an invention not to be lost to society due to excessive secrecy by the inventor.

The most common form of patent in the United States is called a *utility patent.* It may be granted by the U.S. government to inventors (domestic or foreign), individuals, or corporations. Utility patents provide the right to exclude anyone other than the inventor (or those whom the inventor licenses) from making, selling, or using the patented invention for a specific term. This term is normally 17 years, with additional time for certain patents that require Food and Drug Administration (FDA) approval.

The steps in acquiring a patent consist of first filing a patent disclosure and patent application with the U.S. Patent Office. The patent office performs a review and grants or denies the patent. In the case of denial, the applicant may appeal. In most countries, public disclosure of information about the patent before filing the patent disclosure and application invalidates the patent application. In the application, the claim to invention must include the conception of an idea and the reduction of that idea to workable form; that is, an actual device, built and working, or merely a clear description of a workable form in the patent application (Bell, 1984).

In the U.S. system, an invention is patentable only if it satisfies three criteria: utility, novelty, and nonobviousness. For an invention to be useful, the courts have generally followed a common notion of someone using it: "The issue of simple practical usefulness was addressed, and largely settled, in Lowell v. Lewis, heard by Justice Story in 1817 ....Judge Story held...[it] didn't have to be extremely useful or the most useful....It just had to have utility. Clearly the [Perkins pump] was useful; people used it" (Lubar, 1990, p. 11).

Judge Story also added that the patent should not only be useful but also should not hurt society. Also, to be novel, the invention must not simply be a rearrangement of prior art. A revision of the patent law in 1952 added a new standard for patentability—nonobviousness. "It declared that an invention would be unpatentable 'if the differences between the subject matter as a whole would have been obvious at the time the invention was made to a person having ordinary skill in the art to which said subject matter pertains'" (Lubar, 1990, p. 15).

Since this definition was itself nonobvious, a further change in the patent process occurred. "In 1982, to help settle this confusion, a whole new court was created, the Court of Appeals for the Federal Circuit, to handle appeals of patent cases....One of the ways it overcame the problem of defining invention was by putting a greater emphasis on what are called secondary criteria, especially commercial success, in determining the patentability of an invention" (Lubar, 1990, p. 16).

Natural phenomena and scientific laws cannot be patented. However, new materials forms can be patented as products, and the processes to produce them also may be patented. New life forms also can be patented, as well as the biotechnology process to produce them. However, the new biotechnology has created new concerns about patent law; for example (Marshall, 1991, p. 22):

> The confusion over product and process patents has affected the biotechnology industry so much that a few companies [in 1991] have been asking Congress to intervene. Genentech, for one, is cheerleading a reform effort, sponsored by Representative Rick Boucher (D–VA) and Senator Dennis DeConcini (D–AZ). Their bills would change the law to state that "process" and "product" patents can cover the same thing, if the product is truly original. The aim is to make it easier to get patents on genetically engineered proteins, even when the natural analogs may have been well-characterized in the literature. Some think it would be a mistake for Congress to intervene on such a fine point, and the big companies in genetic engineering quietly oppose the bill.

Computer software is also a new area of technology that has been creating changes in patent practices. Initially, software was held to be not patentable but only copyrightable. Then an exception was made such that when software was part of a process it could be patented. This exception has been creating whole new categories of patents (Kahn, 1990, p. 53):

> Last August [1989], Refac International, Ltd, sued six major spreadsheet publishers, including Lotus, Microsoft and Ashton-Tate, claiming they

had infringed on U.S. Patent No. 4,398,249. The patent deals with a technique called "natural order recalc," a common feature of spreadsheet calculations that allows a change in one calculation to reverberate throughout a document....Within the few years, software developers have been surprised to learn that hundreds, even thousands, of patents have been awarded for programming processes ranging from sequences of machine instructions to features of the user interface.

The patent office has distinguished between mathematical algorithms and computer algorithms: "While patents are not awarded for algorithms, which are considered 'laws of nature,' the Patent Office draws a fine distinction between 'computer algorithms' (which are patentable) and 'mathematical algorithms' (which are not)" (Hamilton, 1991, p. 23).

This has been one of the major effects of the new Court of Claims and Patent Appeals established in 1982. However, the U.S. courts still have not completely resolved the issues about software patentability. "Still the CCPA continued to uphold patentability in other cases. Finally...a sharply divided Supreme Court upheld the patentability of a process for curing rubber that included a computer program. The majority concluded that programs that did not preempt all uses of a computer algorithm could be patented—at least when used in a traditional process for physically transforming materials" (Kahn, 1990, p. 55).

In 1991, the issues concerning software were still not clearly resolved. New categories of technology, such as software and recombinantly altered living forms, were presenting new legal challenges to the evolving practices in patent law.

ILLUSTRATION (*Concluded*)
Invention of the Stored-Program Computer

Both functional need and technical opportunity assisted the invention of the computer. (Recall the importance of both the concepts of market pull and technology push to innovation.)

Market pull, however, was originally military need. The military needs included the computation of ballistic trajectories, breaking secrete codes, designing nuclear triggering devices, and radar tracking of airplanes. These military needs provided the funding for the first projects that led to the invention of the computer. The technical opportunities were in electrical circuitry and magnetic data-storage devices.

Let us continue the story of the invention of the computer by going back to John Von Neumann's beginnings. He was born in Hungary in 1903. Von Neumann was the precocious son of a Budapest banker. For example, at the age of 6 he could divide eight-digit numbers in his head and he could talk with his father in ancient Greek. At 8 years of age he was learning calculus. Also, he had a photographic memory. He would take a page of the Budapest phone directory and read it, and then, with eyes closed, he could recite it back.

When it was time for university training, he went to study in Germany under the great mathematician David Hilbert. Hilbert believed that all the diverse topics of mathematics could be established on self-consistent and self-contained intellectual foundations. Hilbert, in a famous address in 1900, expressed his position: "Every mathematical problem can be solved. We are all convinced of that. After all, one of the things that attracts us most when we apply ourselves to a mathematical problem is precisely that within us we always hear the call: here is the problem, search for the solution; one can find it by pure thought, for in mathematics there is no *ignorabimus* (we will not know)" (Heppenheimer, 1990, p. 8).

As a graduate student, Von Neumann worked on the problem of mathematical foundations. Then, in 1931, Kurt Goedel's famous papers were published. Goedel showed that no wholly self-contained foundations of mathematics could be constructed. He showed that if one tried to set a foundation, one could always devise mathematical statements that were formally undecidable—incapable of being proved or disproved purely within the foundational framework. This, of course, disturbed Von Neumann, as it did all other mathematicians of the time.

Goedel's paper also introduced an interesting notation. Goedel supposed that any series of mathematical statements or equations could be encoded as numbers. This later turned out to be a central idea for the stored-program computer—that all mathematical statements, logical expressions as well as data, could be expressed as numerically encoded instructions. However, the first person to take up this new direction was not Von Neumann but Alan Turing.

In 1937, Turing was a 25-year-old graduate student at Cambridge University when he published his seminal paper, "On Computable Numbers." What he proposed was a kind of idealized machine that could do mathematical computations. A series of mathematical steps could be expressed in the form of coded instructions on a long paper tape. The machine would execute these instructions in sequence as the paper tape was read. His idea was later to be called a *Turing machine*.

This supposition, that any mathematical problem could be expressed as a sequence of instructions, in essence, was the description of the concept of a general-purpose programmable computer. Although many people had thought of and devised calculating devices and machines, all previous devices either had to be externally instructed or could only solve a specific type of problem. A machine that could be generally instructed to solve any kind of mathematical problem had not yet been built.

Returning to Von Neumann again, after finishing his graduate studies in Germany in 1930, Von Neumann emigrated to America and joined the faculty of Princeton University. There Turing's and Von Neumann's paths temporarily crossed.

In 1936, Turing came to Princeton to do his graduate work, thinking about the problem of his idealized machine. He worked with Von Neumann, exposing Von Neumann to his ideas. Von Neumann offered Turing a position as an assistant after Turing received his doctorate. Turing chose to return to Cambridge, however, where, in the following year, he published his famous paper.

Thus Turing's ideas were in the back of Von Neumann's mind. In the meanwhile, however, war intervened, starting with the German invasion of Poland in 1939. During the war, Turing went into England's secret code-breaking project and helped construct a large electronic computer to break enemy codes. It was called the "Colossus" and began operating in 1943 (Zorpette, 1987). The Colossus, however, was a single-purpose computational machine.

In America, Von Neumann also became involved in the war effort—in the Manhattan Project to develop the atomic bomb. At Princeton, Von Neumann had been exploring the mathematics of problems in fluid flows. In the design of one of the atomic bombs, it was proposed to place a sphere of dynamite around pie-shaped wedges of plutonium. When the dynamite exploded, the wedges were intended to be blown together into a small sphere (slightly smaller than a soccer ball). This plutonium sphere would then undergo the violent nuclear chain reaction of an atomic explosion. This was called the "Fat Man" version of the atomic bomb. Von Neumann helped design the dynamite trigger for this device.

The trigger problem was precisely what thickness to form the dynamite around the wedges of plutonium so that they would be blown in all together at the same time. To design this, one had to calculate the physical form of the shock waves from the dynamite explosion that would push the plutonium wedges inward. It was a flow-type of mathematical problem, and Von Neumann would be just the one to calculate it. It was a tough problem, however, because of the accuracy required in exactly describing the evolution of the shock waves. To solve it, Von Neumann and his colleague at Los Alamos, Stanislaw Ulam, devised a kind of human computing system. They had one of their colleagues, Stanley Frankel, devise a lengthy sequence of computational steps that could be carried out on mechanical calculating machines made by IBM.

Frankel then had a large number of Army enlistees running these steps on the calculating machines. It was a slow kind of human computer, but it worked. Von Neumann got the solutions he needed to design the explosives for the "Fat Man." The bomb was the one dropped on Nagasaki, killing more than 100,000 Japanese civilians. This technical challenge taught Von Neumann an important lesson—the need for a general-purpose computer. (Afterwards, one of Von Neumann's favorite pasttimes was observing atom bomb tests, and exposure to radiation may have been the cause of his fatal tumor.)

Then came the many crossings of paths in computational destiny. One day in August of 1944, Von Neumann had gone to the Army's Aberdeen Proving Ground in Maryland on a consulting assignment. Afterwards, he waited at the station for a train. On the same platform happened to be a Lt. Herman Goldstine (who before the war had taught mathematics at the University of Michigan). Goldstine recognized Von Neumann, at that time already a world-famous mathematician.

Goldstine introduced himself. Personally, Von Neumann was a warm, pleasant man. He amiably chatted with Goldstine, asking him about his work. "When it became clear to Von Neumann that I was concerned with the development of an electronic computer capable of 333 multiplications per second, the whole atmosphere changed from one of relaxed good humor to one more like the oral examination for a doctor's degree in mathematics" (Heppenheimer, 1990, p. 13). It turned out that Goldstine was one of Mauchly and Eckert's team building the ENIAC, and at this meeting, Von Neumann first heard of the ENIAC project.

The famous ENIAC project had been earlier conceived by John W. Mauchly. Mauchly had been born in Cincinnati, Ohio, in 1907. In 1932, he had received a Ph.D. in physics from Johns Hopkins University. Teaching physics at Ursinus College from 1933 to 1941, Mauchly had begun to experiment with electronic counters while doing research on metrology. In 1941, he attended a summer course in electronics at the Moore School of Electrical Engineering at the University of Pennsylvania. He was then invited to join the faculty there.

In 1941, Mauchly also had heard of some computational work by physicist John Atanasoff at Iowa State University. He visited him, and Atanasoff showed Mauchly an experimental electronic adder that Atanasoff had just completed in 1940. This machine gave Mauchly ideas. In the fall of 1942, Mauchly wrote a memorandum on the idea of using electron vacuum tubes for a calculating machine (Brittain, 1984). The key idea in Mauchly's proposal was Atanasoff's use of flip-flop circuits. Circuits had previously been designed for electron tubes to display binary states. A signal applied to the grid of a tube could turn the tube on, and thereafter it would stay on and conducting. Another signal might then be applied to the tube that would turn it off, and thereafter, it would stay off until a new signal arrived. This was called a *flip-flop circuit* and required a pair of tubes hooked together properly. Mauchly outlined in the memorandum that one could use a set of flip-flop circuits to express a number system. Mauchly's proposal was the use of these circuits for a reconfigurable calculating machine.

Mauchly submitted his proposal to officers of the Army Ordinance Department: Colonel Paul N. Gillon, Colonel Leslie E. Simon, and Major H. H. Goldstine. They approved the idea and gave an R&D contract to the Moore School of Electrical Engineering to build the machine, which Mauchly would call the ENIAC.

For the project, Eckert was the chief engineer, and Mauchly was a research engineer. J. G. Brainerd was the project supervisor, and several others made up the team, including Arthur Burks, Joseph Chedaker, Chuan Chu, James Cummings, Leland Cunningham, John Davis, Harry Gail, Robert Michael, Frank Mural, and Robert Shaw. It was a large undertaking that required a large R&D team (Brainerd and Sharpless, 1984).

When constructed, the ENIAC took up the whole of a large air-conditioned room, whose four walls were covered by cabinets containing electron tube circuits. Altogether, the circuit cabinets weighed 30 tons and drew 174 kW of power.

The major operational problem was tube failure. Mauchly and Eckert had calculated that the 17,468 tubes they were to use in ENIAC were likely to have one failure every 8 minutes, on the average—which would have made ENIAC completely useless. What they did was to run the tubes at less than one-half their rated voltage and one-fourth their rated current. This reduced the failure rate to one tube about every 2 days, long enough to finish a calculation on ENIAC.

It was a clever approach. This problem had been so daunting that no one else had dared try all tubes for an electronic calculator. For example, Howard Aiken at Harvard University (with IBM assistance) was building a special-purpose computer about the same time for the Navy using solenoid relays as much as tubes.

Still the ENIAC was not programmable (nor was Aiken's machine). To set up a new problem for calculation, one had to connect circuits physically with patch cords between jacks and cabling that ran up to a total of 80 feet in length. The patching task itself could take at least 2 days.

In the summer of 1944, on the day when Goldstine (who was the Army's representative on the project) met Von Neumann, the ENIAC was working. The Eckert/Mauchly team already was beginning to think about a successor, which they intended to call EDVAC ("electronic discrete variable automatic computer"). At Goldstine's urging, the Army was considering awarding the Moore School $105,600 to build the EDVAC. "Into this stimulating environment stepped Von Neumann. He joined the ENIAC group as a consultant, with special interest in ideas for EDVAC. He helped secure the EDVAC contract and spent long hours in discussions with Mauchly and Eckert" (Heppenheimer, 1990, p. 13).

The contract was let in October of 1944, and in June of 1945, Von Neumann completed his famous paper, "First Draft of a Report on the EDVAC." This was one of the most influential papers in what was to become "computer science." Goldstine circulated the draft with only Von Neumann's name on the title page. "In a later patent dispute, Von Neumann declined to share credit for his ideas with Mauchly, Eckert, or anyone else. So the 'First Draft' spawned the legend that Von Neumann invented the stored-program computer. He did not, though he made contributions of great importance" (Heppenheimer, 1990, p. 13).

However, neither Von Neumann nor Eckert nor Mauchly built the EDVAC. The ENIAC group broke up. The University of Pennsylvania hired a new director of research, Irvin Travis, who quarreled with Eckert and Mauchly over patent rights. Eckert and Mauchly had a letter from the university's president stating that they could hold the patents on ENIAC. However, Travis told them that they must sign patent releases to the university. Mauchly and Eckert refused and resigned from the university in 1946.

They formed the Electronic Control Company, which became the Eckert-Mauchly Computer Corporation in 1947. It successfully built the first commercial general-purpose electronic computer, the UNIVAC, selling its first product to the U.S. Census Bureau. Running out of funds to build that first machine, however, Eckert and Mauchly sold out to Remington Rand, which in turn later became Sperry Rand in 1955. In 1959, Mauchly left Sperry Rand to form a consulting firm.

In the meantime, Von Neumann had returned to the Institute for Advanced Study at Princeton after the war. He had decided to build a computer, and Herman Goldstine joined him there. Von Neumann also hired Julian Bigelow, who had worked on radar-guided antiaircraft gun control during the war. Goldstine and Bigelow built a 2300 vacuum tube computer for Von Neumann at the institute. This was the first fully automatic stored-program computer.

A patent on this basic invention would have been very valuable. Somewhere between Turing, Von Neumann, Mauchly, and Eckert the inventive idea arose. But Von Neumann published the idea before filing for a patent. In patent law, this invalidated anyone forever from filing for a patent. Thus it happened that there was no basic patent on the stored-program computer.

Von Neumann's famous report had described the logical scheme of the single-instruction, single-processing stored-program computational architecture. The key advance in this scheme was the stored-program instructions, which make the machine a general-purpose computational devise instead of all the preceding special-purpose computing devices. The logic requires several transformational units: input/output, data bus, arithmetical and logical processing unit, and memory divided into data storage and program storage.

This schematic logic depicted the functional form for a calculating device. It required both a program to be inputted into memory and data to be inputted into memory. Then the central processing unit calls the instructions in sequence from the stored program and executes each instruction on the data. After the data have been transformed according to an instruction, the calculated result is temporarily stored and then further transformed by the next instruction in the program. After all instructions in the program have been executed, the final calculated results are both stored in the memory and outputted from the computer.

In this computational activity, the program expresses the logic of the mathematical transformation, and the operation of the computer is to map each instruction of this logic onto a specific configuration of the electronic circuits within the computer. This is a mapping of logic to physical electronic processes within the computer circuit. It is the *highest level of logic* mapped to physical structures of the computer.

At the lowest logical level of computer technology, a logical binary number system also has been mapped into physical circuits of the computer. The instructions of the program and the data on which calculations are performed are expressed in binary numbers. These binary numbers are the *lowest level of logic* mapped into the physical structure of the computer.

Recall that a binary number system is a way of writing integer numbers with only two symbols (0 and 1). Thus, in a binary system,

zero	is symbolized as	0
one	is symbolized as	1
two	is symbolized as	10
three	is symbolized as	11
four	is symbolized as	100
five	is symbolized as	101
six	is symbolized as	110
seven	is symbolized as	111
eight	is symbolized as	1000
nine	is symbolized as	1001
ten	is symbolized as	1010

and so on

Then, addition in the binary system, which in decimal system looks like

$$1 + 2 = 3$$

would in binary system look like

$$1 + 10 = 11$$

Now, in the Von Neumann architecture, one could build a calculating machine using only binary-state electron vacuum tubes, and all calculations in the machine would be performed in the binary number system. The theoretical point here is that this is an example of *mapping of a logic structure onto a physical structure* (the logic of numbers mapped to physical binary states of electron tube flip-flop circuits). In the mathematical operations, calculation is performed as *a coordination of physical processes to logical operations* (arithmetical operations, such as addition, are coordinated to a series of physical processes, such as turning some tubes in circuits off and others on).

What Von Neumann did was to use Turing's idea of devising an algorithm to perform a mathematical calculation in which the algorithm could be expressed as an ordered sequence of operations beginning with the initial data and transforming them to calculated results. The concept of the stored-program computer was to store both the data and the program algorithm in the active memory of the computer architecture.

In the Von Neumann computer architecture, when a program was run, it was timed to an internal clock that governed the following cycle:

1. Initiate the program.
2. Fetch the first instruction from main memory to the program register.
3. Read the instruction and set the appropriate control signals for the various internal units of the computer to execute the instruction.
4. Fetch the data to be operated on by the instruction from main memory to the data register.
5. Execute the first instruction on the data and store the results in a storage register.
6. Fetch the second instruction from the main memory to the program register.
7. Read the instruction and set the appropriate control signals for the various internal units of the computer to execute the instruction.
8. Execute the second instruction on the recently processed data whose result is in the storage register and store the new result in the storage register.
9. Proceed to fetch, read, set, and execute the sequence of program in-

structions, storing the most recent result in the storage register until the complete program has been executed.

10. Transfer the final calculated result from the storage register to the main memory and/or to the output of the computer.

Thus the Von Neumann stored-program computer worked like a Turing machine, in which any mathematical calculation was expressed as a sequence of ordered algorithmic steps that transformed initial data into calculated results (with both the data and program stored in the main memory of the computer).

Functional Morphology

In engineering practice, the two concepts of functional logic and physical form have traditionally been called the *function-form dichotomy*.

Technology invention is the creation of a mapping of natural form to the function in human purpose.

I will call this general concept a *functional morphology*. Physical structures and physical processes exist in nature.

Technological invention consists of imagining the selection or construction of a particular physical structure and the arranging for the sequences of physical process in the structure to correlate with the sequences of a logic set of operations.

Thus the secret of any technology is the invention of a physical structure whose processes can be mapped in a one-to-one manner with a logical operation of a functional transformation. There are two important sides to this mapping:

- *Logic*—a kind of spirit, an intention and purpose, a mind of a creator
- *Morphology*—a kind of mechanism, a kinetics and dynamics, a physical universe, potentials of nature

What is odd about the concept function and form is that humans, too, are a part of the physical universe.

In technology, one part of the physical universe, humanity, is interacting with other parts of the physical universe, nature— using logic as a means.

Metaphysically, this may sound odd, but technologically, it is straightforward.

ILLUSTRATION
Scaling the Mechanistic Picture
of Nature

The concept of morphology embodies the notions of structures and processes of nature. It is based on and made conceivable by the scientific perspective on nature as mechanism. Let us briefly pause to review this mechanistic picture of nature. Although mechanism is a familiar idea, reviewing its main features will help us discuss this fundamental notion of how logic may be used to alter natural structure.

In the modern scientific perspective, *nature is represented in mechanistic terms.* Space and *time* and *matter* and *force* are the primitive terms in scientific representation. All things in the universe and the universe itself are individuated and depicted in a spacial and temporal framework. Two objects in nature are said to be different because they exist at the same time at different points of space.

Space is the individuator of temporally coexisting objects.

Kinematics, motion of objects, is then depicted in science, literally, like the old motion pictures on celluloid film. Each frame of the film shows the configuration of all objects in space at a given time—this is spacial coexistence of objects in the scene. Then the picture frames of space are run past the eye of an observer—this is the sequencing of spacial coexistence. The observer can perceive the motion of objects as their rearrangements in space as the frames are shown.

Kinematics is the temporal sequencing of spacial reconfigurations.

Dynamics expresses the forces between objects and is explained either through the direct spacial contact of objects or through fields of force in the space surrounding objects. These force fields summarize the *potential* of one object to influence the kinematics of a second object *if* the second object enters its surrounding picture frame. Currently, there are four fundamental kinds of force fields known (Huth, 1992):

1. Gravitational force, binding matter together
2. Electromagnetic force, binding atoms and molecules
3. Weak nuclear force, binding neutrons
4. Strong nuclear force, binding nucleons

In quantized field theories, these force fields are explained as virtual field-quantized particles boiling off a charge-source particle and returning (within the very short time of Plank's uncertainty principle to avoid violating the conservation of energy principle). If a second charge-source particle enters the surrounding picture frame of the first charge-source particle, then the virtual field-quantized particles can become real and be exchanged between the charge-source particles—thereby forcing changes in their motions. This is a force-at-a-distance action, as compared with a force of mechanical contact.

The particles of light, photons, are the field-quantized particles of the electromagnetic field between electrically charged particles (such as electrons and protons). The field-quantized particles of the weak nuclear force are the Z and W bosons. The field-quantized particles of the strong nuclear force are conjectured to be gluons. These field-quantized particles of gluons that bind quarks into elementary particles have not yet been directly observed.

Space and time, matter and forces, and kinematics and dynamics together provide the representational terms of physical mechanism.

Physical materials and processes are then constructed in this mechanistic picture from very, very small spaces up toward very, very large spaces. *This is the microscopic-macroscopic explanatory strategy of science.*

In the very smallest space we have to date, the subparticle space, the fundamental particles are made up of smaller particles, leptons, quarks, and gluons. The very, very, very tiny quarks circle about each other and interact with each other through the gluon field to constitute the fundamental particles, such as protons, neutrons, pions, etc. Quarks and gluons have not been observed outside the internal spaces of the fundamental particles, but their effects have been seen in very high energy collisions between fundamental particles and in the spectrum of fundamental particles. *This is the quark-level scale of space.*

In the next size up, the fundamental particles of the universe exist in space: electrons, protons, neutrons, etc. *This is the particle-level scale of space.*

In the next spacial size up, some of the protons and neutrons bind into stable states of atomic nuclei. The nuclei of atoms are composed of neutrons and protons bound together by the gluon-carried strong nuclear force. The weak nuclear force causes the decay of neutrons to protons. *This is the nuclear-level scale of space.*

In the next spacial size up, protons, neutrons, and electrons form atoms. The atom is constructed of negatively charged electrons orbiting a positively charged nucleus. Exchanges of the virtual field photons between the electrons and protons in their electromagnetic fields bind the atom together. The electrons form themselves into shells, with a fixed number per shell (1, 2, 4, 8, 16, etc.). Two electrons cannot exist in the same state (the Pauli exclusion principle). Electrons have an intrinsic spin of one-half (analogous to a charged sphere spinning and creating a dipole magnetic field around it). Yet they also move as waves, and so around the atom, in the same orbit, two electrons can share the same energy and momentum state (analogous to two spiral waves circling in the same orbit but with opposite helical directions). *This is the atomic-level scale of space.* With the invention of the tunneling electron microscope, we can now directly observe this scale of individual atoms instrumentally.

In the next spacial size up, molecules are formed from combinations of atoms that bond together by exchanging outer electrons (valent bonding) or sharing outer electrons (covalent binding). In addition, molecules also stick together through the distant effects of the electromagnetic fields of the atoms (Van der Waals binding). Gases are molecules in a contained space, flying too fast to stick together. Liquids are contained molecules, stuck together but vibrating strongly enough to prevent rigid structural configurations. Microscopic solids are contained molecules, shivering, but stuck together in rigid structural configurations in crystalline (regular order) or glassy forms (irregular order) or polymeric forms (long molecular chains). *This is the molecular-level scale of space.* After the invention and development of nuclear magnetic resonance and imaging technologies, we can now directly observe this scale of molecular structure instrumentally.

In the next spacial size up, atoms or molecules stabilize in liquid or solid configurations as crystalline lattices, domains, or polymeric structures. *This is the domain-level scale of space.* We have been capable of directly observing domain-sized structures after several instrumental inventions, such as the electron microscope and the scanning ion probe.

In the next spacial size up, we find the microscopic level of the organization of matter into aggregates or organisms. *This is the cellular-level scale of space.* We can directly observe this cellular scale with optical microscopes.

We humans exist on a *macro scale of space of organism systems and macroscopic phenomena.* Here nature is directly experienced by our human senses. Gases are experienced as vapors or winds; liquids are experienced as fluids, rain, rivers, lakes, and oceans; and solids are experienced as solidified matter.

At this macro scale, crystalline or glassy macroscopic solids are aggregates of crystalline or glassy domains. Polymeric macroscopic solids are spaghetti-like tangles of polymers. Generally, molecules based on the metallic families constitute inorganic matter, while molecules based on carbon atom chains constitute organic matter. In organic matter, molecules are self-organizing and self-replicating through special organic polymers such as DNA. The self-organizing molecules create organs and living organisms and biologic processes. Evolution of living forms is an explanation of the variety of species as mutations of DNA to create organs and living organisms adapted to a physical environment.

Finally, there are two more scales of space above this macro level in the mechanistic picture of nature, the planetary and cosmic levels. At the *planetary-level scale of space,* nature appears as environmental systems of physical processes and ecologic systems. At the *cosmic-level scale of space,* the physical universe is formed of stellar systems, galaxies, and clusters of galaxies.

All the preceding scales—from quark and particle through atom and molecule to domains, cellular organisms, macroscale, planetary, and cosmic—together constitute the mechanistic representation of nature. It is a very complex but powerful scale of spacial individuation of the things in nature.

The morphological problem in technology is selecting and arranging physical phenomena in the various spacial scales to be used by the human organism as an extension of its human spirit.

Theoretical Concepts of Mechanism

A set of physical objects interacting together through a force is called a *physical object system.* An example of this is a fundamental particle (such as an electron) represented as an object system of quarks interacting through the gluon force. Another example is a nucleus of an atom as an object system of protons and neutrons interacting through the strong nuclear force. A third example is an atom as an object system of an atomic nucleus and a set of electrons interacting through the electromagnetic force. A fourth example is a molecule as an object system of a set of atoms bound together through the electromagnetic forces of the atoms.

The physical dynamics of a physical object system is the temporal representation of the relative positions of the components of the system as determined by the forces between the components.

This notion of dynamic representation of physical object systems is the *second major underlying concept of mechanism.*

A description of a physical system consists of the states of the objects of the system individuated in space and time and other descriptive attributes (e.g., energy, momentum, charge, intrinsic spin, baryon number, etc.). An explanation of a physical system is the law that derives subsequent states of a physical system from initial states. (In introductory mechanical physics, often a description is called the *kinematics* of a system and an explanation the *dynamics* of a system.)

> *A mechanistic representation of a physical phenomenon consists of a description and an explanation of its dynamic evolution over time.*

Technology and Mechanism

Why these concepts are fundamental to technology is that the *morphology* or *structure* of a physical system is the representation of the physical phenomena (the representation being both descriptive and explanatory.)

> *The importance of the representation is that the technologist not only can describe the structure in detail but also can predict the change in structure under manipulation.*

Prediction requires a cause-effect relationship between existing objects. If two events A and B occur, and A precedes B in time, then A is said to cause B if and only if the occurrence of A is both necessary and sufficient to the occurrence of B. By *necessary,* one means that every time B occurred, A occurred earlier. By *sufficient,* one means that every time A occurred, B occurred later.

> *If one can predict the morphologic phenomena, then one can use logic to arrange subsequent states to map out a purposeful transformation—invention.*

Thus morphology is necessary to technology but not sufficient. Functional logic is also necessary for technology.

ILLUSTRATION
First Evidence of Mechanisms in a
Toxic Cause of Cancer

An illustration of how mechanism works as a strategy for explanation in science can be seen in the efforts to understand the connections between environmental toxins and cancer. In the period from 1950 through 1970, many statistical studies had suggested that a correlation existed between cigarette smoking and lung cancer. Yet, because the exact mechanism of how irritants in the smoke caused cells to turn cancerous was not known, cigarette producers maintained that it was not "proven" that cigarettes caused cancer. In fact, the connecting links had not been observed. Finally, however, in 1991, the first of a connecting link between viruses and environmental factors to cause cancer was observed: "Scientists have detected a molecular 'hot spot' that is strongly linked to liver cancer, one of the commonest and most lethal malignancies in the world. The spot is a tiny region of a single gene where toxins that infiltrate the liver seem to home in, sabotaging the gene and touching off cancerous growth" (Angier, 1991, p. A1).

Liver cancer was then one of the top five cancer killers worldwide (although relatively rare in the United States and Europe). The scientists suspected that either the hepatitis B virus or a toxin from a fungal poison called *aflatoxin* could cause the liver cancer. Contaminated grain and other foods improperly stored in warm, moist places grow molds that produce aflatoxin. This was the first instance of a causal link to a cancer.

The Mechanistic Strategy of Explanation

The various levels of spacial scale in the mechanistic picture of nature form the basis for the use of science by technology.

Manipulation of the conditions of nature at a microscopic level can activate potentially existing and desired states of nature at a macroscopic level.

The mechanistic strategy of explanation assumes that the aggregate effects of a microscopic world can be translated into the physical effects of a macroscopic world. In the preceding illustration, cancer was seen as an aggregate effect in the macroscopic world of human health that was explainable at the level of the microscopic physics of cellular growth.

At the level of the macroscopic world, physical concepts can be formulated in terms of direct experience. For example, one can see particle motion with one's own eyes in the motion of billiard balls on a billiard table. Or one can see wave motion directly in bath tubs or lakes and at the seaside. However, at the level of the microscopic world, one can "see" only indirectly through instrumentation, and the concepts constructed there use the concepts from direct experience as analogies.

To reconcile the physics of the macroscopic and microscopic worlds, physical facts are all instrumentally derived (measured) at either level with specified instrumental techniques and under experimental conditions.

Physical terms and variables and laws are abstractions of the generic forms of instrumentally derived facts generalized to conditions of measurement.

Physical *laws* are the abstractions and generalizations of *causal* relationships between physical variables. Physical *theory* is a self-consistent body of terms and laws *necessary and sufficient* to describe and explain phenomena within a specified field. Physical *models* are *causal representations* of generic objects within a phenomenal field.

Physical models are used by technologists to understand and conceive of inventions and then by engineers to design and produce products embodying the inventions. The power of physical models is their predictive capability.

Since a technologist wants to manipulate nature, the ability to predict what nature will do when manipulated is most useful.

This predictive power of science useful to technology is one of the fundamental relationships between science and technology.

ILLUSTRATION
Reductionist Methodology in Biology

This use of the scaling approaches for explanation has philosophically been termed a *reductionist methodology.* Reductionist methods underlie all the scientific disciplines of physics, chemistry, and biology. For example, Douglas Kell and G. Rickey Welch (1991, p. 15) described the reductionist program in molecular biology: "Molecular biology is widely reputed to have uncovered the secret of life, with DNA being heralded as the quintessential component of biological systems....The biologist's problem is the problem of complexity. According to conventional wisdom, the way to study complex systems is by breaking them down into their parts."

This complexity in living organisms is related to spacial scale: The more complex the organic mechanism, the larger is the scale. The strategy of reductionism assumes that living systems are arranged hierarchically. First, molecules make up macromolecules, and macromolecules make up organelles. The organelles make up cells, and cells make up tissues. Then tissues make up organisms, and organisms make up populations and socioecosystems. Each larger organization of life can then be understood by seeking explanations of their properties in terms of behavior at hierarchically lower levels of organization.

Kell and Welch (1991, p. 15) then give an example of such a hierarchical-based explanation: "Hydrolysis of energy-rich molecules causes protein filaments to slide, sliding filaments cause muscles to contract, muscle contraction causes the acquisition of food and the escape from predation, and group behavior centered around common defense and protection of food supply causes the formation of social systems. This analytical approach may be termed 'hierarchical reductionism.'"

Kell and Welch next assert that while this approach provides much insight, in itself it cannot provide a complete program for science. "In biology, as in any science, we must distinguish between the 'method' and the 'philosophy' of reductionism. Methodologically speaking, reductionism is fruitful; it yields much utilitarian detail about observable phenomena. However, reductionism can never be complete; there is always a 'residue' which requires new conjectures or hypotheses for its solution."

What they mean by *reductionism,* as a philosophy as opposed to a methodology, is that reductionism as a philosophy cannot explain all of life. A philosophy of life must include functional as well as reductionist explanation.

Ordering Principles in Science and Technology

Technology (as opposed to science) includes function as an explanation (in addition to mechanistic reductionism as explanation). *Mechanism* is a description of the world of mechanical things related to mechanical things. *Function* is a description of the world as mechanical things related to intelligent things. Intention cannot be reduced to mechanism but only mapped to it. As we saw earlier, the heart of invention is mapping functional explanation to mechanistic explanation.

Mapping intention to mechanism is technology.

The reductionist method cannot provide a complete philosophy of explanation of life because it cannot explain the purpose that is observed in living forms. The problem when some scientists try to make the reductionist method into a reductionist philosophy of life is that such scientists have ignored technology, focusing only on science.

Science and technology are two halves of a whole. *Science* is the discovery and understanding of nature. *Technology* is the manipulation of nature for human purpose. Scientific explanations do not include human purpose as an explanatory mode, but technology does include human purpose as an explanatory mode.

In technology there are two independent representations that must be mapped to each other for invention: the function *of human purpose and the* form *of scientific representations of natural phenomena.*

ILLUSTRATION
Making Molecules by Numbers

As an illustration of the independence of purpose from mechanism, let us review the techniques for trying to correlate chemical properties with molecular structure (Ronvray, 1991, p. 23):

> In principle, chemists can calculate molecular properties from the fundamental theory that underpins all chemistry, known as quantum theory. The Schrodinger wave equation for the molecule provides a mathematical description of its electronic structure. Solving the equation yields reliable values for the energy levels of the electrons and the distribution of the electronic density in comparatively small molecules containing up to about 100 atoms. This approach is valuable for calculating some electronic properties....But it *cannot* predict the more general physical and chemical properties of materials in bulk, which is what the chemical industry is interested in.

There are two problems here. The first is simply the calculational complexity of large molecules with hundreds of atoms and thousands of electrons. This problem will eventually yield more and more complex molecules as calculational capability progresses with better techniques and faster computers.

The second problem is more fundamental and is intractable by reductionist explanation. Many of the properties of materials in bulk ("which is what the chemical industry is interested in") are *functionally defined* from the *technological uses* of the chemical phenomenon. For example, one of the technologically functionally defined properties of molecules is their technical performance as fuel when producing soot: "Another property of hydrocarbons that is difficult to predict is their tendency to produce soot when burnt. Soot, which consists primarily of carbon, represents the final stage in a long and complex combustion process. The 'sooting' tendency of fuels measures how efficiently they burn in air and, therefore, how much pollution they are likely to cause" (Ronvray, 1991, p. 25).

The *potential* of nature here is the residue of unoxidized carbon atoms remaining after a physical process of combustion. However, nature does not care about the amount of that residue as soot; only technologists care. "Soot" is a *functional* concept whose meaning is defined by the efficiency of the combustion process and by the *environmental concerns* of the technologist. "Soot" is the idea in the mind of the *chemist as technologist, not as scientist.* "Carbon atoms" is the mechanistic representation of nature's oxidation process of hydrocarbon molecules in the mind of the chemist as scientist. Soot is a functional property.

In technology, function and morphology are mappings, correspondences—not mathematical derivations of one from the other.

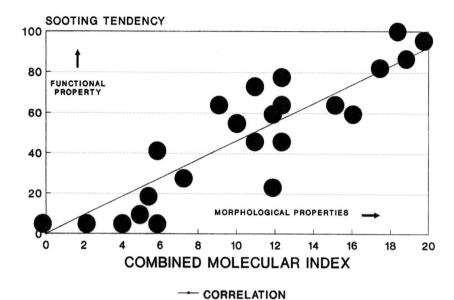

Figure 5.1 Sooting tendency of different hydrocarbon fuel molecules.

In principle, larger spacial scales of mechanistic representations of nature can be calculated up from smaller scales of mechanistic representations of nature. However, functional interpretations of nature cannot be calculated from mechanistic representations—they can only be mapped to one another. For example, Fig. 5.1 shows a correlation plot mapping the topologic configurations of hydrocarbon with the sooting tendency of fuels. The reason the correlation exists is that the geometric topology of how the carbon is arranged in hydrocarbon molecules affects the incompleteness of the oxidation of carbon molecules when the hydrocarbons are burned in air.

Physical structure, morphology, can be correlated with function, but neither morphology nor function can be derived from each other.

The plot of physical topology to functional performance is an *empirical correlation*. It cannot be calculated from physical mechanism alone. The empirical correlation of physical structure with function is a functional mapping.

The fundamental reason for trying to correlate morphology with functionally defined properties is the same as the reason behind invention: technology as the purposeful manipulation of nature. "One of the most rewarding aspects of chemistry is the ability to create new forms of matter that were previously unknown and perhaps even un-suspected. Chemists have now made or at least identified more than 10 million different molecules. Around 400,000 new molecules are de-scribed in chemical research journals every year. Some of them occur naturally, whereas others are synthesized in the laboratory for the first time" (Ronvray, 1991, p. 22).

You see, chemists as technologies do love these "ghosts" of nature—the potential forms of molecules that the chemists may arrange. Moreover, when a chemist arranges these for the first time, he or she invents a new chemical compound. The payoff for this game of arrang-ing nature lies in the functional use of the new arrangement.

How does one map functionally defined properties to chemical struc-tures? One technique has been to classify chemical molecules by topo-logic pattern and to correlate topologic pattern with functional property. "To predict how a chemical will behave in bulk, chemists—particularly those in industry—are starting to use an exciting new method that can estimate these [functional] properties satisfactorily. The method is 'topological,' which means that it considers the ways atoms in a molecule are linked....Based on the pattern of linkage, there are various ways we can derive a number that characterizes the molec-ular structure—the 'topological index'" (Ronvray, 1991, p. 24).

Once a class of molecules has been analyzed according to the linkage patterns and has been assigned a topologic index, then the industrial chemists can run empirical tests to see if that index correlates (maps) to some functional property. "On the biological side [for example], chemical structure has been correlated with odor, taste, carcinogenic-ity and enzyme inhibition. Coleman's Mustard has exploited molecular connectivity indices to predict the taste and smell of ingredients in food. and Harp Lager has even used these indices to improve the taste of its beer" (Ronvray, 1991, p. 25).

In this illustration, one can see that the key to the usefulness of sci-ence to society is the mapping of mechanistic morphologies to func-tional properties in technology.

Technology and Logic

Technology uses not only the ordering principles of science (the method of mechanistic reductionism for explanation) but also the ordering principles of functional logic. Since invention requires structuring mechanistic nature in correspondence with functional logic, let us next review the fundamentals of logic.

Logic is the technique of correct inference. There are three general kinds of inference:

1. Deductive inference
2. Inductive inference
3. Functional inference

Deductive logic is the correct way of deriving a true consequent from a true antecedent. A deductive inference is correct if whenever the precedent p and the consequent q are both true, the inference derives the truth of q from the presence of p. Modern deductive logic is expressed in the mathematical topics of set theory and Boolean algebra. Most of the other topics of modern mathematics (group theory, algebras, vector spaces, calculus, etc.) are specialized techniques and languages for performing deductive inferences (e.g., solving equations).

In contrast, *inductive logic* is the correct way of inferring a general statement from a particular statement. If an exemplar A exists with a specific property p, with what confidence can we say that all the members of the sets of things to which A belongs also share the property p? The modern mathematical topics of probability and statistics summarize techniques for correct inductive inference.

Functional logic is a logic for describing a purposeful transformation. It is fundamental to action and decision making, as well as to technology. Functional logic describes how to get from a beginning situation to a desired situation. For example, in a decision about action, the environment is perceived as forms that suit the desire of the decision maker. All the mechanistic representations of the decision maker's universe are re-sorted into classes of initial states, means, ends, and evaluations. Functional logic is expressed in what is now called *modern decision theory:*

Let the mechanistic configuration of the immediate environment in which a decision maker exists before action be symbolized as C_i.

Let the set of operational means that the decision maker can apply to the initial configuration be symbolized as T_{ji}.

Let the final configuration of the environment of the decision maker after actions be symbolized as O_j.

Then, symbolically, we can write that the effect of action T_{ji} on the initial state C_i is to transform it into a final state O_j:

$$O_j = \text{SUM}\ (T_{ji}C_i)$$

Finally, let us denote the value of the final state to the decision maker as U_j.

Then we can write an expression that Von Neumann and Morgenstern called the "expected utility of an action" $U_j(O_j)$:

$$U_j(O_j) = \text{SUM } U_j T_{ji} C_i$$

What this expression says is that action should be depicted as a set of means leading to a set of ends evaluated on a purpose.

Choices in action can be guided by choosing the means having the highest efficiency of leading to an outcome having the highest utility (value) to the decision maker.

This notation is useful because one can model many managerial decisions, such as production or logistic decisions, in quantitative forms using this logic. An academic specialty was built around this functional logic called *operations research.*

In technology, functional logic is the design of a sequence of transformations to obtain a desired output from inputs.

Functional Logic Contrasted with Prediction in the Mechanistic Picture

The mechanistic representation of nature is built around the concept of *prediction.* The representation of purpose is built around the concept of *prescription.* To prescribe that B is to occur, one must first arrange for A's occurrence. However, the occurrence of A cannot guarantee the occurrence of B if it is insufficient. Other factors (say $C, D, E,$ etc.) also may have to be arranged to make B occur.

A prescriptive relationship is one in which a preceding event A *is necessary to a succeeding event* B *but not sufficient.*

This is a very important point. Although technology utilizes the mechanistic picture, technology forecasting itself is in the functional and therefore prescriptive mode.

- All strategy is in the prescriptive mode.
- All planning is in the prescriptive mode.
- All engineering and all management are in the prescriptive mode.

In any of these situations, the best one can do is to identify all the necessary factors $(A, C, D,$ etc.) for a present event A to create a subsequent event B.

The logic problem of engineering and management is to arrange for both the necessary factor A *and also all the other coproducing factors* (C, D, E, *etc.) to bring* B *about.*

For example, in technology forecasting, one can identify progress in a performance parameter *necessary* to improve performance and lower cost in order to exploit a market. However, the research required to improve the technology, the products designed and manufactured to embed the technology, and the finance and sales required to profit from the technology are all the *C, D, E,* etc. coproducing factors required to bring about innovation.

In technology, then, human spirit is expressed in functional logic—which is the arranging of all the necessary and coproducing factors to attain a desired future from a present state. Only in the mechanistic picture is there ever a cause-effect relationship. And, therefore, only in the mechanistic picture is *prediction* possible. Most human situations, however, are not only mechanistic but also purposeful and functional. In functional situations, relationships between events are either prescriptive (if planned) or accidental (if unplanned), but not predictable.

ILLUSTRATION
Boolean Logic and Turing's Machine

We saw in the computer illustration that the *lowest* level of the computer morphology mapped to logic was the mapping of a set of bistable flip-flop circuits to a binary number system. We also saw that at the *highest* logic level, Von Neumann's stored-program concept, was to devise a transformational cycle that fetched instructions and data from main memory and in the clock cycles of the computer set and executed the ordered sequence of algorithmic instructions on the data and stored the final calculation back into main memory.

Also, between the binary number logic and the algorithmic logic of the computer was an *intermediate* level of logic in the computer

- The mapping of electronic circuitry into Boolean logic
- The combination of Boolean operations to perform elementary mathematical operations

The basic logic operations in Boolean logic are

NOT

AND

OR

Electronic circuits have been devised to map directly on these logic operations. For example, an inverting electronic circuit can receive a signal of one voltage and invert it into one of opposite sign, so the negation logical operation can be mapped as a physical voltage sign inversion.

As another example, the AND logic operation can be mapped as an electronic circuit that does not output a signal unless two proper input signals arrive simultaneously at the inputs to the circuit. Finally, the exclusive OR logic operation can be mapped as an electronic circuit with two inputs that will output a signal only when a proper signal is received either on one input line or on the other.

Next, mathematical operations such as addition, subtraction, multiplication, and division can be represented logically as combinations and sequences of the Boolean logic operations. Mathematical calculations can then be performed by connecting the appropriate circuits to map onto a Boolean expression of the mathematical calculation. Thus, for computers, there are several hierarchical levels of mappings of logics to physical morphologies:

1. Binary numbers mapped to bistable electronic circuits

2. Boolean logic operations mapped to basic electronic circuits

3. Mathematical basic operations mapped through Boolean constructions to similar constructions of the Boolean basic electronic circuits

4. Computer architectures that map a Turing-type algorithmic procedure into temporally sequenced electronic circuits that through Boolean logic combinations create transformations on binary number–mapped bistable electronic memory states.

Transformational Logics of Technologies

The preceding illustration shows that complex functional transformations in inventions may require more than a simple one-level mapping of logic to morphology. The logic of complex transformations requires complex sets of mappings at different levels, such as (1) sequences of mapped transformations, (2) combinations of mapped transformations, (3) hierarchies of levels of mapped transformations, and (4) topologies of sequenced, combinatorial hierarchies of transformations.

Complex inventions are hierarchically arranged systems of morphologies mapped as systems of transformational logics, and these are called technology systems.

This view of technology is fundamental to planning technological progress, as we will examine in later chapters.

The functional logic of any technology can be expressed as a topologic graph connecting the inputs into the transformation through sequential and parallel unit transformations into the output.

A technology system is depicted as a hierarchically ordered topologic graph of unit functional transformations.

If any unit functional transformation within the system is itself complex, then that unit transformation also can be described as a *subsystem* of transformations. The levels of subsystems in unit functional transformations of a technology system provide a hierarchy of functional transformations within the technology system.

Technology and Logic

Now we can more accurately describe what we began metaphorically talking about as the spirit of technology. The "ghosts" of nature in technological transformation and the Promethean seizing of control over nature from the gods—these are the functional logics for ordering the unfolding sequence of nature's expressions, as guided by inventions.

Technological invention is a creation of a means for humans to take the action of altering physical environments from present configurations into desired future configurations.

For example, the technological device known as an automobile functions as a transportation transformation over land, moving passengers from place to place. Similarly, the airplane provides the function of transportation through the air, and ships, on sea. A computer provides computational functions. Even a simple technological device such as a hammer provides a pounding function.

Function is the defining characteristic of a technology as a system, and different morphologies accomplishing the same function provide physical alternatives for a technology system.

Practical Implications

The lesson to be learned in this chapter is an understanding of how technology relates both to nature and to logic.

1. Invention is creating a new mapping of natural form to logical function—in order to manipulate nature for human utility.
2. All inventions can be described as a transformational logic mapped onto systems of physical morphologies.
   a. The total technology system can be depicted as a topology of unit functional transformations.

3. In the mechanistic perspective on nature, a description and explanation is called a *representation* of a physical system.
   a. The *morphology* of a physical system is the representation of the
      physical phenomena.
   b. The mechanistic strategy of explanation assumes that the ag-
      gregate effects of a macroscopic world can be translated into the
      physical effects of a microscopic world.
   c. To reconcile the physics of the macroscopic and microscopic
      worlds, physical facts are all instrumentally derived (measured)
      at either level under specified instrumental techniques and ex-
      perimental conditions.
   d. The importance of the mechanistic representation to technology
      lies in the prediction of changes in physical structure under in-
      strumental manipulation.
   e. Physical models are used by technologists to understand and
      conceive of inventions and then by engineers to design and pro-
      duce products embodying the inventions.
4. Functional logic in technology is the design of a sequence of unit
   functional transformations to obtain a desired output from inputs.
   a. In a decision, the environment is perceived as forms relevant to
      the desire of the decision maker, and all the mechanistic repre-
      sentations are re-sorted into classes of initial states, means,
      ends, and evaluations.
5. Patents protect the commercial exploitation of novel inventions.
   a. In the U.S. system, an invention is patentable only if it satisfies
      the three criteria of utility, novelty, and nonobviousness.
   b. Although natural phenomena and scientific laws cannot be
      patented, new material forms can be patented as well as new
      life forms.
   c. Currently in the United States, software algorithms may not be
      patented except as a part of a transforming process.

For Further Reflection

1. Identify a basic invention that established a new technology. When
   was it invented, and by whom? Describe the history of the circumstances leading up to the invention.
2. Describe in the invention how the natural structures and processes
   (of the phenomenon underlying the technology) map in direct correspondence with the functional transformation of the technology.
3. What was the first application of the invention? How well did it perform? How safe and efficient was the new technology?

6

Implementing Technology in Product and Services

Product (or service) design is the embodiment of the inventions of technology into an economic good (or service). Just as the concepts of morphology and function are central to invention, so too are they essential to design. Technological innovation in a new product/service can alter either the physical structure or the functional logic of the product/service.

The economic uses of technology occur both in the design of products and services and in production. In this chapter we will review the problems and opportunities inherent to embedding new technology in the design of new products or services, and in the next chapter we will examine implementing new technology into manufacturing—since the problems of implementing technology in products and production differ in important ways.

Product design is an initial activity of the business system that is very essential to competitiveness. For example, the previously cited study by the National Research Council Committee on Engineering Design Theory and Methodology (1991) emphasized the importance of product design in industrial competitiveness. "Engineering design...is the fundamental determinant of both the speed and cost with which new and improved products are brought to market and the quality and performance of those products" (NRC, 1991, p. 15).

Product design is also an important determinant of product cost and quality. "It is estimated that 70 percent or more of the life cycle cost of a product is determined during design" (NRC, 1991, p. 1).

Central problems in implementing new technology in new products or services include

1. Making the appropriate performance/cost tradeoff in designing a new-technology product/service

2. Determining the best application focus for a new-technology product/service
3. Determining appropriate requirements and specifications for a new-technology product/service
4. Deciding wherein should lie the proprietary and competitive advantage of a firm's product

Traditionally, technologies for products and technologies for services have been relatively distinct, but with modern computer and communications industries, such technologies are increasingly being shared for information and control in both product and service systems. Also, most modern products now embed control technologies in hardware and software. In this chapter we will consider the issues of implementing new technologies both in hard products and in software.

ILLUSTRATION
Evolution in the Design of Early
General-Purpose Microprocessors

The design of successive technology generations of microprocessors for personal computers offers an illustration of design tradeoffs (between cost and performance in products) when technology is rapidly changing. For the sake of illustration, a comparison will be made of the reasoning behind the designs of the Mostek 6502 microprocessor (introduced in 1978), the Intel 8086 microprocessor (introduced in 1983), and the Motorola 68000 microprocessor (introduced in 1985).

The architecture of microprocessors was the same for general-purpose computers from the 1940s through the 1970s. However, in the evolution of microprocessor designs, compromises frequently were made to address cost constraints. When personal computers were innovated, superior computing technology existed in higher product lines, such as the supercomputer, the mainframe computer, and the minicomputer. The question for the designer of microprocessors was how much computing performance could be designed into a product at a targeted cost. For example, limited by the available transistor density in chip fabrication in 1977, the Mostek 6502 microprocessor used 8-bit word lengths for instructions and data. Available transistor densities in chip fabrication similarly limited the designed performance of the Intel 8086 and Motorola 68000 microprocessor chips in their times.

Design of the Mostek 6502. In the Mostek 6502, the designer was principally limited by the available transistor density of the chip technology of the time—large-scale integration (LSI), which allowed up to about 10,000 transistors on a chip. Then the designer first considered the necessary features for the arithmetic and logic unit (ALU). As a minimum, the ALU should perform at least addition and subtraction of binary numbers.

To perform the steps of addition or subtraction for binary numbers, a designer can list all the steps and find that there are fewer than 256 steps and more than 16 steps. This suggested to the designer of the Mostek 6502 to consider using 1 byte of information to name a particular step in an addition operation, for an 8-bit word length (a byte) could distinguish among 256 names. Thus the 6502 chip designer chose an 8-bit operation code (Wilson, 1988, p. 215).

The next design consideration was the size of the registers in the memory that would hold the binary numbers to be added or subtracted from one another. Memory size then was limited by the number of memory chips that could economically be used with a microprocessor. In the mid-1970s, 16K chips were the largest random access memory (RAM) chips then available. Four of these chips could provide a total of 64K memory. This was then judged economically feasible by the microprocessor designers of the time. To address this size memory, the address registers in the microprocessor would require at least 16 bits for naming all the memory cells. Accordingly, the designer of the 6502 chip thought of instruction lengths in the computer of 24-bit size (8 bits of the operand code plus 16 bits of the memory address code).

This was a problem. However, the 24-bit instruction size would use a lot of the precious 64K memory for programs, leaving little memory for data. Therefore, the designer of the Mostek 6502 thought of a clever way to get around this technical limitation. "The number 24 looks nasty; not only is it not very binary, it's also very big—our programs will use up lots of expensive memory. And besides, we want the machine to do more than loads, adds, and stores—we'd like, for example, to be able to look at each member of a collection of values in memory. We can do this by inventing another operation—an *indirect load*" (Wilson, 1988, p. 215).

The constraints of component technologies (in this illustration, chip density) often require a designer to optimize a design under technological and economic constraints. Yet such inventions to optimize designs under economic constraints also create new problems. "But such an indirect load has two problems. First, it involves two memory accesses, and the first rule of microprocessor design says that it's always easy to build a processor that can do something much faster than affordable memory can cycle" (Wilson, 1988, p. 215).

Today's optimization solution for a technical problem will become tomorrow's performance bottleneck. Thus, in implementing indirect address, a new performance bottleneck was designed into the processor. In addition, playing with the address location in memory meant destroying the values in the microprocessor's accumulator—which requires more loading and storing of information to and from the memory (and this is a major source of slowness in computing). Again, the designer of the 6502 chip thought of a clever solution by giving the processor another register to be used as an address register.

Again, however, there were problems. There were not enough transistors in the large-scale integration chip technology in 1975 (when the 6502 chip was being designed) to *afford* a 16-bit register, so the designer managed with only 8 bits.

Then, in the design of the 6502 microprocessor, one other optimization was made by the designer. Instructions were made 16 bits long, instead of 24 bits, by restricting the address portion of the instruction to 8 bits. This restricted the microprocessor to using the bottom 256 locations in memory as direct values. However, the advantage was that programs could then be shrunk by this by 33 percent, and the computer went faster (with fewer memory cycles to read instructions from memory because the instructions were smaller).

The designer of the 6502 also added a program register (of 16-bit size) to indicate the next instruction to execute. Some instructions also were added to manipulate this register (so that a program could jump from one instruction to another rather than simply sequencing instructions). The designer also added the capability of performing subroutines by adding another register (plus some instructions) that provided for a "call." "A *call* saves the current program counter in memory whose address is specified by the new register, and then increments that register, while a *return* decrements the register and reloads the program counter from the indicated memory location" (Wilson, 1985, p. 217).

Thus both the limitations of computer memory and the cost of memory motivated the designer of the Mostek 6502 to optimize his design—with restrictions on instruction word length and register size. These restrictions then created new bottlenecks: limited addressable memory and slowness of performance (due to the ALU accessing memory through indirect addressing).

Still, given the economics and technical limitations of LSI chip technology in 1975, the Mostek 6502 chip was a very good design. "The resulting machine [described above] is very close to the 6502, originally designed by Mostek and used in the Apple II and the Commodore PET....It used fewer than 10,000 transistors" (Wilson, 1985, p. 217).

Next-Generation Technology Products

The activity of designing a product requires an understanding of the technologies that will be embedded in the product as well as decisions on the tradeoffs between desirable aspects of a product design and the constraints of economics and the limitations of current technology. Product design requirements are determined by the class of customers and the applications (for which the customers will use the product). In technology innovation, the most easily approached first market is an established market in which the new technology can substitute, since some knowledge of customer applications and product requirements already exist for that market (this will be described in more detail in Chap. 10). Products/services with embedded new functionality create brand new markets; improved-performance products/services with existing functionality substitute in existing markets.

For radically new products, the first critical design decision is identifying the customers' applications for the product (and a substitution market that can provide guidelines).

Still, it is difficult for a new-technology product design (or new-technology service design) to establish the customer requirements for the product/service. What does the customer want? Seldom is this question simply answered by asking the customer. Of course, it helps to ask. However, the customer may not know until he or she tries a new product.

The more radical the technical innovation in a product/service, the less likely the customer can know product/service requirements before actual experience in using the product/service.

Moreover, the more radical the functional capability of a new-technology product/service, the more unlikely the initial substitution market will prove to be the most important and largest market.

A critical design decision is identifying the necessary performance criteria and essential features of the product/service for the application.

Usually, the cost of an innovative product will be higher than the price of existing products for which it is to substitute.

A third critical design decision is estimating the acceptable sales price to a customer of the product and the required production cost for profitability at such a sales price.

As technology progresses, the products embedding the technology substitute for existing applications and start new applications. The product requirements and specifications are determined by the applications, and as the application systems develop, product requirements change. This is a kind of "chicken and the egg" problem. Which comes first in technical progress, product specifications or application system requirements? The answer is both, since they are interactive.

When radically new technology products are innovated, the immediate applications that are found for them soon indicate the performance limits of the technology. This provides incentive to improve the performance of the technology guided by the applications. (In the case of the early microprocessors for personal computers, it was primarily memory address limitations that applications highlighted.) As performance in the product improves and costs decline, then new applications open up for the product.

In a new technology, products often go through several generations of next-generation technology products. Yet in each generation there will be tradeoffs between desired performance and cost.

As technology progresses, product design must focus on optimizing product performance within the constraints of currently limiting technological capabilities and economics.

ILLUSTRATION (*Continued*)
Design of the Intel 8086

Our next example in the technology generations of microprocessors is the design of the Intel 8086, which occurred in 1980. By then, more transistors were available on a chip to improve the microprocessor design. One of the first design choices made by the designer of the 8086 was to increase all register sizes to 16 bits. By then, denser memory chips had been produced, and memory was cheaper. However, the 16-bit registers could still only address 640K of memory. The ideal, then, would have been 32-bit address registers, but again, the economics and current technology prohibited such a choice for the designer of the 8086.

Therefore, another clever design trick was used. Some bits were added to the 16-bit address to create a larger bit address. Then the processor could use a small program to write into this additional register. This allowed one to skip around in a big memory (of, say, 1 million bytes).

However, this still meant that one could not "see" more than a 640K size of that larger memory at any one time. "So we'll provide one magic register for addressing code and one for data. And sometimes the code will want to play with two different collections of data, so we'll add a third. That's not very binary, so we'll give the machine four such base registers....That machine, of course, is the 8086 design....The 8086 uses about 30,000 transistors" (Wilson, 1985, pp. 218–219). In the 8086 design, these four additional base registers were each 16-bit size, and a megabyte of memory was addressable—but only in 640K segments.

A base address was shifted four places left and then added to the machine's 16-bit address. Yet even with this clever design, new performance bottlenecks were introduced. "What are the bottlenecks in [the] design? First, the machine is using that memory too often....those loads and stores are crippling performance....Second, that megabyte of memory has people writing more ambitious programs, so they write in a high-level language to finish in a reasonable time....programs are getting bigger...using procedures all over the place" (Wilson, 1985, pp. 219–220).

Technical Progress and Design

To optimize performance under cost constraints, a designer in a new technology may invent new techniques that solve a particular problem at the time of the design. Every clever solution for optimizing a design, however, to get around some economic or technical limitation, usually also will create a new bottleneck for technical performance.

The performance/cost optimizations in one generation of design usually become technical bottlenecks to be solved by the next generation of products for which the performance/cost ratio is higher.

This is what makes design evolution possible for products from one generation of technology to the next when cost is one principal constrainer of design. Product generations will continue as long as the embedded technology systems continue to make large, discontinuous leaps in performance.

ILLUSTRATION (*Concluded*)
Design of the Motorola 68000

Continuing progress in transistor chip density provided the designer of the Motorola 68000 chip with new opportunities. First, the designer of the 68000 attacked the memory-access bottleneck by giving the new processor several data registers on the same chip so that they could be accessed quickly.

However, each of these new registers required instructions for use. "Rather than our one-address instructions (one-address because the instructions literally specified just one operand, the other being the accumulator), we'll have two-address instructions that say things like 'take 13 and add it to register 8'....We've gained performance because we no longer need to go messing around putting things into an accumulator, adding stuff into that and copying it out again" (Wilson, 1985, p. 220).

These new registers also could be expanded by having more transistors available for making the new registers of 32-bit length. Moreover, these registers could be divided into address registers and data registers. Now doing these neat new things for increased performance also created now complexities in the design. "Trouble is, we can now call for some pretty complex operations in one instruction. Things that used to be done by a whole sequence of instructions now fit into one. So let's implement the thing as an interpreter. We'll build a much simpler machine whose instructions are unbelievably crude but very quick to execute, and have a tiny program in a ROM on-chip....The little program is called *microcode*" (Wilson, 1985, p. 222).

The designer of the Motorola 68000 innovated a new feature in microprocessors, microcode. This was an innovation in logic in the form/function of the microprocessor. The chip-processing technology of very-large-scale integration (VLSI) in 1985 provided more transistor capability, making possible more functional capability in the design.

The microcode capability with the new registers allowed complex sequences of operations performed swiftly as a single but complex instruction. "We can have several different instruction sizes, with the longer ones specifying a complex sequence of operations involving multiple steps (e.g., add this constant to an address register, look into memory, take the value, use that as an address, read from memory, and multiply register 7 by that value). Real power at last! This machine...is 68000-flavored. A 68000 needs around 70,000 transistors (Wilson, 1985, p. 222).

Principles of Design

Similar to invention, design activity is also a creative mapping of the logic of a function onto a physical form. For example, Nam Suh (1990, p. 26) described designing as follows: "...the objective of design is always stated in the functional domain, whereas the physical solution is always generated in the physical domain. The design procedure involves interlinking these two domains at every hierarchical level of the design process. These two domains are inherently independent of each other. What relates these two domains is the design."

In the functional domain, Suh called the specification of the logic requirements for a design the "functional requirements." In the physical domain, Suh called the specification of the requirements of the physical forms for a design the "design parameters." The design then maps functional requirements into design parameters.

Technical innovation can be embedded in a design in either (1) new functional logic in the design, (2) new physical forms (morphologies), or (3) new mappings of functional requirements to physical forms.

Innovative software design is an example of new functional logic. Software can be used for conceptualization and communication, information handling and processing, and transactions and system control. Innovations in service technologies are frequently dependent on new software design.

Innovative hardware design is an example of new physical forms or of new mappings of functions to physical forms. For example, a metal part may be shaped by machining, cutting, or grinding. Each process is a different physical form for the same shaping function.

Some technical progress in a new technology will be accomplished by scaling up previous techniques, such as increasing the numbers of transistors on a chip. However, increasing scale also may introduce increased functional complexity. New functional logics may be invented to handle increased morphologic scale.

Technical progress implemented in designed products/service can progress not only by improvements in physical form (morphology) but also by new functional logics (to handle increased functional complexity made possible by progress in morphology).

New logics are important sources of innovation in service technologies. For example, in the software industry, new classes of functional logics create new product lines, such as word processing software, database software, spreadsheet software, computer-aided design software, etc. Also, innovation in combining logical functionality creates new generations of products. For example, the combining of database and spreadsheet logics produced a next generation of spreadsheet software product. Generations in service technologies are often created by functional integration.

Decomposition of a complex design is one way for a designer to handle a complex function. Design decomposition is a hierarchical ordering of the logic of the transforming operations of a designed product (or process or service). For example, Suh (1990, p. 37) provided an illustration of the functional logic of a metal removal device. At a sublevel, this device requires the functional operations of power supply, workpiece rotation source, speed-changing devices, workpiece support and tool-holder, support structure, and tool positioner. And below each of these may be a lower functional level, such as, for example, the workpiece support and toolholder requires subfunctions of a toolholder, a positioner, and a support structure.

Innovation in a product system may occur in any functional subsystem or at any level of the functional hierarchy.

Design constraints are specifications of limits within which the design mapping must operate. Constraints arise both from the design specifications and from the relationships of physical parts within a morphology.

Innovation in a product system also may occur from innovative ways of removing design constraints that limit performance.

For the practice of good engineering design, Suh (1990, p. 47) also proposed two general design axioms:

1. *The independence axiom:* Maintain the independence of functional requirements in the design.
2. *The information axiom:* Design the functions to require minimal information from prior functions or from the outside to perform the function.

Suh argued that when designs require the same part to provide multifunctions, the overall design suffers from increased complexity, more catastrophic failures, more difficult maintenance, and unanticipated functional interactions. Suh also argued that when designs require information from prior functions or from the outside, the reliability of that function decreases as a result of its outside information requirements.

Innovation in design also can improve product quality through improving the design's functional independence or through improving the functional information autonomy of subsystems.

ILLUSTRATION
Design Shoot-Out for the U.S. Air Force
ATF Fighter

In April of 1991, a design choice was made for the next generation of fighter airplanes, the advanced tactical fighter (ATF). The announcement came just after in the U.N.–Iraq war of 1990–1991, which demonstrated the importance of electronics, "smart" bombs, and stealth fighters as a new generation of weapons. "The U.S. Air Force chose the Lockheed F-22 aircraft as its advance tactical fighter and the Pratt &

Whitney F119 turbo-fan engine as the powerplant in an at tempt to attain 21st century air superiority at minimum risk and lowest cost" (Bond, 1991, p. 20).

Then the tactical fighters of the U.S. Air Force were the F-15 and F-16 planes, which had been upgraded in performance and in electronics as technology changed. However, they were becoming obsolete in terms of performance. The technical capabilities of radar evasion, i.e., stealth capabilities, had become important as radar became the major way of sighting aircraft at long distance. "'Since the Persian Gulf war, there has been renewed understanding of the importance of stealth,' [Kent] Kresa [Northrop chairman] said, 'because it has been shown to save lives and be more cost-effective by permitting operations to be conducted without large numbers of support aircraft'" (Bond, 1991, p. 29).

However, the emphasis on fighter agility also remained important during tactical dogfighting after the enemy has sighted the fighter. "This aircraft, thanks to its...agility, will be just as lethal...within visual range as it is beyond visual range..."(Bond, 1991, p. 22).

Lockheed, Boeing, and General Dynamics teamed to design and produce the airframe, and Northrop and McDonnell Douglas were teamed as their competitor. Pratt & Whitney and General Electric were competing to design and produce the engine. Two copies each of engineering prototypes were built: (1) the YF-22 with Lockheed–Boeing–General Dynamics airframe and Pratt & Whitney engine and (2) the YF-23 with Northrop–McDonnell Douglas airframe and GE engine. Both the YF-22 and the YF-23 were flight tested to demonstrate that the companies had reduced the risk of applying the advanced technologies required to satisfy Air Force objectives.

The technical risk is very high when designing a next generation of a technology. Thus the comparison between the prototypes was made not only on specific performance measures but also on whether the prototypes attained desired specifications. Even though both prototypes were designed to the same set of customer (Air Force) requirements, there was still room for choice in the two different design approaches (Bond, 1991, p. 22):

> Advanced tactical fighter designs submitted by Lockheed and Northrop–led teams displayed fundamental differences in air combat philosophies....Both contractor teams were free to interpret the Air Force's ATF requirements....Lockheed...opted for relatively conventional-looking "balanced design."...The F-22, they decided, must balance low observable (stealth) and supercruise qualities against agility, reliability and maintainability...."Our emphasis was on developing a fighter that had unprecedented agility combined with unprecedented stealthiness. But we never allowed stealth or supercruise [objectives] to make us back away from the agility issue," Al Pruden of Lockheed said.

At that time, the Air Force preferred the Lockheed-led team's design balance (of agility with stealth and supercruise).

Markets and Performance/Cost Tradeoffs in Design

The preceding illustration highlights another problem at the heart of product design—balancing functional performance capabilities. Even with one set of customer requirements, two different design teams can provide different product designs.

In response to the same customer specifications, innovative designs can still differ in tradeoff balance between the conflicting customer demands on the system.

In the preceding illustration, the two design teams chose different balances between the criteria for cruise speed and stealth and agility. (We also saw two different airframe designs with two different engine designs for four different product design alternatives.)

Different designs can achieve different balances, and different customers may prefer different balances.

Products require several different functional performance attributes because such products may be required to perform under different conditions or in different environments. Also, different markets for new technology may value the performance/cost tradeoff differently:

1. In the military market, performance usually will be valued over cost.
2. In the industrial market, performance usually will be valued at a cost that can be justified as an appropriate return on investment.
3. In the business market, performance usually will be valued within the upper boundaries of the business practices of monthly expenses (leasing will be used for large-ticket items).
4. In the hobbyist market, performance usually will be valued within the upper boundary of disposable recreational income.
5. In the consumer market, performance will usually be valued within price ranges of comparable categories of consumer purchases.

Therefore, for each market, the design performance/cost tradeoff must be made differently.

ILLUSTRATION (*Concluded*)
The U.S. Air Force ATF Fighter

Each of the two airframe copies, from the two teams of Lockheed and Northrop, were supplied with alternate engines from Pratt & Whitney and General Electric: "Lockheed's two YF-22s, one powered by a General Electric YF120 engine and the other by a Pratt & Whitney YF119, flew 74 sorties over a 91-day period last year. The team fired AMRAAM and Sidewinder missiles and demonstrated an ability to operate with thrust vectoring at [a] 60 degree angle of attack. YF-22s accumulated more that 24 hours of supersonic cruise time at speeds of about Mach 1.4–1.6. Lockheed test pilots said the YF-22 team demonstrated full-sick aileron rolls at 50-degree AOA, and the prototype engines performed well at those angles of attack. The fighter could maneuver at high Mach 'as well as other aircraft do subsonically'" (Bond, 1991, p. 23).

The airframe of the Lockheed plane used only 23 percent of the new composite materials to help give it stealth characteristics, with the 64 percent being metals such as titanium and aluminum to give it the high-speed performance. The Pratt & Whitney engine design had an innovation called a "variable cycle capability," which enabled the engine to operate as a conventional turbojet at supersonic speeds and as a turbofan at subsonic speeds. This innovation improved the efficiency of the engine over the different ranges of its performance.

The combination of the Lockheed airframe and Pratt & Whitney engine became the preferred combination. "F-22 will have a larger speed/altitude operating envelope in military power than the F-15C (the current fighter) has in afterburner" (Bond, 1991, p. 23).

The purpose of the prototype demonstrations was both to demonstrate performance and to reduce the risk of introducing new technology into a production plane. "...officials estimated that the [Air Force] service will have spent $98 billion on the F-22 by the time it pays for its final order (of 648 aircraft) in 2012....The 648 aircraft correspond to the equivalent of 5.5 tactical fighter wings" (Bond, 1991, p. 20).

The commercial benefits to the Lockheed team were to be substantial (Bond, 1991, p. 28):

> The Lockheed–General Dynamics–Boeing team that won the advanced tactical fighter competition will not reap a financial windfall, but will see a steady buildup in funding and resources that will peak during the late 1990s and continue until at least the year 2015....the win will provide the partners with a strategic boost in technology and resources, placing them in the forefront of the tactical fighter arena for decades.

Systems Analysis in Design

Embedding a technology system into a product design concept requires a system analysis of product functioning. The product must be viewed as an engineered system with a functional logic mapped into a product morphology—function/form in design. Systems analysis of a product design is a systematic depiction of the functional transformations required for product performance.

> *A systems analysis for a product / service design lays out (in a flow diagram) the logical transformations as a series of both parallel and sequential operations of the product functioning.*

In this analysis one can identify the locations of technological innovation within the system as

- Boundary
- Architecture
- Subsystems
- Components
- Connections
- Structural materials
- Power sources
- Control

Accordingly, targets for implementing new technology into the product system should be expressed in terms of an engineering systems analysis of the product system, identifying loci of innovation and estimating their effects on the functional performance parameters of the product system. This analysis establishes the location and results of technical improvement in the product system.

The system questions for technical innovation in product design include

1. What are the loci in the product system for embedding the technical progress?
2. What improvement in functionality and/or performance will the technical advance provide?
3. How will the customer recognize value in the applications for this improvement in functionality/performance?
4. How important will this be to the customer?
5. What are the technical risks of innovating such a technical advance in the product system?

6. What is the appropriate tradeoff between technical advance and cost of such innovation that will be appreciated by the customer?

The important commercial question is whether a given innovation is visible and important to customers of the product.

Technical innovations in a product that are invisible to or unused by a customer produce no competitive advantage.

In innovative designs, major components in the product morphology (e.g., airframe and engines) can be matched differently to expand the design choices.

ILLUSTRATION
Systems Configuration of an Airplane

In the preceding ATF airplane illustration, the principal technology system morphology was partitioned into airframe and engine. Figure 6.1 depicts a decision flow diagram for this process for an airplane design. The configuration specifications state operating parameters for the plane system: load capacity, top speed, fuel efficiency, rate of climb, range, acrobatic stress limits, damage tolerance, maintainability, etc. These configuration specifications then provide both targets and constraints for (1) the airframe design and (2) the engine design.

In the airframe design, choices are available to the designer about the shape of the airframe, its structural reinforcement, and the packaging of its exterior surfaces. In the ATF fighter illustration, the two teams had choices about shapes for stealth and materials for stealth and strength and high-speed cruise. The Lockheed design chose to use fewer new composite materials than the Northrop design. Although the composite materials in 1991 provided better stealth capability, they were more limited than titanium to temperature heating due to high cruise speeds.

In the engine design, choices are available to the designer about power, weight, features, and dependability. In the ATF fighter illustration, the Pratt & Whitney engine produced 35,000 pounds of thrust. Its special features were vectoring nozzles that steered the exhaust/thrust gases. An external flap at the nozzle ensured that the exhaust flow lines were tailored to meet the low-observable requirements. An internal flap controlled the direction of engine exhaust by several degrees to increase agility.

Moreover, the Pratt & Whitney engine had 63 percent fewer rotating parts and 25 percent fewer parts overall than the General Electric en-

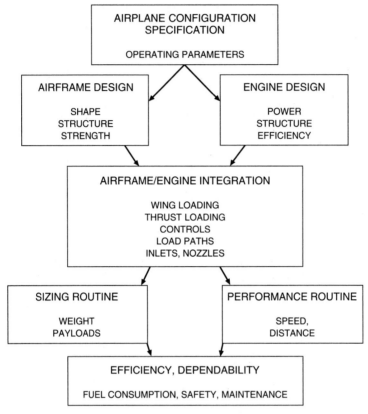

Figure 6.1 Decision flow in airplane design.

gine. These additional parts did give the General Electric engine a special feature, an advanced variable cycle technology. But the Air Force rejected this feature in preference for the fewer parts of the Pratt & Whitney engine (for lower-cost maintenance and repair). Design choices are tradeoffs.

Once the airframe configuration and engine design have been selected, then the next systems design problem is integrating the two, adding controls. This integration may lead to revisions in design for final requirements for sizing and weight or for performance (payload, speed, agility, stealth, etc.).

Service Technologies and Software Design

Technologies for the service industries (or for internal services within a firm) use products and devices from manufacturers but in a procedural system that requires information and communications. Therefore, what is unique in a service application is the information and communication and control procedures.

The design of software for information handling, transactions, communications, and control is essential to services.

The differences between technology and product and service are that *technology* is a *knowledge* of the way to manipulate nature, *product* is a *designed embodiment* of technologies in a good, process, or service, and *service* is a procedural application of technology. The central concepts in the logic of technology are invention, research and development, and innovation. The central concepts in a product are design, form and function, and prototypes. The central concepts in a service are application, procedures, and control.

With the advent of computer technology, the role of information handling and communications in applications has increased. In service technology, information, control, and communications are increasingly embedded within and facilitated by software.

Software design uses a systems approach to envision the functional transformation and specifications for the design of the software application. The differences between software design and material product design are (1) there are no form/function mappings in software design and (2) incompleteness of specification of the application impacts the quality of software design more than in product design.

In designing complex logics for software applications, large teams of developers are often necessary. For example, Robert Rowen (1990, p. 10) has emphasized the problem of management in software development:

> Large software development projects confront two major obstacles. The first is that projects this large normally are not performed by one or two people. The existence of multiple levels of personnel and large expenditures...makes a management process that controls the project mandatory. The second hurdle is overall systems complexity. The complexity inherent in such systems places a burden on all written documentation in that it be useful as an unambiguous vehicle for communication.

Software design is the creation of a logic for a specific application. The elements of the logic are conceptual primitives such as terms and relations and algorithms. Moreover, logic is the place wherein software development projects most often fail.

The three major design sources of commercial failure in software are (1) incompleteness of understanding the boundaries of the software application, (2) lack of insightfulness about the key logic features of the application, and (3) failure to rid the software of bugs.

ILLUSTRATION
Crash of the First Prototype of the F-22
Advanced Tactical Fighter

After Lockheed won the design competition in 1991 for the ATF fighter, Lockheed continued development of the product design. A year later, in April of 1992, flight testing of a prototype of the ATF (then designated the F-22) ran into trouble when the prototype plane crashed. "'A computer software problem probably caused the weekend crash that destroyed the only flying prototype of the F-22 Advanced Tactical Fighter,' the Air Force's top general said yesterday" (Gellman, 1992, p. A3).

The ATF was one of the first U.S. Air Force fighter planes in which the controls were wholly electronically connected. Previous planes had the pilot's control instrumentation physically connected to the control surfaces of the airplane system by wires or hydraulic systems (even in addition to any electronic connections). In the ATF, the pilot's instructions are translated by a computer, which issues electronic data to control the plane's flying. Thus the computational software that implemented control was critical to the performance and stability of the aircraft.

On Saturday, April 25, a Lockheed test pilot, Tom Morganfeld, had taken the prototype plane up for supersonic flight testing. However, before going supersonic, he was told that the telemetry link between the test instruments on the plane and the ground receiving stations was not working. Thus any data collected in the testing would be lost. Accordingly, he had to turn back and land the plane.

Then his problems continued: "Already airborne, the F-22 was too heavily laden to land safely, and so Morganfeld began a series of high-speed, low-altitude passes over the runway to burn excess fuel. On the second pass, Morgenfeld lost control....The F-22 crashed, burst into flame and slid 8,000 feet—well over a mile—before stopping. Morganfeld escaped with minor injuries" (Gellman, 1992, p. A3).

What happened was that Morganfeld retracted his landing gear and ignited his afterburners at about the same time to continue to hold altitude and burn fuel. The resulting burst of speed from both actions required the on-board computer to react quickly to control the plane's stability: "...on modern 'fly-by-wire' aircraft, a pilot has no direct control of the physical movement of the [airplane's] flaps. Morganfeld's commands were interpreted by a computer-controlled servo-motor that continuously made thousands of calculations to adjust the controls, much as anti-lock brakes do on late model automobiles" (Gellman, 1992, p. A3).

Although, at the time, the investigation was not complete, Air Force General Merrill A. McPeak, testifying to the House Armed Services Committee, said that preliminary indications were that the fault lay in the control software and not in the airplanes's aeronautical configuration. "McPeak said he believed that 'something in the logic of the fly-by-wire flight controls system' failed to move the control surfaces far or fast enough to keep up with the pilot's commands" (Gellman, 1992, p. A3).

This illustration underscores the importance of both software and hardware implementations of new technologies and the "bugs" that occur in new designs.

Product Innovation and Design Bugs

Because of the complexity of technology systems, whether product or service technologies, all new innovative designs will encounter problems—bugs.

It is important to manage the implementation of new technology in product/service design with anticipation of technical problems and bugs.

For example, Douglas Clark (1990, pp. 26–27) described the designing of the Digital VAX 8800 family of midrange computers announced in 1986 as follows: "We simulated the VAX 8800 design extensively before building the first prototype....Engineers simply marked a centrally posted sheet with a tick-mark for each bug they found....We regarded finding and fixing hardware design bugs as our most important activity....over 96 percent of the total reported design bugs were found before the prototype was built."

Douglas Clark also emphasized the importance of including bug identification and fixing bugs as a formal process within the design process, arguing that *debugging* (the tracking down and fixing bugs) is a normal and necessary part of a design process. Software designers have added the jargon of *bug* to our lexicon to indicate any problem with a new product that was not anticipated and correctly avoided in the design. In software, bugs are logical errors in the design. In hardware, bugs can either be logical errors or physical problems. Debugging is important to both software and hardware design.

Design bugs arise from the complexity of a technological and product system. The more complex the system, the more difficult it is to fully think out and consider all the aspects of the system and the interactions within the system. Therefore, bugs surprise the designer because the full implications of a design for a complex system may only be understood in the testing of that system. This is particularly true of designs that embody new technology.

The more radically new the technology, the more likely there will be bugs—not from lack of judgment but from lack of experience with the new technology.

Therefore, bugs are inherent in new technology, and when innovating new technology into a product design, the design procedures should include debugging as normal to design and not simply regarded as a "failure" of a designer.

The important thing in managing technological innovation in product design is to find the bugs early.

The later in the product-development process that a bug is discovered and fixed, the more costly its solution will be and the slower will be the development of the software.

In order to identify bugs early, it is useful to design and test products in a modular fashion. Also, the ability to simulate a product design in a computer is very useful to early identification of design bugs. Areas in the design that are most likely to produce bugs include (1) areas where complexity is greatest, (2) interfaces in subsystems when subsystems have been designed by different people or groups, and (3) parts of the design that are more rarely used in the product functioning.

The procedures for product testing are also important to the early identification of bugs. The traditional approach to product testing has been to make a list of tests that test each subfunction of the product and then to perform these tests in sequence. And this is a proper way to begin testing, but it may be insufficient, particularly for a complex system.

Clark (1990) has suggested complementing testing design with a "bug-centered approach" on the part of the product designers. For example, the product design team may set aside some periods (a day, a week, or a month perhaps) as deliberate bug-finding days. The idea is that finding bugs is a positive contribution to design, not a negative contribution. Prizes might even be given to the team member discovering the most or most-challenging bugs.

Keeping a chart of the number of bugs discovered per week during the design is also a useful procedure to managing innovation in design. Confidence in the design should be determined not by the number of bugs found but by the *rate* of finding bugs. There will never be a guarantee that all design bugs have been found in a complex product system. However, when the rate of finding bugs decreases toward zero, particularly in a design process that deliberately and continuously seeks to find bugs, then the design is approaching robustness and will be ready for final testing.

It is also helpful to review the types of bugs being found in the design. For example, is a bug indicative of a more general problem with the design? Why did the test methods not catch the bug in the first place? Could the method used to find that bug find more bugs? And so on. In summary, making bug finding an active, deliberate, and focused part of the design process as a complement to the formal testing of the completed design improves one's ability to innovate new technology in a new product design.

Risks in the Design of Technologically Innovative Products

For any product, the principal functional transformation capability of the product must first be designed into a principal technological morphology for the product. This constitutes an *embedding* of a technology system concept into the product system concept. Recall that the product system is a broader concept than the technology system because the former also will contain subsystems and features that are necessary for the product application and are additions to the principal technology system.

Innovative product designs can therefore fail commercially in two ways:

1. The product design may fail as its expression of a principal technology system performs functionally less well than a competing product.

2. The product design also may fail in its balance of performance and features as perceived by the customer (even if technologically performing as well as a competing product).

As described in terms of the linear innovation process in Chap. 1, product design for a product embedding new technology (either radical innovation or next-generation technology innovation) should take two distinct steps: (1) as an engineering prototype of the new product and (2) as a manufacturing prototype of the new product. The reason for this is that in radical innovation the engineering prototype is a great challenge to design a functioning product within the size, shape, and cost constraints. Then this design must be modified substantially to make it manufacturable in large quantities with consistent quality and at low cost.

The engineering prototype design of a new product should establish five criteria for the innovative product:

1. *Performance* focused on and adequate for the product application
2. *Robustness* of performance under the field conditions of the application
3. *Dependability* of operation over the intended life cycle of the product
4. *Safety* of operation of the product by the customer
5. *Cost* reasonably low so that the product can be sold at the intended price

Innovation in technology systems to be embedded in the product system can be thus evaluated on the contribution of the new technology to performance, robustness, dependability, safety, and cost.

The manufacturing prototype design of a new product should establish two additional criteria:

1. *Manufacturability* such that the product can be produced in adequate volumes in reasonable throughput times
2. *Quality* such that the product can be produced with acceptable yields and no defects

Also, it is important to appreciate that the product system itself may be only one component in a larger technology application system of a customer. Thus the first problem in designing a new product, i.e., determining customer requirements, can be complicated. If a product fully provides a complete application system to a customer, then the problem of determining customer requirements is easier—since the application requirements would be in one-to-one correspondence with the product specifications. However, since the product from the customer's view is incomplete for an application, this incompleteness creates ambiguity about the product specifications. When the product is radically innovative and creates a wholly new market, then even the customers do not know their requirements, since applications are new and often remain to be invented.

The more radically innovative a new product in technological function, the more uncertain are its principal markets and customer requirements.

There are other complications that make developing the technical specifications of a product difficult. When customers use a product for several applications, no one set of product specifications will be optimal for all applications. This can give rise to product model variations and differences in competing products. Which customer application should the product design engineer optimize in the product specifications? This is a puzzle for the designer and another risk.

Designing a product to specifications most demanding for customer applications will likely produce the most robust design.

Another complication is that the technical performance of a product is often in proportion to the cost of the product—more performance for a higher price. However, if the product is priced too high, a customer may not buy it. Misjudging the right ratio of performance versus price for a class of customers makes commercial failure.

So how much performance should one trade off for cost? This creates more ambiguity about customer requirements in product design. Generally, there will be an upper limit to the pricing of an innovative product, and this ceiling becomes more rigid as one moves from the military to the industrial to the business to the consumer markets.

Another complication is that all products contain technologies other than the principal transforming technology that the product embodies. They contain other technologies in the parts of the product, in materials in the product, in connections in the product, and controls in the product. And they also depend on technologies in peripheral components to complete the product. The technology in the product design will be limited by the technical limitations and costs of all these other technologies. Again, this produces limitations in product design and often engineering ingenuity to design around these limitations—more sources of variation and ambiguity in product design.

There are several sources of risk in product design, including uncertainty in establishing customer requirements, tradeoffs between product performance and cost, and incompleteness of the product as an application system for a customer.

Therefore, the design process for next-generation technology products should include an application testing of engineering prototypes with alternate design balances. This is particularly important when a combination of technological innovations is to be incorporated within the product system, and their interactive impact on performance should be tested.

Thus being a technology leader in the introduction of new-technology products has some competitive advantages and some disadvantages. The advantages center around temporarily having a unique product, and the disadvantages center around the initial limitations of an innovative product. Being a technology follower has the opposite disadvantages of competing with a "me too" product but with the advantages of an improved and better-focused product.

New technology embedded in a new product carries the risk that the technology may be premature in terms of needed performance, dependability, and safety. It is also costly. Recall that the newer the technology, the more clumsy and costly it is. The technology leader may gain an advantage in being first in a new technology but at the risk of having a product quickly outmoded by a technology follower's competing product with improved technology and at lower cost.

Differing organizations within a firm, from research to product development, will have different visions of a next-generation product design, and it will never be clear whose vision is correct (except in retrospect). When the first model of a next-generation technology product has been marketed, it usually has been too expensive and not properly refined and focused for a business/consumer market. Since the market will decide, it is important to be prepared to quickly alter the first product design.

> *Product strategy for a next-generation product should include the intention and capability for a rapidly redesigned and remarketed product.*

In summary, several kinds of risks occur in innovating products that embed new technology:

1. Selecting the application focus of the product and arranging its completion into an application system
2. Judging the appropriate tradeoff between performance and design
3. The make/buy decisions and competitive advantage
4. The need to coordinate product design with manufacturing process design
5. The relative competitive advantages of continuity or discontinuity in the design of next-generation technology products/services

Practical Implications

The lessons to be learned in this chapter are about how to facilitate the transfer of new technology into the design of new products or new services.

1. Product design is the first step of the product innovation process, with the majority of the total cost of a product determined in the design phase.
    a. Product design may span a spectrum of innovativeness from the design of a radically new product line, to the emergence of a design standard in a new technology, to the redesign of a product line, to a next-generation technology design of a product line, and product generations will continue as long as the embedded technology systems continue to make large, discontinuous leaps in performance.

   b. The performance/cost optimizations in one generation of designs usually become technical bottlenecks to be solved by the next generation of products for which the performance/cost ratio is higher.

2. The activity of designing a product (or service) is a creative mapping of function into a physical form.
      a. A systems analysis for a product/service design lays out in a flow diagram the logical transformations.
      b. Technical innovation can be embedded in a design in either a new functional logic or a new physical form or both.

3. Problems in implementing new technology in product (or services) design include
      a. Determining the best application focus.
      b. Determining the product specifications.
      c. Making the appropriate performance/cost tradeoffs.
      d. Deciding the competitive benefits of product continuity or discontinuity in the design of next-generation technology products.

4. The proper design of a product for a customer's application will always involve uncertainty.
      a. The more innovative the technology in the product, the greater is the uncertainty in product design specifications.
      b. Different markets for new technology will value the performance/cost tradeoffs differently for a product design.
      c. Innovative product designs can still fail against poorer existing products when the innovate product is not complete for an application.
      d. For any innovative product to succeed in the market, the value perceived by the customer must exceed the value the customer is asked to pay.
      e. Designing a product for manufacturing is the most important factor in the cost of the product.
      f. It is important to manage the implementation of new technology in product/service design with anticipation of technical problems and bugs.

For Further Reflection

1. Identify the key product of a major technology which has a history of several generations of products.

2. In each generation of products, describe the improvements in per-

formance, features, and/or cost that distinguished the generations. What was the major technical bottleneck overcome at each generation?

3. Identify which companies and their products dominated the markets during a generation. Was it the same company or different companies?

4. Did the types and ranges of applications expand over these generations?

5. In the current generation, how has the key product forked into different major product lines? What are the principal customers and applications for each product line, and how do design balances between features/performance/cost differ in each line?

Implementing Technology in Production

In addition to improving products, technology also translates into economic utility through improving manufacturing. However, there are important differences between implementing new technology into manufacturing and implementing it into products and services. In implementing technology into products, one aims for large changes in product performance to provide added value to the customer, whereas in implementing technology into production, one aims for small, continuous advances to minimize disruption in the production processes.

Manufacturing requires *repeated* acts of production with *each and every* detail under *tight* control. Tightly controlled repetition of details means that implementing new technology into manufacturing will disturb a process whose main goal is to *avoid* disturbances.

Innovation in a manufacturing system should be incremental or modular to minimize disruption of the system for continually improving it.

ILLUSTRATION
General Electric's Refrigerator War

In the 1970s and 1980s, one American manufacturing business after another was assailed by foreign competitors producing equivalent products at lower cost and higher quality. General Electric's (GE) consumer appliance business faced such competition. As a result, GE gave up its small appliance business but chose to battle it out in their large appliance business.

Ira C. Magaziner and Mark Patinkin (1989, p. 114) describe the challenge GE then faced: "Fifty miles south of Nashville...is one of the

world's most automated factories. Had it not been built, U.S. households might soon have had yet another product—the refrigerator—
stamped 'Made in Japan.' Instead, here in the heartland, General
Electric found a way to build products better and cheaper than those
made by foreign workers paid one-tenth American wages."

The GE manufacturing manager who helped implement the improvement was Tom Blunt (Magaziner and Patinkin, 1989, p. 114):

> Tom Blunt still remembers the day in 1979 that he first stepped into
> Building 4, the plant in Louisville, Kentucky, where compressors for GE's
> refrigerators were made....The plant was a loud, dirty operation built with
> 1950s technology: old grinders, old furnaces, too many people. Finishing a
> single piston took 220 steps. Even the simplest functions had to be done by
> hand. Workers loaded machines, unloaded machines, carried parts from
> one machine to the next. The scrap rate was ten times higher than it should
> have been, 30 percent of everything the plant made was thrown out.

In 1979, this sort of practice was frequent in American manufacturing. The American industrial capacity had aged after World War II. In
the 1970s, the rise in energy prices prompted by the OPEC cartel had
pointed out the obsolescence of American industry, while the accompanying inflation reduced incentives to make the investments necessary
to improve manufacturing technology. While the American manufacturers hesitated, however, the Japanese had not. (For example,
Japanese automobile manufacturing had continued to improve productivity during the 1970s so that by the end of the decade, American manufacturers were only half as efficient.) The subsequent decade of the
1980s became the period when America became aware that its had lost
manufacturing superiority and was rapidly losing manufacturing base
in many industries.

Blunt had only recently joined GE's Major Appliance Business Group
(MABG) as chief manufacturing engineer for ranges. "He was too new
to start pushing for major projects—especially projects in someone
else's department. Besides, he thought management would never pour
huge dollars into redoing a whole factory. He'd come to realize that
MABG preferred cosmetics—bells and whistles—engineering. In
Louisville, the money was all on the marketing side" (Magaziner and
Patinkin, 1989, p. 115).

Things had to change at GE's Louisville plant, and Blunt found that
other managers at GE had begun to worry about the poor state of manufacturing in Louisville The major appliance business group's profits
and market share were declining. One competitor, Matsushita, was
manufacturing better and cheaper compressors in Singapore and was
selling them to the GE appliance subsidiary in Canada. Matsushita also
was experimenting with a different technology for compressors—rotary

compressors. And this was a technology GE had invented (but had used only in their air conditioners). Moreover, Whirlpool was moving its compressor manufacturing to Brazil for lower labor costs. In the fall of 1981, several manufacturers approached GE offering lower-priced compressors than GE could produce. Then GE's managers began talking about the strategy of sourcing rather than producing their compressors.

However, as a manufacturing engineer, Blunt did not like that strategy. The compressor was the core technology of the refrigerator, important to performance and energy efficiency. "'Tom Blunt didn't appreciate that kind of talk. By now, he'd been named head of advanced manufacturing for refrigerators, MABG's biggest product....' John Truscott, MABG's chief engineer, agreed. 'The compressor is the heart of the refrigerator,' he says. 'The refrigerator is the heart of this group. I didn't want to give away our heart'" (Magaziner and Patinkin, 1989, p. 115).

Recall that in any product, some technical aspect may or may not provide a competitive advantage. If a technical aspect does provide a competitive advantage for differentiating the product, one should not source out that aspect. In refrigerators, energy efficiency is one differentiating competitive factor, so it would have made poor competitive strategy to source out the compressor.

In this illustration, we are also seeing a historical situation—although manufacturing is the heart of the business enterprise, it had been neglected by American business. After World War II in America, production was neglected, mismanaged, and undervalued by senior management in many American manufacturing firms. Throughout the 1960s, many executives in America assembled conglomerated firms and attempted to manage them solely on the basis of financial criteria. In the 1970s, energy-based inflation had made the cost of investment in improving manufacturing an apparently poor immediate investment. By the 1980s, American manufacturing capability in the mechanical manufacturing industries had generally fallen behind the best of then international manufacturing capability in terms of quality, cost, and productivity.

Manufacturing Culture

Let us pause in this illustration to appreciate the special perspective of the manufacturing engineer and the manufacturing manager. This is important if we are to understand how best to implement new technology into manufacturing.

To perceive the special aspects of manufacturing activity, one should examine them strategically. For example,

1. Cost, quality, throughput, and safety are the primary values in the manufacturing function.

2. Details of manufacturing processes and materials matter very greatly in manufacturing; they are important factors in competitive quality and cost.

3. In aggregate, the secondary characteristics of materials and processes (their details) are as important as their primary characteristics to manufacturing cost and quality; and in this emphasis on details, manufacturing engineering differs from product-design engineering.

4. In repetitious situations, variation and other problems are endemic, so product characteristics will always vary within the same production process no matter how tightly the process is controlled.

5. A kind of Gresham's law applies especially to manufacturing managers: Whatever can go wrong in a manufacturing system will go wrong—frequently.

6. A corollary to this also affects manufacturing management: Whatever is fixed will go wrong again, for a different reason.

7. There is even a kind of "Gremlin law" in the experience of implementing technical improvements in manufacturing: Any technical change intended to improve a manufacturing system, no matter how carefully planned, will first bring the system to a dead halt.

8. And a corollary also applies to technical implementation: After any improvement, the production system cannot be made to operate properly again without more effort and problem solving.

9. Manufacturing is a sociotechnical system that makes people an essential part of the manufacturing system along with equipment and automation, and people are principally the sources of both errors and improvements in manufacturing procedures.

With precepts such as these, one can see that there can be cultural gulfs between the experience of the manufacturing engineer and manager and those of other business personnel. For industrial researchers and product designers, the essence of experience lies in the principal characteristics of a product. For manufacturing personnel, on the other hand, the essence of experience lies in the secondary characteristics of product production. For finance personnel, the essence of experience is profitability, cashflow, and return on investment. For manufacturing personnel, the essence of experience is continuity, throughput, quality, and safety.

The other business cultures regard the manufacturers as nitpickers about detail, whereas manufacturing engineers regard the designers, researchers, and finance personnel as naive. Senior management rewards production management for output and quality, not down time and rejects. Research personnel will be enthusiastic about offering innovations to improve the manufacturing process, but manufacturing management will be skeptical because it knows that the production system will have to be debugged and retuned before it operates properly again.

In America, these cultural differences between manufacturing and the other functional areas of business have often fostered feelings that senior management lacks appreciation for manufacturing. To understand the culture of manufacturing personnel in America in the post-World-War II era, one must appreciate their very deep feelings of neglect. For example, in 1991, the U.S. National Research Council published a study on research priorities for manufacturing (Manufacturing Studies Board, 1991, p. 18):

> The image of manufacturing as a career [in the United States] has not evolved at the same rate as the manufacturing environment. Despite the advanced manufacturing technology that requires operators with high-level, multidisciplinary skills, manufacturing retains a sweatshop image, characterized by dirty work in noisy environs. Manufacturing receives little promotion from career guidance counselors and the generally low esteem in which manufacturing is held is reinforced by companies' recruitment policies, which often rank manufacturing engineers below [product] design engineers....

To understand how an American manufacturing engineer in the decades of the 1950s to the 1990s felt, try imagining yourself as a god responsible for a universe that is perpetually running down, populated with disobedient people and neurotic machines, and supplied from distant sources that take light years to respond to reasonable requests and then respond with defective materials. Now add the view that, as a manufacturing engineer/manager, you are only a demi-god who, while having all the responsibility for productivity in this universe, yet see all the other gods, particularly those of Finance, Marketing, and R&D, get all the perks and high salaries.

The National Research Council Committee on Manufacturing also pointed out that Americans had not properly trained their manufacturing personnel. "The qualifications of [American] manufacturing management also come into question. Among Japanese users of advanced manufacturing technology, most managers have been trained as engineers. Most U.S. manufacturing managers have graduated from programs that stress financial management and have spent little or no time on the shop floor" (Manufacturing Studies Board, 1991, p. 18).

ILLUSTRATION *(Concluded)*
GE and the Refrigerator War

Now you may understand how GE engineer Tom Blunt probably felt in 1980 and, moreover, how deep the problem was in GE's Louisville plant. In 1981, Ira Magaziner was a manufacturing consultant and hired by GE to provide information on their competitors' refrigerator manufacturing costs. He visited all the compressor manufacturers in the world and then returned to Louisville to give his report to GE's Major Appliance Business Group.

Magaziner had documented a major cost problem for GE. At that time, it cost GE's MABG over $48 to make a compressor, while (in comparison) the costs at Necchi and Mitsubishi for the same compressor were $32 and $38. In addition, other competitors were designing new plants that would further drive costs down—such as Hitachi and Toshiba aiming at a cost of $30 per compressor and Matsushita and Embraco aiming at new plants to produce compressors at $24.

One reason for their competitors' cost advantage was cheaper labor costs—$1.70 per hour in Singapore and $1.40 per hour in Brazil, compared with $17 an hour with benefits in Louisville. Yet another reason was productivity. It took GE 65 minutes of labor to make a compressor, whereas the cheaper labor in other locations performed faster—48 minutes in Singapore, 35 minutes in Brazil, and 25 minutes in Japan. The competitors' new plants and designs were more efficient.

It was obvious to Magaziner that GE's only hope lay in the new design for a rotary compressor because it had fewer parts and could therefore be manufactured more cheaply than a reciprocating compressor. GE had invented the rotary compressor for air conditioners but had not bothered to use it in refrigerators, where it would have to withstand harder use. Magaziner went to see Tom Blunt and asked him if he could design a plan to produce the rotary compressor. Blunt replied that that was precisely what he did for a living, if he got the chance.

Indeed, Blunt would get his chance. Truscott, MABG's chief engineer, assembled a team of product design engineers and came up with a model the team thought could be made cheaply. However, the design required a precision in working parts to fifty-millionths of an inch—more precise tolerance then used in any mass-produced consumer equipment.

Next, Truscott told Blunt to assemble a team to design a factory to produce the compressor. Blunt assembled a team of about 40 people, including engineering product design people and manufacturing people (a concurrent approach). He put the product engineers and manufacturing engineers in offices across the hall from each other. Together they refined the design of the rotary pump to the most automatable model, with less than 20 parts.

However, there was a major technical bottleneck in the production system. Existing production equipment normally could not meet the close tolerances required of the new product. Blunt and his engineers therefore also had to improve the equipment (Magaziner and Patinkin, 1989, pp. 118–119):

> One of Blunt's chief engineers was Dave Heimedinger. Under his direction, a team of engineers began to negotiate with a supplier of grinding and gauging machines. The vendors would look over the plan to mass-produce parts at jet-engine precision, and then they'd shake their heads. "You can't do that," one vendor said. "We think we can," said Heimedinger.... Heimedinger and his team began experimenting with combinations that had never been tried....The first prototyping machines began to deliver parts the manufacturers said they couldn't produce.

Finally, Truscott and Blunt had a design for an inexpensive rotary compressor and an automated factory to produce it. Should GE build it? The investment was high, at least $120 million. Millions more would have to be spent on redesigning GE's refrigerators to fit around the new compressor. The GE engineers would have to sell the project to the finance people in GE's headquarters in Connecticut, who were wary of large capital investments in appliances, since recently they had lost a lot of money in a failed new washing machine plant.

To prepare for their visit to headquarters, Truscott and Blunt asked Magaziner (who was still consulting for them) to check the proposal for the new plant. Magaziner advised them that it was a close call, but he would recommend the investment. Then Truscott sent the proposal to Roger Schipke, then head of MABG.

Schipke, Truscott, and Blunt flew from Kentucky to Connecticut to present the proposal. The CEO of GE then was Jack Welch. He listened to their presentation. Welch wanted to keep major appliances as a core business for GE and understood that it would require a major investment to turn things around in manufacturing. He approved the project. The new plant would be built in Columbia, 200 miles from Louisville.

We see here a strategic commitment to a core business and a management understanding that only a major investment would keep the company in business. If you cannot produce a product cheaper or at least as cheaply as a competitor, you cannot keep a core business.

We also see a major business risk depending on a technological risk. The compressor technology worked. Could the manufacturing technology produce it within the tolerances required? In addition to the technical risks, there also were people risks, since even the most automated production process is still a sociotechnical system. "But if GE were to succeed...—making the new fully automated plant work—it would have to face another challenge just as important as improving its hardware: improving its people" (Magaziner and Patinkin, 1989, p. 121).

GE's MABG management developed a training plan for its work force to staff the new factory and asked the workers to undergo extensive training: "Workers lined up for training. Partly, it was because of the prestige GE gave. Those who made it through got diplomas and graduation dinners. But there was another draw as well,...plants throughout America were closing and...it was a matter of time before distant forces put...[American workers] out of work too" (Magaziner and Patinkin, 1989, p. 121).

Now one might have hoped for a happy fairy-tale kind of ending to this story of management daring in the MABG at General Electric—a kind of ending without a problem. In real life, however, particularly in business, there are always problems that have to be met and overcome. This story is no exception. "The celebration was short-lived. In January 1988, 22 months after the first compressor rolled out of the new factory, a problem surfaced. Some of the larger compressors—those in GE's bigger refrigerators—began to fail. It was only a small percentage of the plant's total production, but for a consumer product like this, reliability is essential and so is customer satisfaction" (Magaziner and Patinkin, 1989, p. 121).

GE's management and engineers addressed the problem. Schipke formed a project team to analyze the problem. They worked incessantly, often through the night, for several weeks. It was a difficult problem because only a small portion of the compressors had yet failed in service. Finally, however, they saw the source of the problem—lubrication had not been properly designed and was allowing one of the compressor's small parts to wear more quickly than the designers had calculated (and this is an example of what we saw in the last chapter that design bugs will likely occur in the most complex part of the product system). In engineering, defining the problem is halfway to a solution. Truscott and other GE engineers improved the compressor design to eliminate the lubrication problem.

Still, the management problem was not over. Schipke first approved a customer-service plan to go ahead and replace immediately at no cost to the customer any compressor that broke down in the refrigerators GE had already produced and sold. However, it would still take several months to redesign the compressor for manufacturing. This meant that production would continue during that time to make a product that would prematurely fail, or alternatively, Schipke could stop production until the redesign was ready. Either way, it would cost a lot of money and lost customers.

Implementing new technology in manufacturing is difficult because even when an engineering problem is solved, the management problem is not yet fixed.

An engineering solution which disrupts production creates costs greater than the direct costs of the solution (even perhaps seriously affecting customer perceptions of quality).

Schipke cut through the Gordian knot of his manufacturing dilemma a hard, direct way. He did not wish to lose GE's reputation for quality by shipping refrigerators that might develop problems. Therefore, he decided to purchase reciprocating compressors from abroad while engineering fixed the problem. It was a tough decision, but it was the only decision by means of which GE could keep major appliances as its core business. The final responsibility of the quality and cost of a product is manufacturing's responsibility.

GE paid the price, and in the 1990s, GE was still in the major appliance business. "The compressor problem has drawn hard criticism from both the competition and the press. Some at GE are embarrassed because they had to source. Others are irritated at how much money the problem has cost....Still others are angry because it was potentially avoidable. Early lab tests showed the compressors would last 20 years, but obviously, the truest test is performance in the field" (Magaziner and Patinkin, 1989, p. 122).

This illustration underscores an important point about innovation from the perspective of manufacturing: "Looking back, many MABG executives see a lesson. When using a brand-new technology, it may be wiser to introduce it gradually, working out the bugs over a few years, than to convert your entire production to it immediately" (Magaziner and Patinkin, 1989, p. 123).

Experience in manufacturing teaches that in any new product design and any new production process (and any new technology used either in product or production), there will be unanticipated bugs. The costs of fixing problems rise exponentially with the risks of new technology.

From the perspective of manufacturing, technological innovation should always be incremental and continuous.

However, if one lets product design and production capability slip so that only desperate measures can possibly save the business, then the costs of radical change in technology must be paid.

This story of American manufacturing did have about as happy an ending as any story of manufacturing can have. In the 1990s, GE still had a core business in major consumer appliances (Magaziner and Patinkin, 1989, p. 123):

> Tom Blunt will always be a figure from an earlier America. "I like the gritty stuff," he says....At his desk in Louisville, he leans over a computer terminal and punches a few keys. A moving diagram appears on the screen. He explains that he is monitoring the Columbia factory. Sitting in Louisville, 200 miles away, he can peer by computer into the guts of any machine he wants—judging how it's working...."Since 7 AM," he says, "we've made 3412 pumps."...
>
> Building 4 still stands a few hundred yards from Tom Blunt's office. Only now, its half empty. Even if Columbia hadn't been built, it would have been dead by now. The choice was simple: it would either be replaced by a foreign plant or by an American plant....Despite the setbacks, GE is proud it chose to go with this side of the ocean.

Quality in Manufacturing

In the GE illustration we saw the importance of product quality to the long-term competitiveness of the company, and we saw the costs of product-quality problems. There are several meanings to the term *quality:*

1. *Quality of product performance*—How well does it do the job for the customer?
2. *Quality of product dependability*—Does it do the job dependably?
3. *Quality of product variability*—Producing the product in volume without defective copies.

Product performance is a primary focus of engineering design. However, in manufacturing, the second and third meanings of quality (dependability and variability) are the primary foci. In U.S. manufacturing until 1980, traditional quality control focused primarily on the third notion of quality—product variability. Then the standard technique was to control quality after production. Batches of products were sampled to determine if product variation was within acceptable specifications. However, newer approaches to quality have emphasized the importance of considering both product dependability and product variability together. In the 1980s, after witnessing Japanese manufacturing leadership in quality, American manufacturing began paying more attention to the issues of manufacturing quality.

Genichi Taguchi has emphasized the total cost of poor quality. "When a product fails, you must replace it or fix it. In either case, you must track it, transport it, and apologize for it. Losses will be much greater than the costs of manufacture, and none of this expense will necessarily recoup the loss to your reputation..." (Taguchi and Clausing, 1990, p. 65).

Taguchi argued that quality should be seen from the perspective of the customer. From the customer's perspective, the important aspect of quality in a product is whether it works immediately on taking it out of the box and into the field, after it has been knocked, dropped, and chipped. Too often the quality engineer has been overly concerned with staying "in spec," forgetting whether simply staying within manufacturing specifications means real quality to the customer.

To use the notion of manufacturing specifications correctly from the customer's perspective, Taguchi emphasized setting the specifications in the design of a product so as to have the product perform under the field conditions of product use, i.e., *product robustness.* (To explain this concept, Taguchi used a metaphor from electronic circuits products, in which one is concerned that the signal the circuit is to process can be identified above the noise level, i.e., the signal-to-noise ratio. In the customer's eyes, the performance of a product in use is like the electrical engineer's detectable signal over background noise, for the design of the product for quality should be robust not only to manufacturing variation but also to field use.) Technological innovation in products improves the quality of performance when the product functions more effectively in the hands of the customer.

Technological innovation in products improves quality when it reduces deviation from target design specifications.

When all parts of a product are on target in terms of field robust specifications, the product's performance in field use becomes independent of field conditions.

ILLUSTRATION
Ford and Mazda Transmissions and
Tolerance Stack-Up in Product Quality

One way to see how the effects of manufacturing variation and field use interact with each other is through the concept of *tolerance stack-up* in manufacturing assembly. If each part is produced within a specified tolerance about a specified target mean value, some parts will be produced with specs near the mean and others with specs near the tolerance limit. Since these variations are random, when parts are assembled into a product, some products will have (1) parts with specs all near their target mean values, (2) parts with some specs near means and some near tolerance limits, and (3) parts with specs all near tolerance limits.

The third case of the parts all near tolerance limits is called *tolerance stack-up.*

While the product with tolerance stack-up will work (since all parts are within tolerance, even though at their limits), the tolerance stack-up will make the product sensitive to field use *and will* increase wear.

Taguchi and Clausing used an example from Ford and Mazda to illustrate the deleterious effects of tolerance stack-up. "Consider the case of Ford vs. Mazda...which unfolded just a few years ago. Ford owns about 25 percent of Mazda and asked the Japanese company to build transmissions for a car it was selling in the United States. Both Ford and Mazda were supposed to build to identical specifications. Ford adopted Zero Defects as its standard. Yet after the cars had been on the road for a while, it became clear that Ford's transmissions were generating far higher warranty costs and many more customer complaints about noise" (Taguchi and Clausing, 1990, p. 67).

The "zero defects" quality standard which Ford used was to have all parts for the transmission produced within tolerances. A defective part would be out of tolerance limits. Thus Ford was later puzzled why its transmissions were having more warranty costs and complaints than Mazda's, since they were identical designs and the parts in Ford's transmissions were all within their tolerances. Ford then disassembled and measured all the parts in samples of transmissions made by both companies. The parts in transmissions made by Ford were all within specifications, but the parts in the Mazda transmissions were all on the mean targets of the specification. The accuracy of production was the reason Mazda transmissions were produced with less production scrap and reworking and had lower warranty costs.

Although the Ford transmission components were all within tolerances, many were near the outer limits of the specified tolerances. These many small deviations tended to stack up. For example, in the Ford transmission, a slightly larger diameter (from its target diameter) gear might come up against another gear that also was slightly larger in diameter from its target, and then when the two gears pressed together in operation, they were pressed slightly harder than designed (and harder than the equivalent two gears in Mazda's transmissions). The Ford-made transmissions were wearing out earlier than were the Mazda-made transmissions. The former's transmissions with more out of target but within spec gears were wearing out faster than the latter's transmissions.

Taguchi's message was that it is not good enough for manufacturing to produce parts within tolerance specifications. Manufacturing must produce parts near the targeted mean of the specifications, for parts near the outer limits of tolerances, while good enough to ship a product, will not be good enough *to please a customer in use.*

Technological innovations in manufacturing which improve the precision of manufacturing processes contribute to improvement of product quality and customer satisfaction through the improved dependability and durability of the product.

Quality-Loss Functions

As we saw in the two illustrations concerning GE and Ford and Mazda, it is important to express the total costs to business of poor-quality products. The poor quality may arise from design flaws (as in the GE illustration) or from manufacturing flaws (as in the Ford-Mazda illustration). In either case, what the customer sees are products that fail in use.

To focus management attention on product quality from the customers' point of view, Taguchi also has advocated use of a "quality-loss function" for communicating the total business costs of poor quality (Taguchi and Clausing, 1990). Taguchi defined a quality-loss function as the total costs C as a function of the deviation of product quality y from targeted quality T:

$$C = f(T - y)$$

The form of such a function would be difficult to derive exactly in practice. For example, it would have been impossible for GE's Schipke to know in advance that the small compressor part would wear out prematurely and how much it would cost GE to rectify the problem because of the added costs of sourcing until a new design could be put into production.

It does not really matter how accurate the quality-loss function is as long as one uses at least a first approximation (to indicate the rough order of magnitude of the cost). For an approximation to a quality-loss function, Taguchi argued that the function needs to be symmetric (as to whether deviations are above or below target value) and needs to rise very rapidly as the deviations lay further from a targeted value. Accordingly, Taguchi has used a simple quadratic term:

$$C = k \, \text{SUM} \, (T - y_i)^2$$

where k is a constant to adjust to appropriate monetary value,
   y is a parameter for the ith quality tolerance.

An important use of the quality-loss function is for financial decisions about implementing new technology into a manufacturing system. When one is considering building a new manufacturing facility, the normal financial analysis compares the investment in the production facility and equipment with the return on investment from manufactured products sold. And this is reasonable. However, when one is deciding whether or not to implement new technology into an ongoing manufacturing process to improve product quality, this financial calculation is incomplete. Moreover, used alone, it is misleading. The reason is that the financial return-on-investment (ROI) calculation expresses only direct opportunity costs, not lost-opportunity costs. When considering investments in technology to improve production capability, one should add to the ROI calculation a lost-opportunity cost calculation—which is what the quality-loss function calculates.

In estimating the relative cost / benefit value of investment in a technological innovation in manufacturing to improve product quality, a quality-loss function should be used to estimate the costs of the lost opportunity of not achieving product quality.

When requesting funding for projects to innovate new technology into manufacturing, it is important that the business costs of not improving product quality be calculated (as well as the return expected on the investment).

Design for Manufacturing and Product Life Cycle

Recall that competitive strategies require a focus on product differentiation or manufacturing leadership in low-cost and high-quality production. Yet these two strategies need not be seen as opposites.

Technology strategy should include not only improvement of product performance but also improvement of product quality and lowering of product costs.

Since all products eventually become "commodity-type" products, product redesign must include not only improvements in the product model but also improvements in the manufacturability of the product. (As we saw in the preceding GE illustration, the choice of the technology for rotary compressors was primarily that they were simpler in numbers of parts and could therefore be manufactured more cheaply than other designs. Since refrigerators were commodity-type products, cost was the primary competitive factor. As we also saw, this redesign of the compressor required much tighter manufacturing tolerances than existing equipment could certify. Much manufacturing innovation had to be made by GE engineers in adding sensors and on-line control to make the machines capable of producing the tighter tolerances.)

Studies on many different kinds of products have reached the conclusion that at least 75 percent of the eventual total cost of a product is determined in the design phase (NRC, 1991).

Accordingly, the notion of designing-in manufacturing quality and cost is very important to manufacturing science and practice. Cost and quality of a product are determined by the design of the product's systems, components, and packaging. Since manufacturing processes inherently introduce variation, design must minimize product variation.

Technological innovations in design and manufacturing of a product which lower its total life-cycle costs will contribute to the customer perception of total quality of the product.

Considerations in the design of products to lower manufacturing costs and improve manufacturing quality have been termed *Design-for-manufacture* (DFM). For example, Kenneth Crow (1989) summarized the goals of DFM as (1) the improvement of product quality, (2) an increase in productivity and capital utilization, (3) reduction of lead time, and (4) product flexibility to adapt to market changes.

DFM concepts focus on design of parts with appropriate tolerances and producibility rules, simplifying parts and numbers of parts, and simplifying fabrication and assembly. Crow (1989, pp. 4–8) listed some rules of thumb that DFM practitioners have collected:

1. Reduce the number of parts, because for each part, there is an opportunity for a defect and an assembly error.

2. Foolproof the assembly design ("poka-yoke") so that the assembly process is unambiguous. Components should be designed so that they can be assembled in only one way.

3. Design verifiability into the product and its components. For example, for mechanical products, verifiability can be achieved with simple go/no-go tools in the form of notches or natural stopping points.

4. Avoid tight tolerances beyond the natural capability of the manufacturing processes. Also avoid tight tolerances on multiple connected parts. Tolerances on connected parts will stack up, making maintenance of overall product tolerance difficult.

5. Design robustness into products to compensate for uncertainty in the product's manufacturing, testing, and use. While a general quality goal is to reduce variance within the production system, "noise" exists outside the system which cannot be controlled. Robustness is intended to offset this external noise as well as the uncontrolled variances within the production system.

6. Design for parts orientation and handling to minimize non-value-added manual effort and ambiguity in orienting and merging parts. Parts must be designed to consistently orient themselves when fed into a process.

7. Design for ease of assembly by utilizing simple patterns of movement and minimizing fastening steps. Complex orientation and assembly movements in various directions should be avoided.

8. Utilize common parts and materials to facilitate design activities, to minimize the amount of inventory in the system, and to standardize handling and assembly operations. Common parts will result in lower inventories, reduced costs, and higher quality.

9. Design module products to facilitate assembly with building block components and subassemblies. This modular design should minimize the number of part variants while allowing for greater product variation.

10. Design for ease of servicing the product. Easy access should be provided to parts which can fail or may need replacement or maintenance.

Technological innovations in design tools or in unit production processes which facilitate design-for-manufacturing will lower product costs and improve product quality.

Loci of Introduction of New Technologies in Manufacturing Systems

The production system of a firm is a series of process subsystems of unit production processes embedded within the sociotechnical system of production. Technological change can therefore occur in any aspect of a production system:

1. The boundary of the production system

2. Unit production processes within the production system

3. Connections between unit processes, such as materials handling technologies

4. Production system organization, communication, and control

Thus technical innovation may occur in any part of a production system:

1. New or improved unit processes of production
2. New or improved tools or equipment for unit processes
3. New or improved control of unit processes
4. New or improved materials handling subsystems for moving workpieces from unit process to unit process
5. New or improved tools and procedures for production scheduling
6. New or improved tools and software for integrating information for design and manufacturing

ILLUSTRATION
Unit Process Concepts in Chemical Engineering

The origin of the engineering discipline of chemical engineering illustrates the concept of production arranged as a set of generic unit processes. Ralph Landau and Nathan Rosenberg (1990, p. 59) describe the importance of chemical engineering to the progress of the American chemical industry: "The essence of all chemical engineering is the cluster of skills needed to design and coordinate chemical-processing equipment. We know now that the acquisition of these skills involves an educational curriculum vastly different from that offered to a chemist or a mechanical engineer. But this fact took many years to be worked out. The development of the discipline happened, to a striking degree, at a single institution: the Massachusetts Institute of Technology."

In 1903, Arthur Noyes, an MIT graduate (with a Ph.D. in chemistry from the University of Leipzig), established a German-style research laboratory in physical chemistry at MIT. It produced the first Ph.D.s from MIT, and in the same year MIT established its Graduate School of Engineering.

Another professor at MIT, William Walker, was a colleague of Noyes, but Walker differed in educational philosophy with Noyes. Walker placed less emphasis on training in basic physical chemistry and more emphasis on having students and faculty work with industry. Walker found an intellectual colleague in Arthur D. Little, then an engineering consultant. They shared a vision that MIT train engineers for building and leading industry through focusing on the application of science.

With their differing focuses on basic science and industrial practice, Noyes and Walker disagreed with each other. "A sharp controversy began between Noyes and Walker that would last for years. He [Walker] held that only through an understanding of actual problems drawn from industry could a student move from theory to practice. He found this particularly important in dealing with problems of scale—matters of materials, costs, markets, safety and the like that only industrial firms had ever dealt with" (Landau and Rosenberg, 1990, p. 59).

Science and engineering have different goals in that science simply wants to understand nature, whereas engineering also wants to manipulate nature. Engineering emphasizes the technological systems required to deliberately realize preselected states of nature. Science has no goals of selecting states of nature but only of discovering and predicting states of nature. Thus the intellectual differences between Noyes and Walker were typical of those seen between most scientists and engineers. Moreover, with industrial experience, engineers usually come to appreciate far more than do scientists the problems of translating basic knowledge into commercial practice.

Noyes and Walker never reconciled their differences. Eventually, Noyes retired, and Walker then changed MIT's chemical engineering program. Walker reorganized the program in industrial chemistry into a new program of chemical engineering. Walker used a new idea from his industrial colleague, Arthur Little. The new idea was to conceive of industrial production as a system composed of unit operations. Any general chemical process could be thought of as a particular selection and configuration of generic unit processes such as distillation, absorption, heat transfer, filtration, and so on. Together Walker and Little developed the basic principles of chemical engineering that would distinguish it from the science of chemistry.

The unit processes of any manufacturing system are logical steps required in the sequences of transformation which together make up the total transformation of resource inputs into product output.

Production engineers become familiar with and conceive of the manufacturing system as a topology of unit production processes in order to abstract past experience in the design of production systems to the design of new production systems. This conceptual approach to chemical manufacturing was used for MIT's chemical engineering training. An engineer trained in unit processes could select and configure these processes as necessary to carry out any general production of a desired chemical. Walker and his colleagues vigorously pursued these ideas as a new engineering discipline. In 1916 they established at MIT a School of Chemical Engineering Practice. Walker focused the discipline of chemical engineering on an overall systems approach to the design of continuous automated processing, first in petroleum refining and later in chemicals.

This approach to thinking about production systems had a major impact on the competitiveness of American industry (Landau and Rosenberg, 1990, pp. 60–61):

America's chemical industry forged ahead even during the Depression. Germany, the original pioneer, lagged behind, partly because of postwar restrictions of the European cartel. German industrialists were locked into the inflexible methods of the nation's dyestuffs industry....many German chemists weren't trained to comprehend process design and its integration with product manufacture....[Later after the war] as the Germans grafted the American approach onto their own strong chemical-research institutions, they became once again a major factor in the industry.

Unit Process Innovation

Technological innovation can be implemented in the unit processes of a manufacturing systems as (1) new unit processes, (2) improvements in existing unit processes, (3) improved understanding and modeling of the physics and chemistry of unit processes, and (4) improved control of unit processes. The invention of a new unit process may be required by (1) a need to process new kinds of materials, (2) a need to process with improved precision (improved purity or at greatly reduced tolerances), (3) a need to process with reduced energy consumption, and (4) a need to speed up throughput of production or to scale up the volume of production.

New materials may need to be processed on new or existing equipment. Improvements in tolerances or quality or the reduction of energy or the scale up in volume production will usually be helped by improved understanding of the physics and chemistry of the processes. This understanding can be used as models for sensing and process control.

Technological innovations in unit processes should be introduced into the production system as discrete modules, previously debugged, into an existing production system.

Unit Process Control

Research and technological innovation can improve unit process control in two ways: (1) real-time control through intelligent sensing and control and (2) experimental design for processes one cannot presently model.

Real-time control of unit processes requires

1. Sensors that can observe the important physical variables in the unit manufacturing process

2. Physical models and decision algorithms that compare sensed data with desired physical performance and prescribe corrective action

3. Physical actuators that alter controllable variables in the physical processes to control the manufacturing process.

Innovations in sensor types, sensor techniques, or signal processing from sensors can improve real-time control. Innovations in physical modeling and/or in decision algorithms are other sources of improvement in real-time control. (For example, in the 1980s and 1990s, the use of expert systems, "fuzzy" logics, and neural networks were used as new techniques for decision algorithms.) Innovations in physical actuators also can improve real-time control.

When real-time control is not economically feasible for a unit manufacturing process, then process variables must be set before running a process and must not be altered during the process. The use of statistical techniques can improve preset control. In the case in which modeling a production process from first principles is not practical, the one can use experimental design in learning about the control of complex production processes.

Technological innovation in control systems for production processes improves quality but should be tested first at a pilot scale before implementation at full production scale.

ILLUSTRATION
Implementing Robot Technology
in Automobile Production:
The Nissan Experience and
the General Motors Experience

Implementing radically new unit process technologies is a time-consuming process. In contrast to product innovation, production innovation tends to be a much more long and drawn out process. This is so because new technology usually must be inserted into an ongoing production process with minimal disruption. An example of this is the introduction of robots into automobile assembly by Nissan. Michael Cusumano (1988, p. 228) indicates that this effort began in the mid-1960s and continued over the next decades: "Nissan began experimenting with fixed-sequence robots in the mid-1960s to automate spot and arc welding and then developed a variable sequence manipulator during 1967 to handle materials."

Nissan first bought Unimates (which were built by Kawaski Heavy Industries on license from Unimation). These were numerically controlled mechanical arms invented in the United States. "Nissan tried a few in Oppama but had difficulty getting the Unimates to do even simple spot-welding jobs properly, forcing company engineers to spend years adjusting and modifying the machines. Four years later Nissan still had just 34 of the programmable robots" (Cusumano, 1988, p. 228).

Nissan focused first on applications that were difficult and unsafe for humans, such as spot-welding in the assembly of the automobile. These welding jobs were used by Nissan for 88 percent of the robot applications. The remaining robots' applications were about 5 percent in painting and 5 percent in assembly and materials handling.

By 1979, Nissan had implemented enough robots to worry their unions. In 1983, Nissan President Ishihara signed a memorandum with the union agreeing to inform the union about plans for introduction of more robots. He agreed not to proceed if the robots would result in dismissal or lay-off of union members. Also, Nissan promised to train employees to operate and maintain new robotic equipment.

Thus, by working with unions and though careful, incremental but steady implementation of robots, Nissan proceeded to introduce robot technology into automobile manufacturing. By 1985, a typical robot application cost about 1.7 times as much 1 year's wages for an experienced worker. However, a robot ran on two shifts, thereby paying for itself in 1 year. In addition, robots worked more precisely than humans, aiding production quality.

Contrast this approach to implementing new robot technology in Nissan with the approach taken by General Motors (GM). In 1981, the Cadillac division of GM began building a modern new plant at Hamtramck in Detroit to replace its aged facilities. "It was at Hamtramck that an extraordinary gamble by GM with new technology spectacularly backfired. The production lines ground to a halt for hours while technicians tried to debug software. When they did work, the robots often began dismembering each other, smashing cars, spraying paint everywhere or even fitting the wrong equipment. Automatic guided vehicles (AGVs), installed to ferry parts around the factory, sometimes simply refused to move. What was meant to be a show case plant turned into a nightmare" (Economist, 1991a, p. 64).

After much pain, the engineers and workers at Hamtramck did eventually get things to work. "At Hamtramck some of the high-tech equipment has been removed and other changes have been made to hand more control to workers....The workers at Hamtramck now try to improve things themselves by visiting suppliers and developing methods and processes to improve quality" (Economist, 1991a, p. 65).

What GM tried to do was to put into place, at the same time, too much new technology (incomplete and still rapidly changing) into a whole new production system. GM senior management apparently did not know or forgot that all new technologies take extraordinary efforts to make them work—particularly for the exacting demands of manufacturing.

In addition, it was not just one isolated manufacturing experiment that GM tried at Hamtramck. The new GM chairman in 1981, Roger Smith, made new technology a major GM policy, but Smith and his team neglected the hard-won experience of manufacturing engineers (Economist, 1991a, p. 64):

> In the 1980s, GM invested a colossal $80 billion modernizing its operations worldwide. That included the $2.5 billion spent buying Electronic Data Systems (EDS), a computer-systems firm, in order to mastermind a computerization drive. Another $5.2 billion was spent buying Hughes Aircraft to try to inject even whizzier technology in the cars themselves.
>
> The result was a disaster. According to some estimates, about 20 percent of GM's spending on new technology was wasted. Sometimes that was because the company picked untried technology which promised the earth and simply did not work.

The computer programming company EDS that Smith purchased simply had no experience with computerizing manufacturing (which depends not only on computer science but also on physical process modeling and control). While high-tech, Hughes Aircraft was experienced only in the defense business, where manufacturing experience is the exact opposite of the consumer business (with defense generally producing low-volume, high-unit-cost products in contrast to the high-volume, low-unit-cost products of the consumer sector).

The differences for the success by Nissan and the failure by GM lay in their relative respect for or lack of respect for the special problems of innovation in manufacturing.

Technological Change and Manufacturing Complexity

The preceding illustration points out the importance of an incremental approach to implementing new technology into manufacturing. When technology is rapidly changing, it is best to start with the simplest versions of the technology and implement them in the areas of manufacturing where there is a clear and immediate benefit. As familiarity, experience, and skill with the technology are developed, a firm can then proceed to implement more advanced versions of the technology into other areas of manufacturing. This incremental approach from simpler to more complex both reduces the technical risk and enhances the experience needed for successful implementation of a new technology.

Production innovation should begin with the simplest application and, as experience and technical skill are gained, grow to more complex applications.

Moreover, manufacturing technology is a sociotechnical system involving workers, technicians, engineers, and managers. Implementation of new technology in manufacturing will require modifications, customization, and debugging to make it work in the high-speed, precise conditions of manufacturing processes.

In implementing innovation in manufacturing, the sources of the ideas for debugging, customization, and problem solving will be the workers, technicians, and engineers who have the most and deepest experience with the particular manufacturing processes.

ILLUSTRATION
Sequence of Automation
in Manufacturing
in Post-World War II Japan

Post-World War II progress in manufacturing automation in Japan illustrates the recent technical changes in mechanical manufacturing technologies. In 1991, Hideo Takanaka was general manager of the Regional Project Development Department at Toyo Engineering Corporation in Tokyo. He summarized the broad outlines of automation in Japanese manufacturing from around 1960 to 1990 as follows: "The first stage of development of modern factory automation in Japan started about 20 years ago through the extensive introduction of unmanned operations of individual machine tools and unit operating machines. Today [in 1991], Japanese factories using these techniques are the most advanced factories in the world" (Takanaka, 1991, p. 29).

The Japanese economy was heavily damaged in World War II, and the subsequent American occupation altered much of the governmental and economic structure. Thus, from 1946 to about 1960, the Japanese economy was primarily rebuilding and serving internal markets. After 1960, the export efforts of Japanese industry began to be successful in foreign markets, and by 1991, Japan was one of the great exporting economic powers in the world. The key to their export dominance came through leap-frogging the rest of the world in manufacturing expertise and capability.

The Japanese investment in manufacturing automation and procedures paralleled their exporting efforts. Takanaka categorized progress in three stages. In the 1960s and 1970s, the focus was on automating the production system with special-purpose processing machines. In the latter part of the 1970s and early 1980s, flexible manufacturing systems (FMS) were emphasized with general-purpose processing machines (flexible manufacturing). In the remainder of the 1980s, the focus was on moving production and marketing into computer-integrated manufacturing (CIM).

In the mechanical manufacturing innovations of the 1960s, technologies were primarily focused on automation of special-purpose production lines to reduce the costs of labor in production. However, these production systems lacked the flexibility to accommodate product model changes.

In the mid-1970s, innovations began focusing on making the production system flexible enough to produce different models of a product but still emphasizing reduction of the costs of direct labor. The key innovation was numerically controlled machine tools and then the flexible manufacturing cell (FMC). The latter was provided with tools to do many different unit machining operations.

Thus, in manufacturing progress in Japan, the new technology of computers was beginning to affect manufacturing. In the first stage, computers provided numerical control of special-purpose machines. In the second stage, computers began to provide flexibility through a variety of numerically controlled cutting tools, tool holders, automatic tool changers with robotic assistance, and automatic materials handling with unmanned vehicles and robots. Then FMC units were tied together with automated materials handling and automated warehouses, which the Japanese called that "factory automation" (FA).

In this same period also, computers began to assist R&D through computer-aided design (CAD). "As automation began to affect the research and development, as well accumulating past experience and general information, the FA moved into Computer Aided Design (CAD) and Computer Aided Manufacturing (CAM), where computers were directly connected to the production system lines to exploit the advantages of the new systems" (Takanaka, 1991, p. 30).

The next step in integrating the uses of the computer into the factory came in tying together these different applications. "Then in the last half of the 1980s, these developments of FMC, FA, CAD, CAM could be tied together with marketing in a third stage. It would be more appropriate to call this development a production activity information system rather than simply factory automation" (Takanaka, 1991, p. 30).

Therefore, beginning with control of machine tools (NC tools) in the production flow and at the other side of the firm providing computerized design tools (CAD) and tying all this together with tooling flexibility and automated materials handing, the automated factory (FA) could then be integrated with marketing computer integrated manufacturing (CIM). This sequence of introducing the computer into manufacturing in the last half of the twentieth century was expected to produce as big an impact on the world economic systems as did the introduction of powered tools in the first industrial revolution. This was another industrial revolution—after mechanical power, computer intelligence was revolutionizing the world's production systems.

Takakaka (1991, p. 32) presented some examples of CIM systems being implemented in Japanese firms in 1990. For example, the Yamazaki Mazak Corporation produced machine tools:

> Information about an order [to Yamazaki Mazak Corp] activates the CAD system, which then imitates the FMS of a stamping operation, as it has an on-line connection with a CAD system terminal and thus, design and manufacturing are directly linked together. As a result the required time from designing to stamping and painting operations has been cut down from 3 months to 5 days. An automated...warehouse was built for assembly operation and the parts processed by FMS, with those procured from outside sources,...are loaded onto a palette which combines...those required for each individual final product. Many unmanned transport vehicles carry these palettes to the work shop where the product is finished automatically on a just-in-time basis....

Another example is seen in Kao Corporation, which then had gone even further in computer integration of the enterprise—starting with a sale and moving backward through distribution to manufacturing. "Kao has operated an information communication network in its marketing department since the 1970s. Every day orders from large retailers are sent to the sales company via a Value Added Network using a Point of Sales terminal. Sales personnel call upon medium size and small retailers and send order-information to the sales company through telephone lines or by car radio using portable terminals. The computer processes the data sent to the sales company, who then send shipping instructions to distribution centers located throughout the country. In the houseware division deliveries can be made within 24 hours after the receipt of an order" (Takanaka, 1991, p. 30).

Not only did the computer schedule the delivery of orders from the warehouses, but production in Kao also was scheduled. "CIM directly connects this sales information and 280 manufacturing plants throughout the country which manufacture individual products as well as intermediate products. At the moment [in 1990], more than 780 major plants are connected to the sales information system with an information network. As a result of the introduction of CIM to Kao, distribution stock decreased from 2.8 months prior to its introduction to

0.7 months, while the turnover doubled without any increase in the number of employees from 4500" (Takanaka, 1991, p. 31).

Manufacturing and Economies of Scale and Scope

Manufacturing strategy as to what to produce and how and what to purchase in producing a product depends on considerations of both profitability and competitive advantage. Maximizing short-term profitability without optimizing competitive advantage leads in the long term to competitive failure through eventual loss of market share. Conversely, maximizing competitive advantage without maintaining profitability leads in the short term to failure through loss of positive cashflow.

In an era when total market demand exceeded total supply production capacity (as in the world markets of the first part of the twentieth century), it made sense to emphasize economies of production and produce large volumes of standard products—since anything produced could be sold. However, in an era when total supply production capacity exceeds total market demand (as in the United States and Europe in the 1980s), it makes sense to emphasize market niches, i.e., middle-volume production of quickly developed products.

This latter perspective on the developed world markets was the view that the Japanese exporters of the 1960s through 1990s adopted. In this perspective, implementing CIM was urgent.

When market demand exceeds supply capacity, one should try to maximize return on investment by optimizing utilization of capacity. Conversely, when supply capacity exceeds market demand, one must be a minimum direct-cost producer, irrespective of investment in capacity (optimizing not return on investment but cashflow and market share).

ILLUSTRATION *(Concluded)*
Economic Reasons for Rapid Implementation of CIM in Japan

Of course, Japanese firms were not alone in using these new approaches and technologies in manufacturing. In fact, most of the technical innovations (computers, numerical machine control, programmable robots, CAD software) originated in the United States and Europe; only in Japan, however, were they systematically, incrementally, and steadily implemented into manufacturing systems. Although the American and European firms had led the world in inventing technical innovations for computerizing manufacturing, Japanese firms led the world in *implementing* computerization in actual manufacturing practice.

One of the reasons that Japanese firms were leaders in implementing CIM was that they seemed to understand better than others how to go about obtaining the benefits of the new technology. For example, Tanaka (1991, p. 30) describes how the Japanese firms then viewed the economic situation:

> In the 1980s consumer behavior, reflecting the more abundant material-istic level, became more individualistic and rapidly changing....As a re-sult...it is increasingly difficult to predict long-term demand for consumer products. Consequently, it is even more essential for a manu-facturer to obtain good consumer information from the market quickly so that it can be fed into the product development and/or production changes. This is reassuring enterprises to attempt to automate the entire enterprise, through greater integration of the marketing, design and pro-duction functions."

With the economic perspective of supply exceeding demand, Japanese management focused on reducing direct and indirect costs, not on maximizing return on facility investment. For example, Takanaka argued that in the Japanese automation experience, the first and second stages had focused on reducing direct costs and estab-lishing efficient small-batch production capability. "The target of fac-tory automation in this [first] stage was the efficiency of mass production with a small number of product models...high in demand; this assisted in reducing the labor cost during a period of rapidly in-creasing wages....The aim of this factory automation [of the second stage] was to achieve efficient small volume production of numerous product types using flexible automation of assembly and processing" (Takanaka, 1991, p. 29).

The third stage of CIM for the Japanese was to become very market responsive. "The objective of the CIM system is to construct an efficient management system in relation to the unified information and servic-ing activities, ranging from assessing user needs to the delivery of a product that reflects these needs..." (Takanaka, 1991, p. 30).

Cost emphasis in the third stage of market responsiveness was on neither facility utilization nor direct costs. "...the purpose of the CIM in-troduction was to reduce the factory indirect costs (design, procurement, process control, etc.). To take a specific example, when Hitachi Limited introduced CIM in 1987, their production line had already reached an advanced level of automation and there was little room to reduce pro-duction costs through rationalization of direct manufacturing activities. Direct manufacturing labor costs had already been reduced to 7 percent of total manufacturing cost..." (Takanaka, 1991, p. 30).

By 1990, this statistic on direct labor costs had become common for many manufactured products. Direct labor costs ran around 10 percent or less of total product cost, material costs around 35 percent, and indirect costs around 55 percent or greater in most manufactured products. Takanaka (1991, p. 30) expressed this pattern: "In order to run the production system efficiently for a small volume with numerous types of products and accommodate ever-changing marketing needs, an ever-increasing number of indirect workers appeared to be required."

Many observers of the Japanese success in manufacturing, such as Cohen and Zysman (1989, p. 112), agreed with this analysis. "Manufacturing [has become]...a competitive weapon. The evidence is overwhelming that low cost has not been the only or the most important advantage of Japanese production innovations. The Japanese did not invent the color television, the video tape recorder, or the semiconductor. But they developed designs and manufacturing systems that created decisive competitive advantage."

As another example, J. D. Goldhar, M. Jelinck, and T. W. Schlie argued that in many markets "The conditions that formally supported economy-of-scale logic [in manufacturing] are gone. In their place, conditions now support and reward factories that exhibit economy-of-scope and strategies that utilize flexibility" (Goldhar et al., 1991, p. 245).

The conditions Goldhar and associates described as favoring economies of scope in manufacturing strategy included

- Shorter product life cycles
- Greater product diversity
- Fragmented markets
- Widespread product alternatives
- More rigorous product quality standards
- Sophisticated customers

This combination of product differentiation and low cost that can result from economies of scope has had a major impact on competitiveness. For example, Cohen and Zysman (1989, p. 112) also saw flexibility in product design and manufacturing as central to Japanese manufacturing success. "Equally important have been their [Japanese] innovations in the organization of production, which permit them to introduce new products rapidly and constantly to improve and adapt the workings of that system. Honda defended its market position in motorcycles in Japan by abruptly introducing an entire new product line. The product cycle from design to production for Honda automobiles is faster than any foreign rival's."

The economic goal of focusing on and adapting production capability to fast-response, efficient, small-lot production has required the use of computer technologies to facilitate integration of research, manufacturing, and marketing. It has also focused on reduction of the overhead costs in product design, inventory, and marketing rather than simply on reducing direct labor costs. In this way, manufacturers can compete in crowded markets, where excess production capability sought market share by making niches in markets and rapid responses to changing customer preferences.

Technological innovation in manufacturing can have an impact on competitiveness through economies of scale or scope, and depending on the economic situation, one or the other may provide more competitive benefit.

Without the proper strategy for the implementation of technology, the economic benefits of technological innovation cannot be gained.

Evaluating the Benefits of Technical Improvements in the Production System

The economic benefits of implementing technical improvements in the production system should be evaluated in terms of improving

1. *Production quality*—reducing rejects and improving production accuracy
2. *Production efficiency*—reducing material wastage and/or energy usage
3. *Production effectiveness*—increasing flexibility and variability of producible products
4. *Production capacity and throughput*—increasing the volume of production per unit time
5. *Production responsiveness*—decreasing the time of changeover for product change
6. *Production cost*—decreasing the overall cost of a unit product

Accordingly, evaluating the benefits of innovation in manufacturing (and selling those ideas to senior management) requires the research unit to express the benefits of the improvements in terms of one or more of the following criteria:

1. Improved production precision

2. Reduced materials waste
3. Improved production flexibility
4. Reduced throughput time
5. Improved production control and scheduling
6. Reduced work-in-process inventories
7. Reduced indirect product costs

Financial evaluation of technical innovation in manufacturing based *solely* on direct-cost return-on-investment criteria misses the impact of quality on competitiveness and profitability. Investments in manufacturing should include considerations of the opportunity cost/benefits as well as the cost of lost opportunities.

> *A cost/benefit analysis of innovation in manufacturing should measure both reduction of the costs of producing and the costs of nonproducing time.*

Peter Drucker has emphasized that manufacturing should be viewed in its total enterprise setting (Drucker, 1990). Drucker argued that new techniques for quality, accounting, and innovation in manufacturing together facilitate defining the performance of the business enterprise as *productivity*. They also facilitate viewing manufacturing as a physical process that adds *economic value* to materials.

> *Implementing new technology in manufacturing should have the dual goals of improving* productivity *for the manufacturer and at the same time improving* economic value *for the customer.*

Practical Implications

The lessons learned in this chapter focus on the special conditions for implementing new technology into manufacturing.

1. Technological innovation in manufacturing can affect competitiveness through economies of scale or scope.
    a. Depending on the economic situation, either economies of scale or scope may provide more competitive benefit.
    b. Evaluating the cost/benefits of innovation in manufacturing requires the research unit to express the benefits of technological innovation in terms of several criteria:
        (1) Improvement of productivity and quality
        (2) Improvement of profitability
        (3) Improvement of the speediness of responsiveness

    (4) Improvement of the capability of the firm to add value for its customers

2. In the manufacturing function, quality, cost, throughput, and safety are the primary values.
    a. Details matter very much in manufacturing, and variation and other problems are endemic to manufacturing.
    b. Quality has three different meanings in terms of product performance, dependability, or variability.
    c. A quality-loss function is a useful way to communicate the costs of not achieving high product quality.
    d. Product redesign for innovative products must include not only improvements in the product model but also improvements in the manufacturability of the product.
    e. Technological innovations in design and manufacturing of a product which lower its total life-cycle costs will contribute to the customer perception of total quality of the product.

3. Technological innovations in manufacturing which improve the precision of manufacturing processes improve product quality.
    a. The loci of technological change in a production system can be in any aspect of the system: materials, product design, unit processes, unit process control, system integration, and control.

4. Innovation in manufacturing should be incremental.
    a. Implementing new technology into manufacturing will always at first disturb manufacturing, whose main goal is to avoid disturbance, and after any improvement, the production system cannot be made to operate properly again without more considerable effort and problem solving.

For Further Reflection

1 Compare the history of manufacturing innovation among four industries (e.g., textiles, petrochemicals, automobiles, and semiconductor chips, pharmaceutical, biotechnology products, etc.).

2 When in each industry did manufacturing capability and innovation become a decisive competitive factor?

3. What roles have economies of scale or scope played in competitive situations in each industry?

4. Compare the industries as to whether manufacturing has been technically of a continuous-process nature or a batch-process nature. What role have computers played in each kind of process?

5. Find the historical costs of major products from each industry, and plot prices over time. Compare them. Plot and compare also their performance/price ratios. In which industries have the performance/price ratios historically changed the most? What role did technological innovation in manufacturing play?

8

Technology and Industrial Structure

We have reviewed invention and how invention is the beginning of technological change. Before we can forecast technology change and its economic impacts, we need a technique to establish which technologies are relevant to a firm.

Technologies underpin a business by providing the knowledge for the design of its products and services as well as for the manufacture and production of its products and services. However, the issue of which technologies are relevant to a business is complicated by the fact that a business ordinarily uses several technologies even in the design and production of one product or one service. Furthermore, a diversified corporation with many businesses uses so many technologies that they must be sorted and categorized (as we saw in Chap. 2) into key strategic technologies, core technologies, and even strategic technology domains. Technology relevance to a firm is further complicated by the fact that the competitive conditions of a business may be affected by technological change in related industries.

To handle such complications, we will use the concepts (1) of the firm as an economic value-adding chain and (2) of the relationships between industrial sectors in economic value-adding chains.

ILLUSTRATION
Vertical Integration
in Automobile Manufacturing

Let us illustrate the value-adding transformations of a firm by examining vertical integration in the automobile industry in the 1980s. For example, Michael Cusumano described the production process of the assembly of automobiles in Toyota in 1984 as sketched in Fig. 8.1.

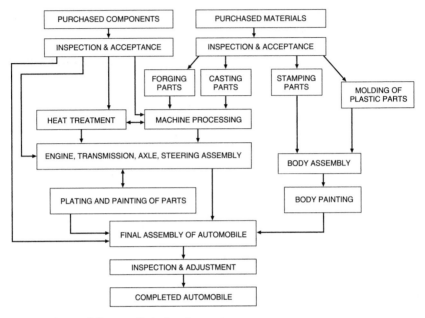

Figure 8.1 Automobile manufacturing flow.

Toyota purchased both manufactured components and materials, such as electrical parts, bearings, glass, radiators, batteries, and so on, for its own manufacturing processes from various suppliers. Toyota also purchased processed materials from other suppliers, such as steel sheets and rolled steel, nonferrous metal products, oils, paint, and so on. Purchased components and materials were subjected to acceptance inspection.

Next, materials went through various production processes to be formed into parts (such as forging, casting, machining, stamping, and plastic molding). In addition, some of the purchased components also went through further processing to be finished as components (such as heat treatment or additional machining).

Materials, components, and parts eventually were all used for three subassembly systems in fabricating the automobile:

1. Power subsystems—engine, axles, transmission, steering assembly, etc.
2. Chassis subsystems—frame, suspension, brakes, etc.
3. Body subsystem—body, seating, windows, doors, etc.

Various plating and painting processes prepared the power and chassis subsystems for final assembly, and the body was painted for final assembly. Then, finally, the three major fabrication subsystems were attached together as an automobile. After adjustments and inspection, the product emerged as a completed car.

Cusumano (1985, p. 262) notes the level of production at Toyota: "During the early 1980s, a dozen assembly plants turned out Toyota automobiles at the combined rate of more than 800 per hour." At this production rate, a great deal of coordination with suppliers was required. "Toyota managers were responsible for coordinating deliveries of components and subassembly manufacturing with the schedules of final assembly lines, where workers quickly joined engines, transmissions, steering components and frames with body shells....To manufacture a small car, from basic components (excluding raw-materials processing) through final assembly, Toyota and its subcontractors took approximately 120 labor hours..." (Cusumano, 1985, p. 262).

In the 1980s, one of the differences between Japanese and American automobile manufacturers was the much lower vertical integration of the Japanese automobile industry (Cusumano, 1985, p. 187):

> ...Japan's 10 major automakers were more like a collection of manufacturing and assembly plants for bodies, engines, transmissions and other key components than they were comprehensive automobile producers. From the mid-1970s through the early 1980s, Nissan and Toyota accounted for only 30 percent of the manufacturing costs for each car sold under their nameplates; they paid the rest to outside contractors, including subsidiaries,...loosely affiliated companies,...and non-affiliated suppliers.

In manufacturing, there usually is a choice of which components and parts to purchase from a supplier or to produce internally (in-house). This is called the *degree of vertical integration* in the manufacturing. Cusumano (1985, p. 192) presented figures that compared vertical integration between certain companies in 1979:

Nissan then produced 26 percent in-house.

Toyota 29 percent in-house.

GM 43 percent in-house.

Ford 36 percent in-house.

Chrysler 32 percent in-house.

The reason that Japanese managers chose low vertical integration was partly historical. "Subcontracting to subsidiaries or other firms reached these high levels in the Japanese automobile industry after demand expanded rapidly beginning around 1955. Managers decided that it was cheaper, safer and faster to recruit suppliers rather than to hire more employees or invest directly in additional equipment for making components" (Cusumano, 1985, p. 192).

A decision to subcontract and purchase components for a product requires a balance of several factors, namely, quality, price, investment, and proprietary competitive edge. While purchased components do not provide any competitive edge in terms of exclusive performance, they may provide equal or lower cost or quality, and they also require less capital investment.

What we see in this illustration is that most manufactured products will be produced from a combination of suppliers and in-house manufacturing and assembly processes. In economic jargon, this notion is expressed as an *economic value chain*.

The Concept of the Firm
as an Economic Value Chain

In thinking about competitive strategy, Michael Porter argued the importance of viewing the firm as a kind of "open system" that transforms inputs into outputs, adding "economic value" in that transformation (Porter, 1985).

The concept of a firm's economic value chain can be schematically depicted as in Fig. 8.2. One sees the following:

1. The firm's inbound logistics *purchasing* resources as materials, components, supplies, and energy from the market environment of the firm.

2. Through *production* operations, these are transformed into the products of the firm.

3. Where from inventory, these are *distributed* to customers.

4. After *sales*.

This transformational sequence connects the firm with its customers in adding economic value to the resources the firm procures.

Above this transformational chain, several other functions of the firm provide *overhead* support to the firm, including the activities of

1. Technology development

2. Human resource management

3. Firm infrastructure

These overhead activities also can be regarded as internal services within the firm.

Three kinds of technology are used by any economic value-adding firm: (1) product technologies in the design of the firm's products,

FIRM OVERHEAD FUNCTIONS

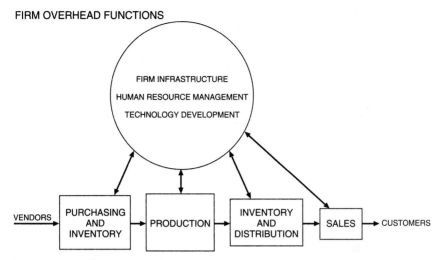

FIRM DIRECT PRODUCTION FUNCTIONS

Figure 8.2 Firm depicted as an economic value-adding chain.

(2) production technologies in the production of the firm's products, and (3) service technologies in overhead functions that serve the direct productive activities of the firm and provide external services to customers.

There are technology systems for inbound logistics in materials, components, supplies, and energy, and changes in any of these technologies can have an impact on the cost, quality, or availability of resources to the firm (and hence the cost, quality, and quantity of products of the firm). Also, changes in the technology systems embodied in the product designs or in the production processes can have an impact on function, performance, cost, quality, and timeliness of the products produced by the firm.

Downstream of the firm, changes in the technologies of applications systems and marketing and sales (usually information technologies) can have an impact on the access of the firm to customers. And changes that affect product maintenance and repair can affect the service a firm provides its customers. In the overhead functions of the firm changes can occur in technologies that affect the profitability and competitiveness of the firm (such as technology development, human resources, and infrastructure of control and financial procedures).

The importance of viewing the firm as a economic value chain is that it highlights the many and different technologies that can affect a firm's competitiveness.

Industrial-Level Economic Value Chains

Complementary to the description of the firm as an open system of economic value-adding transformations is the description of the relationship of industrial sectors as a sequence of economic value-adding transformations. Firms can be classified into industrial sectors whose relations are as a kind of chain of value-adding transformations—beginning with the extraction and processing of natural resources through to the production of products and applications for final customers of the chain of transformations. This had been called an "industrial-sectors economic value chain" (Porter, 1985; Steele, 1989).

As shown in Fig. 8.3, any industrial value chain can be depicted as (1) the extraction of raw materials from nature, (2) which next are processed into materials products, (3) which are supplied to parts and component and subsystem manufacturers, (4) which then go to major device system manufacturers, (5) which may then go to an applications systems integrator, or (6) be distributed by a wholesale and retail sector, (7) to be purchased by a final customer, while (8) all the prior producing sectors are serviced by a tool, machine, and equipment sector.

Interacting industrial sectors provide the necessary transformations for adding value from natural resources to final product.

To understand a complete industrial system, one should envision the topology of technology systems as a flow of materials from natural resources through to customer applications.

Competitive conditions will differ in each sector of an industrial value chain, and also the importance of different kinds of technology for competitiveness in each sector will differ.

The competitive conditions of firms within an industrial sector are directly or indirectly affected by changes in technologies in all parts of their value chain.

Lowell Steele (1989) argued that the industrial value chain influences the basic questions of a firm's competitive strategy, such as

1. In what business is the firm?
2. How does the firm compete in the business?
3. Does the firm supply tools to the customer to help solve the customer's problems, or does the firm solve the customer's problems?

These questions have different answers according to the place of the firm in the industrial value chain.

Innovation in the economic value chain of production includes innovations in the industrial sectors supplying the chain, as well as the innovations in the technology of the immediate production processes. The location of a firm within an industrial system of economic value chains distinguishes (1) the kinds of research and technological development in which the firm is likely to focus and (2) the kinds of competitive factors that are likely to be central to competition within an industrial sector.

ILLUSTRATION
Innovation in the Iron and Steel
Industry System

As an illustration of innovations in an industrial value chain, Bela Gold, William Peirce, Gerhard Rosegger, and Mark Perlman described the industrial system of iron ore and coal acquisition through the refining and shaping of steel products (Gold et al., 1984). Iron ore is mixed with metal alloys and scrap iron and steel along with sinter cider scale to produce iron and steel. In the refining, fuels (such as coke or electricity), fluxes (chemicals), and oxygen are used. Most of the steel produced is then shaped into ingots and products, and some of the scrap from this shaping is recycled back into steel production.

Two kinds of power have been used to heat the iron ore in refining: coal and electricity. After mining, coal must first be coked (partially burned) to remove chemical impurities that impede iron refining. Iron ore is mined or imported and transported to the refineries. Iron and steel scrap is recovered from the production process and also reclaimed from junked products that use large amounts of steel (such as automobiles). In addition, oxygen, flux chemicals, and alloy metals are used in iron and steel production, and these are produced by different industrial sectors in chemicals and metals.

Each of the industrial sectors has its own technology systems for its own process transformations.

Mining technology for coal.   Consider the innovations in one of the supply industries to iron- and steelmaking—the technologies of coal mining. Coal mining technologies divide basically into (1) underground mining and (2) surface strip mining. Underground mining technology was an early process historically, and underground mining of any ore was restricted to depths no deeper than the water table. In the eighteenth century, the innovation of the steam-driven water pump allowed mining to go deeper than the water table. In the nineteenth century, the innovation of dynamite allowed miners to fragment rock and ore more efficiently.

From then until early in the twentieth century, the sociotechnical system of coal mining required coal miners to descend on foot (or in lifts) into the depths of the earth, blast their way in tunnels to coal

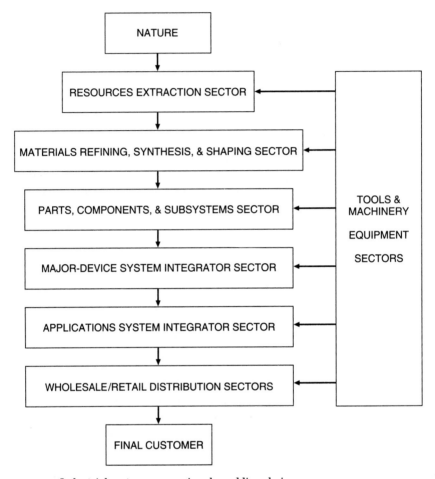

Figure 8.3 Industrial systems economic value-adding chain.

seams, and then pick and blast coal into chunks for loading into cars on rails, pulled by men (or mules or electrical trains) to the surface. The hazards to people working in the coal mines included mine cave-ins and explosions, machinery accidents, partial deafness from machine noise, and black lung disease from coal dust.

Innovations in the coal mining technology included drills to fragment the ore, hoists and conveyors to lift the ore, and machines to rip the face of the ore.

Later innovations in other industries (notably the construction machinery industry) made possible an alternate technology of open strip mining. Open strip mining involves the removal of surface soil with large tractors and shovels and then the systematic removal of ore. This process reshapes the landscape, leveling and terracing hills and digging large holes in the earth. The landscape after strip mining requires extensive reconstruction for environmental recovery. By 1963, the ratio of ore output to labor was more than a factor of 4 times higher for open pit mines versus underground mines (Gold et al., 1984, p. 339).

Coking and ore benefication technologies.   Both coal and iron ore received a processing step before going into iron refining. For coal, this was coking—a partial burning to rid the coal of impurities such as sulfur that interfered with iron refining. For iron ore, this was benefication (processes for fragmenting ore to smaller particle size and concentrating their iron content and agglomerating these concentrated particles into ore pellets).

Coking of coal was the basic innovation developed in England in the eighteenth century that made steelmaking with coal possible. The process was called "beehive coking" and occurred at the mine head. Historically, all previous steelmaking had used wood charcoal, which, along with ship building and sheep grazing, was responsible for the extensive deforestation of the English landscape. Steelmaking using coke and the steam engine and textile machinery innovations were the three basic innovations that created the eighteenth-century industrial revolution in England.

Toward the end to the nineteenth century, the waste product of coking, coal gas, began to have commercial value as a chemical by-product. In 1987, a new innovation, the Koppers oven, was used to capture the waste gases. Koppers ovens were boxlike structures in which the coal was partially burned to coke. Vent pipes captured the coal gas and delivered it to recovery plants for use in electricity and chemical production. Later, petroleum and natural gas substituted for coal gas as chemical feedstocks.

Sometimes a stimulus for technological innovation in one part of an industrial value chain occurs when other industrial value chains find uses for the by-products of the first value chain.

Iron ore benefication made possible the use of ores from different sources in the same refining process and reduced the cost of transporting iron ore from the source to the refiner. Innovations in concentration included washing, magnetic concentration, gravity, and froth methods.

Sometimes innovations in the technologies of both processing and using natural resources are required in order for a new form of a natural resource to find economic utility.

Ore transportation technology. In iron- and steelmaking, the transportation of iron ore from the mines to the furnaces required loading the ore on railroad cars and into ore ships. Incremental innovations in the transport ships focused then on loading and unloading procedures. Unloading equipment was built into the ship—such as hoppers emptying onto conveyor belts. This technology was particularly useful for loading and unloading beneficiated and pelletized iron ore.

In the acquisition and transportation of natural resources, the quality and location of the resources provide technical challenges and stimulate technical change. Once a technical solution has been found (and until better solutions are invented), the technological system will be stable. As the quality and quantity of natural resources decline, technological innovation will be again required.

Technology and Competitive Factors in the Resource-Extraction Sectors

Generally in any resource-extraction industry, sources of raw materials are located and collected, mined, or extracted from nature. For example, the asset value of oil production firms can be measured in the estimated barrels of oil in their reserves, and the most proprietary knowledge they have is oil exploration techniques. One competes by finding and owning the rights to natural resources and by efficiently producing raw materials from these natural resources. The extraction industry firms provide raw materials to their customers, and the availability, quality, and cost of these materials influence sales to customers.

> *For resource-extraction industries, the most important assets are access to sources of raw materials and technologies for extracting raw materials.*

The competitive factors that especially discriminate among firms producing similar products within a resource-extraction industrial sector include

1. The effectiveness of resource discovery techniques
2. The magnitude and quality of their discovered resources
3. The efficiency of the extraction technologies
4. The capacity for extraction and utilization of capacity
5. The cost and efficiency of transportation for moving resources from extraction to refinement

For firms in the resource-extraction sectors, technological R&D focus on improving (1) techniques for resource deposits discovery, (2) techniques for resource extraction, and (3) techniques for transportation or modifying the resource for transportation.

The customers of the resource-extraction sectors are the firms in the different materials refining, synthesis, and shaping sectors.

ILLUSTRATION (Continued)
Iron Refining in the Iron and Steel Industry

Let us look next at innovation in the firms of the industrial economic value chain of iron and steel production. Steel making is a two-step process. First, iron is refined from iron ore and/or scrap, and then, steel is made from the iron. In both processes, the technological heart of the transformation is the furnace technologies.

In ironmaking, the blast furnace was the transformation technology that takes in-plant inventories of raw materials and transforms them into pig iron. A blast furnace plant is a technology system consisting of (1) raw materials handling and storage facilities, (2) equipment for moving materials into the furnace, (3) the blast furnace, (4) facilities for heating and blowing air into the furnace, (5) equipment for drawing the molten iron and slag from the furnace, (6) equipment for removing waste gas from the furnace top and reducing emissions into the atmosphere, and (7) cooling equipment requiring large quantities of water.

Although the blast furnace as a basic technology was unchanged for nearly 200 years, the details of the blast furnace plant system were changed continuously by incremental innovations so that output per manhour increased six times over the 60 years from 1900 to 1960. Performance improvements came both from the incremental innovations and from economies of scale in production capacity.

For example, many of the incremental innovations in the blast furnace system centered on improving control of the operation. And control of the furnace system was very important for both productivity and safety. For example, a stoppage of the air blast to the furnace immediately stops the operation. When the furnace is shut down, it will be out of production for a long period. If there is a delay in drawing off the molten metal or slag, the materials may pile up beyond the melting zone of the furnace, seriously damaging the furnace.

Innovations also were made in other parts of the technology system. One important example is the mechanization of materials handling and charging of the furnace with materials. At first, raw materials were shoveled by manual labor into wheel barrows and taken to the top of the platform and manually dumped into a hopper (which then discharged the materials into the furnace). The innovation of a skip hoist allowed materials to be conveyed directly from the loading dock to the furnace. This mechanization of operations standardized the composition of the charge into the furnace, permitting better quality control.

Another part of the technology system is the means of removing the iron from the furnace. Hot iron is tapped and removed from the top of the furnace, and innovations in drilling through the clay plug in the tap hole (an automatic high-speed drill) and in removal of final product also improved performance.

These innovations in the mechanization of loading and removing materials from the finance next permitted enlargement of the furnace, introducing the economic benefits of economies of scale. It was, in fact, the scaling up of the sizes of furnaces that provided major gains in profitability, by improving energy efficiency and decreasing labor requirements. Further innovations in furnace design also improved performance and productivity. For example, furnace diameters increased from the 13- to 15-ft size at the turn of the century to over 30 ft by 1935. From 1900 through 1960, the number of blast furnaces in the United States declined by a factor of 5, while the output increased by a factor of 3. During the 1930s, efficiency also was increased by enriching the air that was blast into the furnace with oxygen. Furnace lining materials were improved to withstand hotter temperatures. Waste products were recovered and reused. The physical output of iron product in blast furnaces grew from 1899 to about 1950 and then leveled off.

Improvements in many parts of a technology system also may make possible economic benefits from economies of scale.

Primary steelmaking process. Steelmaking is a second process performed after iron is refined. The carbon content of iron must be adjusted into the right range to transform the iron into the stronger structure of steel. First, hot iron from the blast furnaces, cold iron from scrap, alloying metals, and fluxes are put into the steelmaking furnace. The steelmaking furnace then transforms these into steel. The quality of the steel varies according to the proportions of the materials put into the furnace and the details of the process of transformation within the furnace.

In the technology systems of the industrial sectors described so far, mining and ironmaking, the basic technologies have been relatively unchanging (except for open strip mining) and technological change was incremental in improving the basic technologies. However, in steelmaking over the same period (late 1890s through 1970), there were three different basic technologies: the Bessemer furnace, the open-hearth furnace, and the basic-oxygen-process furnace.

Prior to the middle of the nineteenth century, steel had always been made in crucibles, but a new process was invented by Sir Henry Bessemer. The basic equipment was a large pear-shaped vessel which could be tilted along a horizontal axis. Hot iron was placed in the vessel, and air was blown into it through tuyeres in the bottom of the vessel. This blowing of air into the iron was the inventive idea that distinguished the new technology from the older crucible method. It provided the first technique for the mass production of steel to specific metallurgical standards.

Yet even while the Bessemer process was being used, it was at the same time being displaced by another process—the open hearth. The open hearth was first proposed by C. W. Siemens in 1861. Pierre and Emil Martin obtained a license and began production in 1864. The open hearth used heat in the waste gases from the furnace to preheat air and gas fuels. This enabled the furnace to use inputs of scrap and other cold metal in addition to the molten iron. The furnace consisted of a rectangular container and two sets of brick chambers through which the waste gases were alternately blown in order to heat them. The air and fuels were blown into the hearth through the preheated chambers.

In contrast, the Bessemer process depended entirely on the heat generated by the reaction of metaloids with the blast of air. This gave this process several technical limitations, including the inability to process inputs of various mixes. Also, it was difficult to control the transformation in the Bessemer technology because the conversion was short and violent. This resulted in ingots of poor quality that included gas bubbles. It also resulted in a costly rapid deterioration of the furnace lining. It was because of these technical limitations that the open-hearth process eventually replaced the Bessemer process.

Steels produced in the open hearth were uniformly stronger and of higher quality, and maintenance was easier. Thus there were two competing processes for a core technology transformation of the industry instead of one. And there also was a third competing basic process, the electric furnace.

The electric furnace was actually thought of as early as 1889, but the high cost of electricity then made it unfeasible (in comparison with the Bessemer and open-hearth processes). Later, the electric furnace was used for production of calcium carbide, aluminum, and precious metals. During and after World War II, it began to be used for steel making—taking advantage of the availability of steel scrap.

In terms of the utilization of these different processes from 1900 to 1960, there was a steady decline in Bessemer capacity and an increase in open-hearth capacity. After 1935, electric furnace capacity began and grew rapidly (Fig. 8.4).

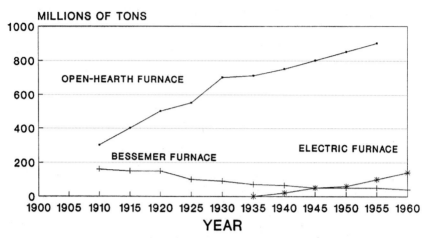

Figure 8.4 Relative U.S. production capacities of different steel furnace technologies.

Technological progress in the open-hearth process was incremental, focused around the use of oil as an alternate fuel and improvement of the furnace lining (doubling lining life from about 200 heatings in the 1920s to as many as 500 heatings in the 1940s). The war production of steel in the early 1940s also stimulated further progress in the use of thinner linings and higher furnace walls. New spectrographic methods for measuring temperature led to improved process control.

In the 1920s, another technological innovation in another industry, the Linde-Frankl process, provided increased supplies of oxygen and decreased its price. Injecting oxygen with air into the open hearth improved the steelmaking process. However, it was not until the mid-1950s (Fig. 8.5) that the new process began replacing the older processes. Experimentation to learn to control the process began in Europe in the late 1940s. Innovators had to solve problems involving rapid destruction of the furnace linings. They learned to blow the oxygen onto the center surface of the hot metal bath to avoid contact of the oxygen with the furnace linings.

The oxygen converter consists of a pear-shaped vessel that could be tilted in two directions. The vessel is charged by introducing hot metal and scrap when it is in a horizontal position, and then it is turned upright. Next a water-cooled lance is introduced through an opening at the top, through which oxygen is blown off the center surface of the charge. The heat times vary from ½ to 1 hour. And then the vessel is tipped and the steel is poured off. The first plant in the United States using this process was built in 1954. Its attractiveness was lower cost for equipment and operating savings over other processes.

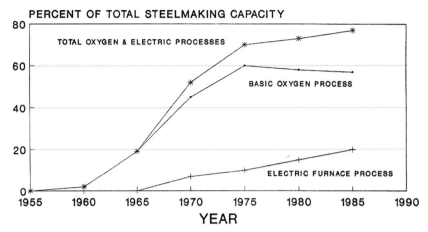

Figure 8.5 Technology substitution of oxygen and electric steelmaking processes in the United States.

In an industrial value chain, technical progress can occur not only incrementally but also as discontinuous, new basic processes.

Thus we see that four different basic inventions were used in the steel furnace technologies. The Bessemer process was eventually completely replaced by the open-hearth process as a result of the latter's improved technical performance. The basic oxygen process was beginning to replace the open-hearth process, also due to improved technical performance. The electric furnace provided a competing alternate process to the basic oxygen process, depending on the availability of scrap steel and the relative costs of electricity versus other fuels, as well as the price of oxygen.

The basic innovation of the Bessemer furnace in its first decade grew to nearly 100 percent substitution over older processes because it was the only technology for producing the quality of steel required by the new railroads. In contrast, the open-hearth furnace (innovated only 5 years later and which eventually replaced the Bessemer furnace) did so only after a long period of time. This was due to the fact that the economic benefits of the open-hearth furnace really came only after many incremental innovations and had to substantially outweigh the capital costs already sunk into the Bessemer furnace before managers invested in the new technology system.

In competing technologies for the same functional transformation, the basic invention with inherently better technical perfor-

*mance will eventually replace the others, but the rate of replace-
ment may be slowed by economic factors.*

Technology and Competition in the Materials
Refining, Synthesis, and Shaping Industries

The preceding illustration concerned a specific industry in the generic
value-adding segment of the materials refining, synthesis, and shaping
industries. Here materials products are produced from raw materials.
The materials products are processed and shaped into materials products that can be used in parts manufacture or fabrication.

*For the materials refining, synthesis, and shaping industries, the
most important assets are plant and synthesis/processing technologies and new-application technologies.*

Competitive factors for firms in the materials refining, synthesis,
and shaping industries include (1) patents and proprietary knowledge
about the creation of materials products, (2) patents and proprietary
knowledge about the processes and control in production processes, (3)
quality of produced materials, (4) cost of produced materials, (5) timely
delivery of produced materials, and (6) assisting customers in developing applications for material products.

In the materials refining, synthesis, and shaping industries, firms
focus R&D on (1) invention of new materials products, (2) modification
and variation of materials products, (3) invention of new unit production processes for producing materials products, (4) improvement and
integration of unit production processes, and (5) development of new
applications for materials products.

The customers of the materials refining, synthesis, and shaping industries include firms in the parts and components, major device system integrator, applications system integrator, and tools, machinery,
and equipment industries.

ILLUSTRATION (*Concluded*)
Final Mechanical Shaping
of Steel Products

As the next step in the illustration of the economic value chain of steel
products, let us describe the mechanical working of steel into final
product forms. After molten iron has been converted to molten steel

using one of the preceding processes, the next transformation step is to form the steel into its product forms, which include girders, rails, pipes, bars, wire, and sheets.

There are alternate processing paths for raw steel into steel product. First, a blooming or slabbing mill shapes the steel into blooms, billets, or slabs. From these, a hot rolling process produces the final shapes:

- From blooms are produced structural shapes, rails, and tube rounds (with the tube rounds subsequently shaped into seamless pipe).

- From billets are produced bars and wire rods, which may then be further shaped into cold-drawn rods (by cold drawing) or wire (in a wire mill).

- From slabs are produced sheets and strip steel, plates, or skelp (which is further processed in a pipe mill to become welded pipe or tubes).

The total market growth for steel products in the United States stopped after 1950 as a result of the substitution of other materials such as nonferrous metals, plastics, and composites. This substitution of other materials for steel products is an example of an incomplete substitution of one technology-based product for a product of another technology (which will be discussed in the next chapter).

The basic technologies for shaping steel are rolling processes, wire-drawing processes, or die and tube extrusion processes. These basic technologies have not changed, but many incremental improvements have provided significant progress in them. These improvements included size of the equipment, power sources, control, and sensing.

Innovation in an Industrial Value Chain

Major changes occur in an industrial value chain when basic inventions are discovered and introduced into a sector of the industrial value chain. In addition, incremental improvements in different sectors over time add up to large improvements in the economics of the whole industrial value chain.

The rate of innovation of either a basic or incremental invention in the industrial value chain depends on both the degree of improvement of the technical performance and the economic criteria of benefit.

For example, processes that diffused rapidly in the iron and steel industry (such as continuous cold rolling of sheets, electrolytic tinplating, continuous hot-strip milling) did so because their uniqueness (or very superior performance advantages) over other processes and because capital investment costs could be recovered quickly.

Other innovations in this industry diffused slowly, because they recovered capital costs more slowly and only provided niche markets against competing processes (such as, washing coking coal, machine loading of coal, by-product coking, machine cutting of coal, and strip mining of coal).

Processes in the iron and steel industry that diffused moderately rapidly (such as continuous cold rolling of strip, basic oxygen furnace, pelletizing, and continuous mining) did so because they provided relatively immediate economic benefits but still required very substantial capital investments.

In an industrial sector, there are different kinds of innovations, basic and incremental, having relatively different economic values and requiring different scales of investment.

The rates of diffusion of an innovation in an industrial sector depend not only on its effect on technical performance but also on the economic conditions and benefits possible due to the present condition of the sector in the industrial value chain.

As illustrated in Fig. 8.6, the rates of innovation also depend on management's perception of (1) immediate value to the managers, (2) long-term value to the firm, (3) pressures of competition, (4) willingness and

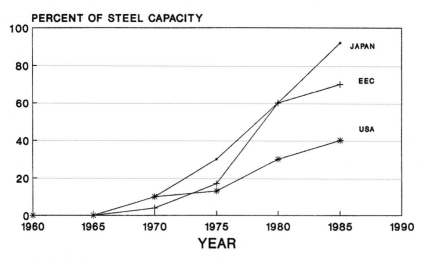

Figure 8.6 Technology substitution in continuous casting of steel.

ability to raise capital, (5) availability and cost of capital, and (6) average profitability within an industry. Technological innovation can be evaluated on its estimated contribution to profitability, and yet the actual profitability also may be determined by other factors, including the competitive pressures, financial and economic climates, etc.

Within an industrial system,

1. Some technological innovations may reduce costs of production, materials, or labor.

2. Others may not reduce costs but may improve quality.

3. Others may neither reduce cost nor improve quality but may increase capacity, increase throughput, or improve customer responsiveness.

4. Others may decrease cost and/or improve quality and/or speed throughput in some combination.

For long-term survivability of a firm in an industrial sector, both profitability and market share are necessary. Quality, capacity, throughput, and responsiveness contribute indirectly to profitability through market share.

ILLUSTRATION
Competition and Technical Progress in Laptop Computers

Traditionally, the depiction of the iron and steel industrial value chain has ended with the materials refining and shaping sector. Accordingly, for a convenient illustration of technological innovation in the parts and components industrial sector, let us look to a different industrial value chain. A recent illustration of the technical progress in and competitive impact of parts is seen in the evolution of the laptop computer product line.

The laptop computer evolved from technical progress in computer systems and illustrates how industrial structure also was altered by technical progress and subsequent market opportunities. The decreasing size of the computer components and increasing power made possible several product lines of computers, the smallest being the laptop computer.

The first laptop computer was marketed by Tandy as the Model 100 and was designed and produced by Kyocera, without a disk drive, weighing 3 pounds. A new company called Grid produced a laptop with disk drives in 1982, weighing 10 pounds. In 1986, Zenith won a government contract to supply 15,000 laptop computers to the Internal

Revenue Service. In the same year, IBM introduced a 12-pound laptop (which was unsuccessful and was withdrawn in 1989) and Toshiba introduced a 10-pound laptop (which was soon redesigned into a 7-pound laptop that became very successful).

By 1990, the laptop market had grown to 832,000 units annually, with projections toward 2.5 million by 1993. However, neither of the two leading personal computer manufacturers, IBM and Apple, was early in leading in the laptop product line. "So far, the roar of the laptop revolution has produced only a pounding headache at IBM and Apple, America's No. 1 and No. 2 PC makers. In the 1980s, both companies struggled in vain to develop competitive homegrown laptops. IBM's PC Convertible, introduced in 1986, was quickly overtaken by laptops that were smaller, lighter, brighter, and cheaper. Apple's 16-pound Macintosh Portable, introduced in September, 1989, has never sold well, despite two big price cuts. In 1990, both companies are desperately trying to get back into laptops before the market passes them by" (Depke et al., 1991, p. 119).

What happened was that technology strategy at IBM and Apple at first had not given enough attention to anticipating the changes in the downstream application systems of PC customers and to the upstream technologies of component manufacturers for reducing the size of computer components and for developing the technology of liquid-crystal display. Accordingly, IBM and Apple introduced laptop products late in comparison with competitors and with poorer features and specifications. However, because of Apple's proprietary operating system advantage, in 1991, it was able to redesign its laptops and begin successfully selling them.

Accordingly, by 1991, Japanese firms attained about 40 percent of the total world market in DOS-based laptop computers and had surpassed the share of U.S. firms—their first real success in the world's computer industry.

For effective progress to make laptop computers possible, progress in the components supplier industry centered on (1) liquid-crystal displays, (2) miniaturization of disk drives, and (3) low-powered transistorized IC chips.

Technology and Competitive Factors in the Parts, Components, and Subsystems Sector

Generally in the parts, components, and subsystems industry, the parts and components or subsystem assemblies are manufactured and fabricated for the producers of major device systems.

For the parts, components, and subsystems industry, the most important assets are (1) proprietary knowledge of part design, (2) production control, and (3) unique equipment for part production.

Competitive factors for firms in the parts, components, and subsystems industry include (1) patents and proprietary knowledge about the design of products and components, (2) patents and proprietary knowledge about the processes and control in production processes, (3) quality of produced materials, (4) cost of produced materials, (5) timely delivery of produced materials, (6) assisting customers in developing applications for material products, and (7) concurrent design capability with customers.

Firms in the parts, components, and subsystems industry focus R&D on (1) invention and design of improved parts, components, and subsystems, (2) modification, variation, and redesign of product lines and new generations of parts, components, and subsystems, (3) invention of new unit production processes for producing the parts, components, and subsystems, and (4) improvement and integration of unit production processes.

The parts and components industry comprises the greatest numbers of firms in the manufacturing sectors of industry and the greatest diversity. Generally, these firms divide into electronics and mechanical-type production. Electronic parts suppliers include IC chip manufacturers and other electronic parts, printed circuit board manufacturers and assemblers, and electronic subsystem designer and assemblers. Mechanical parts suppliers include mold and die makers, fabricators, and subsystem designers and assemblers.

Parts and components tend to be unique in technology or not unique as a commodity. In the former superior performance and in the latter costs and quality of production provide the principal means of competition. The parts, components, and subsystems industry principally supplies "tools" and not solutions to their customers, since product system design distinguishes competitors.

Technology and Competitive Factors in the Major-Device Systems Integrator Industry

In the major-device systems integrator industry, the major device for a generic class of customer applications is designed and assembled. (An example was the Toyota automobile company described earlier.)

*In the major-device systems integrator industry, the most impor-
tant assets are (1) proprietary knowledge of design and of production control, (2) production facilities, and (3) brand name and
access to market distribution channels.*

Brands and distribution distinguish the asset value of the device systems industry from that of the parts industry. Competition early in the technology life cycle of an industry usually depends on proprietary technology, but as the industry matures, firms compete predominantly by brand recognition, production quality, price, luxury features, fashion, and customer service and cost to the customer of device replacement. The device systems industry principally supplies "tools" and not solutions to the customer, since major devices are usually used by customers for many different applications.

Competitive factors for firms in the major-device systems integrator industry include (1) patents and proprietary knowledge about the design of the major-device system and about key competitive components and subsystems, (2) patents and proprietary knowledge about the processes and control in production processes, (3) performance and features of major-device systems, (4) costs of purchase and maintenance of major-device systems, (5) dependability and cost of repair of major-device systems, (6) availability and cost of distribution channels and timely delivery, (7) availability and nature of peripherals to complete a major-device system into an applications system, and (8) assisting customers in developing applications systems around the major-device system.

Firms in the major-device systems integrator industry focus R&D on (1) the modification, variation, and redesign of lines and generations of major-device systems, (2) modification, variation, and redesign of key competitive components and subsystems of the major-device systems, (3) invention of new unit production processes for producing the major-device systems, (4) improvement and integration of unit production processes, and (5) assisting customers in developing applications systems around a major-device system.

Firms in the major-device systems integrator industry include automobile manufacturers, airplane manufacturers, ship builders, building construction industries, weapon systems manufacturers, computer manufacturers, etc. They are basically product system integrators, utilizing parts, components, and subsystems from the prior industry. These firms can vertically integrate backwards into parts and materials to gain cost and competitive advantages; but on the other hand, they can sometimes deverticalize to gain cost and competitive advantages. Vertical integration as a competitive advantage is extraordinarily sensitive to how it is managed, to the costs of capital, and to the technologies of communication and information.

Technology and Competitive Factors in the Tool, Machine, and Equipment Industry

In the tool, machine, and equipment industry, the equipment and supplies that are produced are essential for all the production sectors of an industry.

Competition between firms in the tool, machine, and equipment industry focuses on performance capability and capacity of the equipment, price, and production system integratability.

Competitive factors for firms in the tool, machine, and equipment industry include (1) patents and proprietary knowledge about the design of the production equipment and tools, (2) proprietary knowledge about control in production processes, (3) performance and features of equipment and tools, (4) costs of purchase and maintenance of equipment and tools, (5) dependability and cost of repair of equipment and tools, (6) availability and cost of distribution channels and timely delivery, (7) availability and nature of peripherals to complete production equipment and tools into a production process system, and (8) assisting customers in automating and controlling production processes.

Firms in the tool, machine, and equipment industry focus R&D on (1) modification, variation, and redesign of product lines and new generations of production equipment and tools, (2) understanding and control of industrial processes for which a piece of equipment is central, (3) development of specialized equipment customized to a particular customer's production process, (4) development of software for planning, scheduling, and controlling production processes, and (5) development of software for generic product design and integration with production control.

The tool, machine, and equipment industry supplies production equipment for all producing sectors of an industrial system economic value chain. Firms specialize in types of equipment for different functional value chains, such as equipments for the ferrous materials industry, nonferrous materials industry, chemical industry, electronics industry, etc. Examples are the mechanical machine tool industry, the industrial control industry, electronics equipment industry, chemical equipment industry, etc.

ILLUSTRATION
The Service Industrial Sectors

Let us turn to application systems integrator and wholesale/retail distributor sectors. Service sectors are particularly important manifestations of application systems utilizers. For example, James Brian Quinn, Jordan J. Baruch, and Penny C. Paquette (1988, p. 45) have called attention to the importance of the growing service sector portion of the U.S. economy and have emphasized the interaction between the manufacturing and services sectors: "...the service sector has grown steadily in its contributions to the U.S. gross national product and now [1988] accounts for 71 percent of the country's GNP and 75 percent of its employment. Far from being a negative development, today's service industries create major markets for consumer goods, lower virtually all manufacturer's costs, provide strong, stable markets for capital goods producers, and enhance overall U.S. competitiveness."

Quinn and associates emphasized several points about the service industries:

1. They are capital intensive.

2. Service companies are often of large scale.

3. Some service technologies improve manufacturing systems.

They also pointed out that service technologies improved the responsiveness of manufacturing to customer needs. "Manufacturing success today often requires more rapid feedback from the marketplace, better customized products and more accurate delivery in shorter cycle times—all of which are dependent on service integration....Some manufacturers are closely tied to retailers that...daily aggregate sales data electronically from their entire retail networks, transmit restocking orders by telecommunications to their manufacturer-suppliers all over the world, and expect accurate deliveries within days...to their specific distribution points" (Quinn et al., 1988, p. 48).

Quinn and associates also emphasized that internal services to manufacturing provide value to the manufacturing operation. Generally, at least 75 percent of all costs in manufacturing are from internal service activities. They also pointed out that on a global scale, the economic value of international exchange transactions has become larger than the economic value of trade in goods and services. "Perhaps the most important structural change in international manufacturing competition stems from the continuing integration (through electronics) of the world's financial centers into a single world financial marketplace. Financial flows have already become disconnected from trade flows. Although world trade in goods and services aggregates only $3 to $4 trillion annually, financial transactions by CHIPS (Clearing House for International Payments) alone totaled $105 trillion in 1986..." (Quinn et al., 1988, p. 52).

Their conclusion was that in viewing the industrial systems of the world, it is also important to focus on the manufacturing-services interface.

The depiction of the industrial value chain should include and take into account the interaction of the manufacturing and service sectors.

It is important to anticipate technological change in service technologies that are either internal to the operations of a firm or external in the applications sectors.

Technology and Competitive Factors in the Applications Systems Integrator Industry

In the applications systems integrator industry, the applications systems for a customer are designed and/or acquired for that system developed. Competitive factors for firms in the applications systems integrator industry include (1) proprietary tools for systems design, (2) patents and proprietary interconnect hardware and software, (3) patents and proprietary sensing and control hardware and software, and (4) professional expertise.

Firms in the applications systems integrator industry focus R&D on tools (1) to improve applications system design capability, (2) to interconnect hardware and software, (3) to sense and control hardware and software, (4) to improve consulting services capability, and (5) to improve logistics and scheduling.

Firms in the applications systems integrator industry divide principally into service delivery firms, information and control integration firms, and professional services firms. Service delivery firms include airlines, bus, and railroad lines; transportation firms; telephone and communication firms; etc. Information and control integration firms include engineering design firms, information integration firms, etc. The major professional services firms are medical and legal businesses and engineering consultant businesses.

Technology and Competitive Factors in the Wholesale and Retail Distribution Industry

Competitive factors for firms in the wholesale and retail distribution and service industry include (1) location, (2) brand franchises, (3) inventory control and logistics capability, (4) price-sensitive advertising, (5) after-sales service, (6) point-of-sale information systems, and (7) computerized communication with customers.

Competition between firms focuses on location, prices, brand product lines, and customer service.

Research and development in these firms usually focus on (1) development of customer service applications and (2) development of logistics and scheduling technologies.

Firms in the wholesale and retail distribution and service industry include appliance, food, apparel, household furnishings, automobile dealerships, etc. Generally, each sector specializes around a functional group of products (e.g., food, apparel, automobiles, etc). There is some grouping of products in general retail establishments, such as department stores and large grocery stores.

Firms in the wholesale and retail distribution and service industry have traditionally done little R&D of their own but have been heavily affected by service technologies. A few do product specification and testing. Advances in service technologies such as logistics, computerized information, and communications greatly affect their competitive capability.

Industry/Technology Matrices

We have finished describing the generic types of industrial sectors of industrial economic value chains. We examined their relative emphases on technology and competitive factors in order to understand and appreciate the various kinds of technologies that affect businesses and the conditions of competition.

There are two dimensions in technology strategy, the products and the technologies involved in the products. For example, Rod Coombs and Albert Richards (1991) emphasized that two foci, one on products and one on technologies, are both necessary for technology strategy.

We can now use the concepts of economic value chains to identify all the relevant technologies with which a given firm should be concerned:

1. Describe the industrial sectors value chains for the businesses of the firms.

2. Also identify any actual or potential substituting products from other industrial value chains.

3. List all the products and production and services in the industrial value chains.

4. Identify all the relevant technologies to these products, production processes, and services.

5. Organize all these products, processes, services, and technologies in a product/process/service technology matrix.

6. Enter into each cell of the matrix an anticipation of the rate of change of each technology as slow, moderate, or rapid.

A matrix listing all the product lines (or businesses) of a firm against technologies can show the relevance of different technologies to the different product lines (or businesses) (Fig. 8.7). Product/technology matrices should be constructed for each industrial sector of the industrial value chains in which the businesses of a firm lie. In producing the products and components for the products, different technologies also will be used in the production. A production/technology matrix also should be constructed for each industrial sector. A service/technology matrix should be constructed for each sector that identifies internal service technologies and service technologies linking producers and customers.

For the whole industrial value chain, these different product/technology, production/technology, and service/technology matrices can be added together (eliminating redundant technologies) for a combined industrial sector/technology matrix.

When this also includes substituting products, processes, services, and their relevant technologies, such a combined matrix is a summary of all the relevant technologies to a firm.

After anticipation of the rates of change in technologies has been made, the subset of technologies of rapid change are the first to attend to in technology strategy, with the subsets of moderate and slow change next.

Core Technologies and Strategic Technologies in Diversified Corporations

Building on work by D. Teece, Coombs and Richards (1991) also argued that it is important to distinguish within the technology portfolio those technologies which are critical to a company's existing competitive position or merely enable it as it did competitors. We now define *strategic technologies* for a firm as those technologies relevant in an industrial value chain that are distinctive and central to a given firm's competitive position.

Strategic technologies *for an industrial sector are those relevant technologies in the industrial value chain which differentiate competitiveness among competitor's in the industrial sector.*

Note that this definition of strategic technologies includes all relevant technologies whether or not they are produced by the firms of an industrial sector.

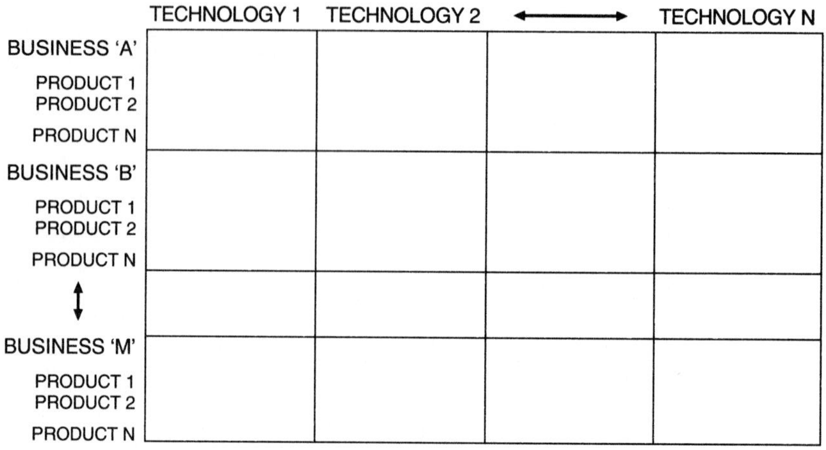

Figure 8.7 Product/technology matrix.

In comparison, we can also define *enabling technologies* as technologies in the industrial value chain that are essential to the value-adding transformations of the chain but do not provide competitive differentiation among firms in the chain of sectors. Strategic technologies will be proprietary to a firm, whereas enabling technologies will not be proprietary to a given firm.

However, what is an enabling technology to one sector of an economic value chain may be a strategic technology to a prior supplying sector of the chain.

In the technology strategy of a firm, enabling technologies should be purchased or acquired because they offer no competitive advantage but are essential. Strategic technologies should be the focus of the company's R&D and internal development of technology because they do offer competitive advantage. However, one should be very careful about using this distinction. When change is rapid in what has been an enabling technology to a firm, it might become a strategic technology to that firm if the firm can move faster with the change than its competitors. Thus the distinction between strategic and enabling technologies is seldom firm, is never set in concrete, and is a critical judgment in technology strategy.

Finally, recall from Chap. 2 that a set of strategic technologies taken together to create a definitive product or production capability for a firm is called a *core technology* for a firm.

Competitive Benchmarking

It is important in determining the relevant technologies that may be strategic also to project a competitor's technological strengths and weaknesses. Groups made up of technical and marketing people in the firm can make judgments about a competitor's technology capabilities. A technique for summarizing this kind of information is called a *competitor/technology matrix,* as illustrated in Fig. 8.8.

On this matrix, one lists both the products and the technologies involved in designing and producing the products. The first column compares the products according to technical performance and cost. The remaining columns compare the relative technological sophistication among competitors on the technologies involved in designing and producing the products.

Products may be compared directly. One can purchase competitors' products and take them apart to "reverse engineer" them. However, information on the technologies underlying these products and their production must be inferred not only from this but also from other sources. Attendance at trade shows at which competitors are showing products is helpful, as well as obtaining competitors' technical literature (particularly service manuals). One also can attend technical and professional society meetings to hear the papers presented by competitors' researchers. And one can read papers published by competitors' researchers in the technical literature. One also should monitor patent

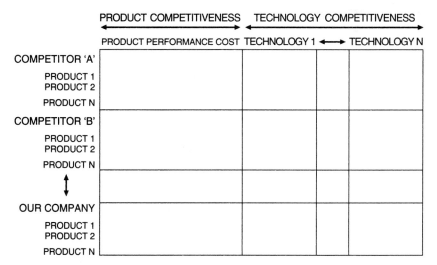

Figure 8.8 Competitive benchmarking—competitor/technology matrix.

disclosures and collect information on the patents held by competitors. One also can hire former employees of competitors.

ILLUSTRATION
Core Technology of 3M Company

The 3M Company is a famous example of a company with a core technology in surfaces and adhesives. Most of its products involve coatings and bondings, on which the company traditionally has had a strong and innovative research program.

From this research strength, 3M in 1980 was producing about 45,000 different products (counting product variations in size and shape) organized within 40 business divisions—with about 95 percent of these products derived from their core coating and bonding technologies. Moreover, 3M's corporate goal at that time was always to have about 25 percent of their products innovated within the last 10 years (Smith, 1980).

Technical Progress and
Industrial Organization

Consider now the issue of technological obsolescence. The obsolescence of a given technology must be evaluated within the setting of the societal function served by industrial value chains. Industrial-sector value chains begin and grow in order to provide goods and services for the basic functional needs of society, e.g., materials (in the case of the iron and steel industry) or transportation (in the case of the automobile industry).

Technologies may become obsolete, but the basic societal needs never end.

The basic needs of society are eternal—food, shelter, clothing, transportation, public safety, education, national defense, recreation, etc. All people in all societies at all times need basic functions fulfilled by some technology. What changes historically in industrial organization are the technologies by means of which industries (and governments) provide goods and services to the public for the basic functional needs of society. There are no obsolete industrial functions, only obsolete industrial technologies.

Dying industrial sectors are those whose technologies no longer provide the most effective and efficient goods and services for societal functions.

ILLUSTRATION
Technical Progress and Industrial
Segmentation in the Computer Industry

Let us turn next to the impact of technological change on the structure of an industrial value chain, and as an illustration, let us review the evolution of industrial sectors as the computer industry evolved from 1950 through 1990.

Recall from Chap. 7 that the general-purpose stored-program computer was invented in the 1940s and innovated as a commercial product in the 1950s, first by Sperry/Rand and then by IBM. The critical components in the first computers were electronic tubes for computational circuits, ferrite memories for active memory, and paper punch cards, electric typewriters, and magnetic tape for data input/output and storage.

Electronic tubes were produced by electronics firms and purchased by the new computer companies. IBM participated in the development of the early commercial ferrite core memories and pioneered production processes for them. Thus IBM's early technical and cost leadership centered on memory, and IBM produced these components internally.

The next major component innovation was the invention of the germanium transistor by Bell Labs. Transistors were produced by new start-up firms (such as Fairchild) and by vertically integrated computer systems companies (such as IBM). The silicon transistor was invented by Texas Instruments. The next generation of general-purpose computers replaced electronic vacuum tubes with transistors (and still used ferrite-core magnetic memories).

The major-device computer systems companies thus focused internal R&D on architecture and control improvements (e.g., IBM invented Fortran as a higher-level control language). IBM also pioneered development of peripheral devices for the computer system, such as the hard disk drive for data storage and retrieval.

Innovation in the component parts of the computer system still occurred upstream in the parts supplier industry. The next basic innovation centered on the invention of the integrated circuit (IC) semiconductor chip in 1959 independently at Fairchild and at Texas Instruments. This component innovation (and subsequent improvements) led both to next generations of computer products and to bifurcations of computers into different product lines and different industrial sectors in the computer industry to produce these lines.

By 1970, progress in IC chips by the component electronics industrial sector made possible the innovation of a minicomputer, and the computer industry bifurcated into mainframe producers and new minicomputer manufacturers (since the mainframe manufacturers at first ignored the newly evolving market for minicomputers). And a new application for IC chips, dynamic random access memory storage (DRAMS), replaced ferrite-core storage memories, facilitating the emergence of minicomputers.

By 1980, more progress in IC chips by the component electronics industrial sector made possible fabricating the central processing unit (CPU) subsystem of a computer on a single chip. Again, the industrial sectors of the mainframe and minicomputer industry at first ignored the market implication of these technical advances, and new firms started a new segment of computer makers—the personal computer.

By 1985, continuing progress in IC chips, the reduced instruction set chip (RISC), made possible the engineering workstation. Again, new firms (such as Apollo and Sun) began to take advantage of this new market. Also, by 1985, progress in using CPU chips in new architectures, parallel processing, made possible mini-supercomputers. Again, new firms (such as Convex and Sequent) were founded to take advantage of a new market.

In 1986, the innovation of liquid-crystal displays (substituting for cathode-ray displays) made possible a new market in personal computers, the laptop computer. Also, by 1986, the boundaries of what constitutes the computational system began to alter as computer networks began tying together the various computer product lines into distributed processing computer networks.

Thus, during the history of the computer industry, technical progress in components, transistors, and integrated semiconductor circuit chips made possible different product lines: (1) mainframe and supercomputers, (2) minicomputers, (3) personal computers, (4) engineering workstations, (5) mini-supercomputers, and (6) laptop computers. And the innovations of these products lines were made by start-up firms that altered the structure of the major-device systems integrator industrial sector for computers: (1) supercomputer manufacturers, (2) mainframe computer manufacturers, (3) minicomputer manufacturers, (4) personal computer manufacturers, (5) workstation manufacturers, (6) mini-supercomputers manufacturers, and (7) laptop computers manufacturers.

As technical progress made possible new computer product lines, new companies arose to form new niches in the industrial sector of major-device integrator firms for computers. Thus the technical progress led periodically to restructuring of the industrial sector.

The technical progress primarily occurred upstream in the components, parts, and subsystems producers of CPU chips, memory chips, and data storage and display devices. Yet this upstream technical progress influenced restructuring of the downstream major-device integrator producer sector. Finally, as new product-line niches in the major-device integrator sector overlapped the applications of the older product line (minicomputers and mainframe computers), dominant companies there found themselves in obsoleted technology life-cycle situations.

Industrial Reorganization Scenarios

One can construct an industrial reorganization scenario in which dramatic changes in a strategic technology could alter the organization of an industrial value chain by (1) altering vertical integration in the chain, (2) creating new product-line variations in segments of the chain, (3) obsoleting product lines in segments of the chain, (4) providing substituting-technology products in segments of the chain, (5) fusing two different industrial value chains together, or (6) obsoleting an entire industrial value chain with a substituting value chain.

Recall that a technology system is a system of functional transformations. Applications of products and processes embodying technological systems also can be viewed as a kind of system.

An application system is a system of purposive activities.

A technology system perspective emphasizes the technical performance of a technology. An applications system perspective emphasizes the purposes in using a technology. Within an industrial value chain, the upstream sectors are viewed as suppliers and the downstream sectors are viewed as customers.

From the perspective of a given industrial sector, the technologies of the upstream sectors are viewed as technology systems; whereas the technologies of downstream sectors are viewed as applications systems.

Upstream technical change can affect downstream applications systems in several ways:

1. Lower cost to move applications into lower-priced market niches
2. Improve quality to improve substitution into current applications systems
3. Improve performance to increase substitutions

4. Simultaneously improve cost, quality, and performance to increase substitutability and move into new market niches

5. Lower cost to provide multiple-copy ownership of products in a market niche

6. Improve performance to adapt technology to new application systems (and new markets)

Therefore, integrating technology strategy with market strategy requires forecasting the effects on downstream applications systems as changing markets for upstream technology systems.

This shift in perspective between technology systems and applications systems is important in an industrial value chain to understand the impacts of technical change on the organization of industry.

Evaluating the possible reorganization impact of technical change upstream on downstream sectors requires asking the question: Can the downstream sector clearly recognize the technical upstream change as affecting purposes?

If the answer is definitely yes, then upstream technical change may affect either the competitive factors or the organization of downstream sectors. If the answer is maybe or not clearly, then upstream technical change will primarily affect the competitive factors and the organization only within upstream sectors.

Procedure for Forecasting Industrial Structural Changes

1. For each business in which the company is currently engaged, construct the industrial value chain that locates the business.

2. For each sector in each value chain, identify and list the core technologies.

3. For each core technology, evaluate how rapidly technology change has been and is forecasted to occur (stable, slow, moderately, rapidly).

4. For each core technology, identify any substituting technology or materials, and evaluate how rapidly technology is forecasted to occur in the substituting technology or material (stable, slow, moderately, rapidly).

5. Construct a *technology/business matrix*
    a. By listing the business along the columns.
    b. Listing the core technology/substituting technologies for the business and for its upstream and downstream industrial sectors.

    c.   Entering in each box of the matrix the forecasted rate of technical change (stable, slow, moderate, rapid).

6. Construct a *strategic technology matrix* by taking all the entries that have moderately or rapidly forecasted change.

7. The strategic technology matrix provides a comprehensive listing of all technologies in the current businesses of the company upon which to form technology strategy.

8. Construct *industrial reorganization scenarios* in which dramatic changes in a strategic technology could alter the organization of an industrial value chain by
    a.   Altering vertical integration in the chain
    b.   Creating new product-line variations in segments of the chain
    c.   Obsoleting product lines in segments of the chain
    d.   Providing substituting-technology products in segments of the chain
    e.   Fusing two different industrial value chains together or obsoleting an entire industrial value chain with a substituting value chain.

Practical Implications

The lessons to be learned in this chapter are how the concepts of economic value-adding transformations of a firm or an industry can be used in techniques to identify the technologies relevant to a given firm.

1. A firm should be conceived of as an economic value-adding transformation through representing the operations of a firm as an open system with inbound logistics, production operations, outbound logistics, marketing, sales, and service, and overhead functions such as research, personnel, etc.
    a.   Three kinds of technologies are used by any economic value-adding business: product, production, and service technologies.

2. Related industries also should be conceived of as a connected set of economic value-adding transformations through representing the serial transformations from nature to customer:
    a.   Beginning with the industrial sector which extracts raw materials from nature
    b.   To the industrial sector which processes the materials into basic materials products
    c.   To the industrial sector of parts, components, and subsystem manufacturers that form the materials into parts

    d. To the industrial sector of major-device manufacturers which assemble parts in major product systems

    e. To the industrial sector of applications systems integrators who connect products into an applications system for a customer

    f. To the industrial sector of wholesale and retail stores which distribute products to customers

    g. All of whom may be served by the industrial sector of tool, machine, and equipment manufacturers.

3. Since the competitive conditions differ in each sector of an industrial value chain, the importance of technology for competitiveness also will differ.

    a. Also, the competitive conditions of firms within an industrial sector are directly or indirectly affected by changes in technologies in all parts of their value chain.

4. One can construct scenarios of industrial reorganization under the impact of technological advances by:

    a. Altering vertical integration in the chain

    b. Creating new product-line variations in segments of the chain

    c. Obsoleting product lines in segments of the chain

    d. Providing substituting-technology products in segments of the chain

    e. Fusing two different industrial value chains together or obsoleting an entire industrial value chain with a substituting value chain.

For Further Reflection

1. Choose a large firm that underwent a severe competitive challenge from new technology that substantially altered industrial organization.

2. Depict the industrial value chain for this firm before and after the technology change.

3. What were the strategic technologies to which the firm did or should have paid attention?

4. What were the economic factors that the technology change affected that facilitated industrial reorganization at that time?

Chapter
9

Technology Substitution

As we saw in the iron and steel industry illustration in the preceding chapter, the early applications of many new technologies are substitutions for older technology in products or processes. This has been called *technology substitution.* Technology substitution proceeds over time and must occur in the opposition to continuing improvements in existing technologies. Sometimes, such substitution does not occur because improvements and pricing keep the new technology out of a given application.

The rate of market penetration of a new technology into a market has been called *technology diffusion.* This rate depends on many factors in addition to functional performance, such as price and user familiarity and product completion and production capability. For business planning, this rate of diffusion is a critical factor in estimating the rate of return on investment in the new technology.

Technological Diffusion
in Substitution Markets

New technologies substitute in existing markets when they provide similar functionality but improved performance and/or reduced cost. The keys, then, are (1) rate of performance improvement and (2) rate of cost reduction.

In contrast, new technologies create new markets when they provide new functionality. Then the rate of diffusion coincides with the rate of market growth, which primarily depends on the rate of improvement of technology performance to open new applications and the rate of cost reduction to extend new markets. The keys are (1) application focus and time to an application completion and (2) the rate of improvement of performance/price ratio.

A technology substitution for a product in an existing market specifies the performance and cost that a new-technology product innovation should meet. The pattern of the growth of substitution of technology and new-technology products/processes is generally mathematically of a sinusoidal form. Many analytical models have been offered to describe these patterns, and they have been called *technology diffusion models*.

J. C. Fisher and R. H. Pry (1971, p. 291) suggested one of the first technology diffusion models, making the following assumptions:

1. Many technological advances can be considered as competitive substitutions of one method of satisfying a need for another.
2. If a substitution has progressed as far as a few percent of the total consumption, it will proceed to completion.
3. The fractional rate of fractional substitution of new for old is proportional to the remaining amount of the old left to be substituted.

With these assumptions, they suggested that the rate (df/dt) at which the fraction of the market (f) attained by a substituting technology will be proportional to the fraction attained and decreased eventually by the square of the fraction attained:

$$df/dt = af - af^2$$

where a is a constant of growth.

The form of this equation is sinusoidal, beginning exponentially and transitioning to linear growth and then transitioning to a final asymptotic plateau, as sketched in Fig. 9.1.

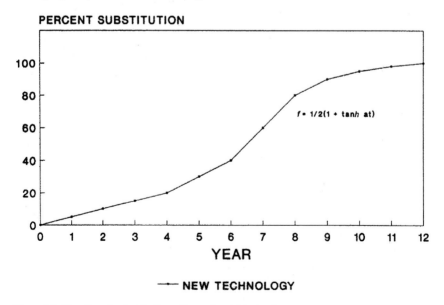

PERCENT SUBSTITUTION

$f = 1/2(1 + \tanh at)$

YEAR

—•— NEW TECHNOLOGY

Figure 9.1 Fractional substitution of new for old technology.

One convenient form of this function is

$$f = \frac{1}{2(1 + \tan h \, a(t - t_0))}$$

where t_0 is the initial time of innovation and $\tan h$ is the trigonometric hyperbolic tangent function.

Other diffusion models based on estimating a differential rate of growth have appeared in the literature to include complications of practice. One example was presented by V. B. Lal, Karmeshu, and S. Kaucker, who argued that the time between when an adopter learns of a new technology and the time when the adopter actually implements the new technology is an important time lag that should be included in the model. "In most models of diffusion of innovation considered hitherto, a basic limitation has been the assumption of instantaneous acceptance of a new innovation by a potential adopter. However, in reality there is a finite time lag between the moment a potential adopter comes to know of a new innovation and the time he [or she] actually adopts it" (Lal, Karmeshu, and Kaucker, 1989, p. 104).

Starting with a linear and quadric set of terms in $n(t)$, where $n(t)$ represents the number of adopters and $N(t)$ is the total population of potential adopters, then a variation of the Fisher-Pry differential model [since $f(t) = n(t)/N(t)$] can be written as

$$\frac{dn}{dt} = a(N - n) - bn(N - n)$$

Lal, Karmeshu, and Kaucker proposed that the most general form for including time lags into the differential equation of diffusion is to add "memory kernels" $K(t)$ and $K'(t)$ to the linear and quadratic terms:

$$\frac{dn}{dt} = Ka(N - n) - K'bn(N - n)$$

Thus one can add complications to the differential model to accommodate complications in practice. However, since one is trying to forecast diffusion, the problem of estimating these new functions is very difficult.

It is often more useful in practice to use a simplified model with good estimates on parameters than a sophisticated model with poor estimates on critical parameters.

All models will result in some sinusoidal form, but the critical differences will be (1) when the curve begins, (2) the inflection point for linear growth, and (3) the inflection point toward saturation.

The origin and the two inflection points are the critical points for fitting parameters to any form of a sinusoidal diffusion model.

Since origin and inflection points in a diffusion model depend on specific applications, one needs to partition potential markets into different applications.

ILLUSTRATION
Diffusion of Industrial Process Innovations

George Ray and colleagues (1989) studied other examples of the diffusion of industrial process innovations such as tunnel kilns in brickmaking, shuttleless looms in the textile industry, the float-glass process in glassmaking, and numerically controlled machine (NMR) tool technology.

Technology substitution of the tunnel kiln process. The tunnel kiln replaced many of the traditional types of kilns. "…In the large continental countries—Germany, France and Italy—some 90 percent of the clay brick output passed through tunnel kilns by about 1980. The more traditional types of kilns, among them mainly the Hoffmann kilns, accounted for a gradually shrinking part of national output" (Ray, 1989, p. 5).

Tunnel kilns provided a savings in the energy costs of brickmaking. Figure 9.2 depicts the technology substitution curves, combined for Germany, France, and Italy and separately for the United Kingdom. One sees in the figure that the technology substitution for tunnel kilns

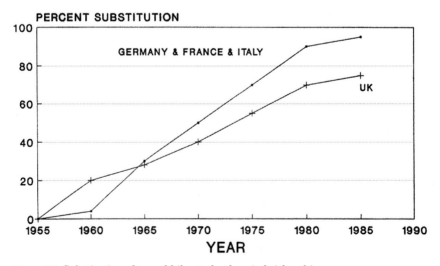

Figure 9.2 Substitution of tunnel kilns technology in brickmaking.

in the United Kingdom peaked at about 78 percent. "...the situation in the U.K. deserves attention....a large part of brick production in the U.K. consists of 'fletton' bricks made from a special type of ('lower Oxford') clay with a high carbon content that makes fire control difficult and vastly reduces the savings in energy costs....Since energy savings is one of the major advantages of the tunnel kiln, its higher capital cost is not justified for using fletton bricks. For this and other reasons the tunnel kiln has always been treated as a process suited only to the non-fletton brick sector in the U.K." (Ray, 1989, p. 5).

In this example we see that if incremental innovations do not occur to make the industrial process efficient for all source materials, then the substitution may stop short of 100 percent.

The savings of a substituting technology for an industrial process must justify the capital costs of substitution.

Technology substitution of the float-glass process.   Another major innovation in industrial history was the float glass process, which substituted for earlier glass sheet–making processes that pulled ribbons of molten glass from a melting pot. In the float-glass process, molten glass was poured out onto a bed of molten tin, with the glass floating on the tin as a thin, flat sheet. Since gravity made both the liquid tin and floating liquid glass almost perfectly flat, the float-glass process eliminated the need for mechanically grinding the glass plate to obtain optically smooth surfaces. This elimination of the mechanical grinding step produced enormous costs savings in the production of plate glass.

It was a revolutionary process innovation. "This [float glass] was one of the most revolutionary technologies in postwar decades. It introduced an entirely new concept of flat glass production which proved vastly superior to earlier methods. It has become the preferred and universal system throughout the world for making transparent flat glass..." (Ray, 1989, p. 5).

Figure 9.3 shows the diffusion of the float-glass process in the OECD countries from 1960 to 1987. By 1987, the float-glass process had replaced 83 percent of the older plate and sheet processes.

The superior process may not wholly replace other processes in a given time period as long as sunk capital investment in older equipment for small markets can still be utilized for a longer time.

Technology substitution of shuttleless looms.   The diffusion of shuttleless loom technology as a substitution for shuttle looms only began around 1980 and offers an example of the early stages of technology substitution. "Since their introduction in the late 1950s each of the four main

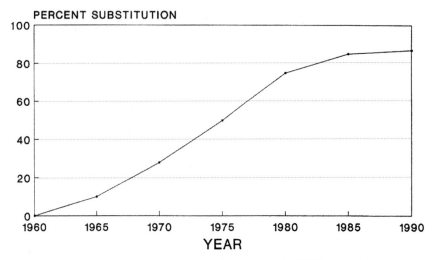

Figure 9.3 Technology substitution of float-glass process in OECD countries.

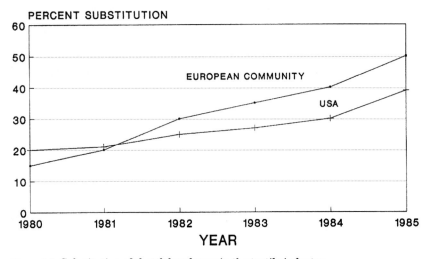

Figure 9.4 Substitution of shuttleless looms in the textile industry.

types of shuttleless loom—the projectile, the rapier, the air- and water-jet looms—have been significantly further developed....These developments were aimed at widening the looms, further reducing their noise level, increasing their speed and enabling them to handle more colors, as well as the more intricate design and structure of the fabric, apart from generally raising loom efficiency" (Ray, 1989, p. 7).

Figure 9.4 shows the percent substitution of shuttleless looms in the textile industries of the United States and the EC-9 countries of Europe. And one sees a slower adoption in the U.S. textile industry.

Again, as in most process innovations, many incremental innovations are required to adapt the new process to most conditions of production. "...Whilst for most fabrics there is at least one suitable SL [shuttleless loom] type, any one type would not necessarily serve all possible technical requirements of the great variety of end products...[yet] the most modern types of shuttleless looms are capable of weaving an increasing array of...specialties. As a result of these refinements, the dissemination of SLs continued rapidly in the 1980s" (Ray, 1989, p. 7).

Although innovated in the 1950s, it was only after two decades of refinement that substantial technology substitution began in the 1980s—in the EC-9, going from 15 to 50 percent from 1980 to 1985.

Often, until a process innovation is improved to handle a variety of products, technology substitution will not proceed to 100 percent replacement of older processes.

Technology substitution of numerically controlled machine tools. Numerical control of machine tools was an innovative technique involving the use of computer-generated code to control the cutting, boring, milling, and grinding machining operations in metal shaping industrial processes. Technical progress in developing machine controllers, software, and computer integration developed steadily but relatively slowly from the 1950s through the 1980s.

However, the innovation consisted not of one radical process innovation (as in the preceding float-glass process) but of the development of many incremental improvements to the basic idea of computer-controlled machine tools. "From the small beginnings in the 1950s and 1960s, NCMTs [numerically controlled machine tools] have come a long way....Originally, NCMTs were considered to be primarily suited to performing certain operations on certain batch sizes, smaller batches and one-off pieces....The improved NCMTs made inroads in both directions (larger batches and moderate volume pieces)....The development of NCMTs has taken other directions also: originally the technology was mainly used for drilling, milling, polishing and turning, but later it was extended to practically all metal cutting and forming operations" (Ray, 1989, p. 9).

Accordingly, the diffusion of NCMTs has proceeded slowly, waiting for technology improvements to make the technology more useful. Figure 9.5 is not a substitution chart but the numbers of numerically controlled machine tools produced by country. Notice that production did not really begin in volume until after 1977, with Japan leading the way.

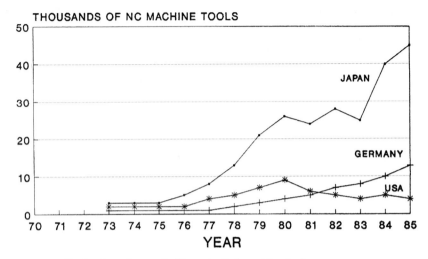

Figure 9.5 Production of numerically controlled machine tools.

When a process innovation is predominantly a set of many incremental improvements (rather than one dominant change), the technology substitution will begin late and only after a high enough level of progress has accumulated.

Stochastic Models of Technological Diffusion

Nigel Meade (1989) proposed stochastic techniques to handle specific information about a market's likelihood of adoption of a new technology, and these techniques do not presuppose a constant diffusion rate (as in the differential growth models of Fisher-Pry and others). Meade assumed that at a time t one might distinguish the following types of consumers for a given substitution market:

$O:$ Decision makers who use the old technology

$N:$ Decision makers who use the new technology

$N':$ Decision makers who are potential users of the new technology but not users of the old technology

If f represents the fraction of technology users having substituted to the new technology at time t, then

$$f = \frac{N}{O + N + N'}$$

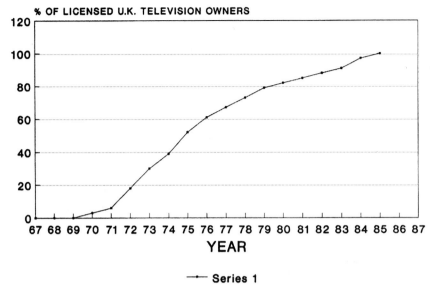

Figure 9.6 Forecast of the percentage of television users in the United Kingdom obtaining color televisions.

Meade next assumed that at time t, n consumers converted to the new technology (from $O + N'$), so that

$$df = \frac{n}{O + N + N'}$$

If one can estimate the probability p of a decision maker converting to the technology at a time t, then the estimated number of conversions would be

$$n = p(O + N')$$

so

$$df = \frac{p(O + N')}{(O + N + N')}$$

Meade then used stochastic methods to create conditional probabilities for estimating p. Figure 9.6 plots the historical data for the substitution in the United Kingdom of color televisions for monochrome televisions. One notes that at the inflection points of 1974 and 1979, Meade's change of the conditional probabilities alters the slopes of the curves.

The major difference that Meade saw between the two approaches of a differential model and a probablistic model was that "...in a substitution process, extra information is available describing the pool from which most of the adopters of the new technology will come. Thus the main practical difference is that data available in a substitution process are bivariate (N = population of adopters and O = population of old technology), whereas in a diffusion process they are univariate (N = population of adopters) (Meade, 1989, p. 394).

To the extent that one has information about the population who have adopted and those who have not adopted and to the extent that one can estimate the probability of an adoption, stochastic bivariate techniques can provide forms for utilizing additional information.

ILLUSTRATION
Introduction of Industrial Robots

Edwin Mansfield (1989) studied the rates of introduction of robots into several industries in two countries, the United States and Japan. He measured the time it took for half the firms of a given industry to adopt some robots. Mansfield used 1961 as the starting point for innovation, which was the year of the first sale of an industrial robot based on George C. Dovol's 1954 patent. Mansfield also used the definition of a robot employed by the Robotics Industry Association, which defines a robot as a reprogrammable, multifunctional manipulator.

Table 9.1 summarized the years from 1961 to when half the firms in an industry had begun using robots. One can see that there is wide variation, from 3 to 20 years in the United States and from 2 to 15 years in Japan.

Mansfield also examined the factor of the "rate of return" as a deci-

TABLE 9.1   Number of Years Until Half the Potential Users in an Industry Introduced Robots

Industry	United States	Japan
Autos and trucks	15	6
Auto parts	8	—
Electrical equipment	17	2
Appliances	19	—
Nonferrous metals	20	9
Steel	3	—
Machinery	16	15
Aerospace	7	—

sion variable in adoption of a new technology. He argued that one reason for the pattern of the more rapid adoption of the robot in Japanese
firms as compared with American firms was the willingness of the
Japanese firms to adopt robots with a lower expected rate of return on
their production investment.

*Decision criteria for adoption of new industrial technology varies
according to financial practices of the countries and industries
involved.*

Procedure for Forecasting
with Diffusion Models

All models for forecasting require measurements of the existing market, potential market, rate of adoption, and probability of adoption.
From the perspective of historical hindsight, such data are easily
found, but for forecasting, they can only be estimated. Therefore, the
problem with using diffusion models for forecasting is guessing what
the values may be for a new substituting technology.

*Models that treat all markets indiscriminately are apt to produce
poorer forecasts than models which are constructed for each possible market.*

Thus it is very important to construct diffusion models on a market-
by-market basis. For an innovator, the most important decision must
be which market on which to focus in order to estimate the rate of adoption of the technology.

Decision factors for an industrial market will generally fall into three
categories:

1. Willingness to adopt a new and incompletely perfected technology
2. Ability to adopt a new technology as part of an existing system of
   production or required reorganization of production
3. Required rate of return for adoption of a new production technology.

*One can use knowledge about the practices and past adoption of
technologies by a given market as analogies for identifying and
estimating the important parameters in a market's likely adoption of a new technology.*

One way to do this is as follows:

1. For the intended class of customers to which the technology is

expected to be marketed, try to find a historical example of another innovation that diffused into that particular market.

2. Fit a Fisher-Pry function to the historical data, and determine parameters for that historical example (or, alternatively, use stochastic techniques).

3. Assume that the new substitution differential rate will be analogous to the historical example, because the substitution is into the same consumer market and therefore may involve similar parameter estimates.

The critical factor in forecasting technology substitution is in estimating when a large-volume market will be penetrated in order to bring the costs of the technology to a level that is competitive with the costs of older technology.

Forecasting the Economics of Incomplete Technological Substitution

Technology diffusion does not always proceed to complete substitution because (1) the substituting technology may not offer absolutely superior performance in all attributes of technical performance and (2) marginally poorer performance may be traded off for a cost advantage. Materials are an example where alternate technologies may be used for the same application—trading off different performance parameters against each other and against materials costs. The market growth and substitution for ferrous metals by nonferrous metals and plastics are an illustration of technology diffusion resulting in distinct market niches.

The dynamics of substitution depend on combinations of factors: unique performance parameters and tradeoffs in price for performance. H. M. Cameron and J. S. Metcalfe (1987, p. 150) looked at the conditions of supply and demand in the asymptotic region of the diffusion curve when the new technology substitutes only partially (and not wholly) for an older technology. In that asymptotic region, Cameron and Metcalfe argued that a combined supply and demand curve for the products of the two technologies (old and new) will find equilibrium based on the relative economic utilities of the two technologies.

They used the standard supply-demand curve of economic theory, where the first assumption is that demand for a product will vary inversely with price:

$$x = c - ap$$

where c and a are appropriate constants for a particular product. a expresses the marginal elasticity of demand volume x with unit change in price p. c expresses the ultimate demand volume x for the product if the price of the product approaches zero (free good, $p = 0$).

The second standard assumption was that the price of the supply p will increase as more demand x is generated and will decrease with less demand. Then the supply equation would read

$$p = h - bx$$

where h and b are appropriate constants for a particular product. b expresses the marginal elasticity of price p with a unit change in supply x. h expresses the ultimate scarcity price of the product as the supply approaches zero $(x = 0)$.

In the asymptotic region of the partial substitution, the industrial capacities and prices will be relatively stable, and the law of supply and demand will determine equilibrium, so the price of demanded product will equal the price of supplied product. Price will adjust to the balance of supply and demand in a market.

Solving the preceding two equations simultaneously will yield the free-market price p' and the free-market volume x' in terms of the constants $(c,\ a,\ h,$ and $b)$:

$$p' = \frac{h + bc}{1 + ba}$$

$$x' = \frac{c - ah}{1 + ab}$$

These are the price p' and volume x' the market will settle on when supply exactly matches demand.

Of course, in practice, these equations are difficult to use for a forecast. It is difficult to determine the constants experimentally (a the marginal elasticity of demand volume, c the ultimate volume of demand for the product if the price of the product approached zero, b the marginal elasticity of price with a unit change in supply x, and h the ultimate scarcity price of the product as the supply approaches zero). However, in retrospect, after a free-market equilibrium point (p' and x') has been reached, one can usually calculate the constants for a product from historical data. Nevertheless, even if the equations are not useful for forecasting, they are helpful to think theoretically about the conditions of substitution.

The question Cameron and Metcalfe specifically addressed was for materials substitution, when the relative prices determined relative substitution. Suppose that the long-run supply-demand situation results in equilibrium volumes and prices, that is, x' and p' and y' and q', for the two materials, old versus new. Then what is the relationship between the two equilibrium prices of the two materials p' and q'?

Cameron and Metcalfe (1990, p. 153) assumed that the ratios of the two equilibrium prices p' and q' can be expressed as a new constant e:

> Since the new material is technically superior to the old, the relative prices of the old and new, in competitive equilibrium, must satisfy
>
> $$p'/q' = e$$
>
> The new [material] commands a price premium dependent upon the qualitative superiority in supplying users with production services.

Now, of course, the forecasting problem is to estimate just what that ratio e will become. Again, this is difficult, but the concept can be used to play "what if" scenarios.

The central assumption in Cameron and Metcalfe's analysis (1990, p. 151) was the following:

> We shall consider substitution between two technologies, each of which produces a nondurable "material" employed as an input in subsequent stages of the production process. The technologies of the competing materials differ with respect to both process of production and product quality. The quality dimension is conceptually the most difficult to incorporate, but it will suffice to assume that e units of a new material are equivalent from the users viewpoint to one unit of the old material (where the magnitude of e is less then one).

Now the only sense in which two technologies may be compared is on the functional transformation in which the substitution is aimed, and the comparison must then be on the effectiveness for this transformation. Therefore, Cameron and Metcalfe's *substitution ratio e must be defined as the ratio of the technology performance parameters for the two technologies of the same functional transformation.*

For example, two land transportation technologies (e.g., different classes of automobiles) might be compared in terms of their fuel efficiency ratios ($e = mpg_1/mpg_2$) or their average cruising speeds (v_1/v_2) or their relative carrying capacities, and so forth. The critical notion here is that the substitution ratio e should be directly translatable into the relative prices or relative numbers of units purchased.

Under what conditions is this equivalence possible? It is possible only when the quantity of the product purchased is directly proportional to the value to the customer of its technology parameter. In the materials example that Cameron and Metcalfe used to formulate this theory, this is sometimes possible. Under this assumption, they then write a new demand equation for a market being supplied by both the old and new materials (with the new material meeting y/e units of demand, as measured in the units for x):

$$x + y/e = c - ap$$

And at equilibrium, the supply equation for the old material will be

$$p = h + bx$$

The simultaneous solution (x'') of the preceding two equations for the new equilibrium market niche for the old material x yields

$$x'' + y/e = c - a(h + bx'')$$

Rewriting this equation yields

$$x'' - abx'' = c - ah - y/e$$

And further rewriting yields

$$x'' = \frac{c - ah}{1 - ab} - \frac{y/e}{1 - ab}$$

Now we can insert the expression for the equilibrium supply x' of the old material before the technology substitution into the preceding equation for the equilibrium supply x'' after the technology substitution and get

$$x'' = x' - \frac{y/e}{(1 - ab)}$$

From this we can see that after technology substitution, the new market volume for the old technology product will find a market niche at a lesser volume than before substitution.

Now, of course, there are complications to this simple model. First, especially in materials, the new technology is never wholly substitutable for the old technology because the performance characteristics of any two materials are never completely identical. Therefore, there may be market niches for old materials because of their unique performance characteristics and independent of their relative price in comparison with substitution materials.

Second, in the substitution, the rate at which the new material prices decline is important in determining the relative market volumes because improvements in older materials at a lower cost may actually prevent the new material from reaching a volume application that would bring its market price down to where the elasticity of supply would apply. Thus the final equilibrium point in substitution materials may not be independent of the path to equilibrium, as Cameron and Mitchell supposed. In forecasting the market volume and prices of a new technology substitution, we must take into account (1) the supply-demand equilibriums of the old and new technologies after substitution, (2) any special market niches due to unique performance characteristics, and (3) any problems in the paths of diffusion that might affect the nature of the final equilibrium markets of the old and new technologies.

Practical Implications

The lessons to be learned in this chapter are techniques for estimating the rates at which a new technology may penetrate markets by substituting for products and processes based on older technologies.

1. The rate of market penetration (technology diffusion) of a substituting technology depends on many factors.
   a. Since new technologies that provide completely new functionality create new markets, the rate of diffusion of a functionally new technology is primarily dependent on the rate of improvement of technology performance to open new applications and the rate of cost reduction to extend new markets.
   b. Since new technologies for existing functionality substitute in existing markets, their rates of diffusion depend on the rates of improved performance/cost ratios.
   c. The rate of diffusion of industrial technology innovations differs between those which require the construction of a wholly new manufacturing plant and those which can be implemented within an existing plant.
   d. Also, the rates of diffusion for industrial process innovations differ between those that are predominantly a major change with subsequent incremental improvements and those that are always a set of many incremental improvements which together and over time will add up to a major change.

2. The pattern of growth of technology substitution and new-technology products/processes is generally mathematically of a sinusoidal form.
   a. One can use knowledge about the practices and past adoption of technologies by a given market as *analogies* for identifying and estimating the important parameters in a market's likely adoption of a new technology.
   b. For the intended class of customers to whom the technology is expected to be marketed, it is useful to find a historical example of another innovation that diffused into that particular market.
   c. To the extent that one has information about the customers who have adopted and those who have not adopted and to the extent

that one can estimate the probability of an adoption, stochastical bivariate techniques also can provide mathematical forms for estimating the rate of substitution.

d. For incomplete substitution which will result in market niches for competing technologies, supply-demand models can be used to play "what if" kinds of scenarios in forecasting.

For Further Reflection

1. Identify an industrial market and a consumer market in which at least two innovations in each were made.

2. Fit historical data on these four innovations, and compare the patterns. Explain similarities and differences.

Forecasting
Market Conditions

It is important not only to forecast technological change but also to anticipate its impact on markets. This chapter examines techniques for forecasting changes in markets and in competitive conditions as a result of technological innovation.

Central to market changes resulting from technological progress are the effects of the different cycles of innovation and obsolescence; and Devevdra Sahal (1984) emphasized that the different cycles involve different levels of industrial organization:

Cyclic phenomena	Organizational level affected
Product life cycle	Division level
Product line life cycle	Firm level
Technology life cycle	Industry level
Kondratieff life cycle	National level

ILLUSTRATION
Product Life Cycle of the DC-2

As an illustration of a product life cycle that had a major impact on the commercial airplane industry, let us examine the design and production of the Douglas DC-2 in 1933. The infant commercial airline industry in the United States was bolstered by the decision of the U.S. government to contract with private firms for the delivery of mail by air. With this market, the early airline firms offered passenger travel on the same planes that carried the mail. By 1930, new airline firms were offering passenger service on various routes, such as United Airlines, Transcontinental and Western Air, Inc. (TWA), and Pan American Airlines. Their long-distance service utilized principally Fokker and Ford trimotor airplanes. The German-built Fokker had an open cockpit and an overhead wing, and it carried 8 to 10 passengers in a boxy wooden fuselage.

United Airlines ordered a new metal plane from Boeing that was intended to fly faster and need fewer refueling stops. And this was not good news for its competitor, TWA. "With its rounded, streamlined aluminum-alloy body, the Boeing 247 would be the first modern passenger plane, and TWA knew it. What's more, TWA couldn't have it. Boeing and United Airlines were sister companies under the same holding company....TWA was going to have to wait until United had received all sixty it had ordered. That wouldn't be for several years, and by then TWA might be out of business" (Allen, 1988, p. 7).

Moreover, TWA had other troubles: "...on March 31, 1931, a TWA flown Fokker crashed into a Kansas pasture killing seven people, one of whom was the legendary Notre Dame football coach, Knute Rockne. An investigation revealed that the insides of the wooden wing had rotted away where it had joined the fuselage, and the way the plane was constructed, routine inspection could not even have detected the problem. The experience marked the beginning of the end not only of the Fokker but for all wooden aircraft. TWA was going to have to find a new metal plane" (Allen, 1988, p. 7).

Jack Frye was then vice president of Transcontinental and Western Air, Inc. (TWA). He knew what TWA needed in a new plane. He wanted an all-metal plane with three motors and a single wing. It should fly 1080 miles without refueling, cruising at 150 mi/h. It should land at under 65 mi/h and be capable of taking off from any TWA airport with one engine out. It should not weigh more than 14,200 pounds and should seat 12 passengers comfortably. Frye sent these specifications to airplane manufacturers: "A brief letter mailed on August 2, 1932, led to the emergence of modern airline travel. It was a form letter sent to airplane manufacturers and was just three sentences long, with a postscript and accompanying single-page list of general specifications" (Allen, 1988, p. 7).

Note that a *user* of the technology, *a customer knowing the application,* could carefully specify the performance and features required by the new-technology product. There had been enough early experience in the application for a clear specification of application requirements.

One of the recipients of the letter was Donald W. Douglas, whose new company, Douglas Aircraft Corporation, had first designed and sold a torpedo bomber to the Navy. Thereafter, Douglas had built Army observation planes, postal mail planes, and flying boats. However, he had not yet built an airliner, and he seized the opportunity when he received Frye's letter. "When Jack Frye's letter arrived, in August 1932, Douglas's company occupied an old movie studio in Santa Monica, where a dozen young airplane designers worked in shirt sleeves and knickers and all knocked off for picnic lunches together. As soon as Douglas received the letter, he and his engineers began looking for ways they could design a plane to outdo the specifications and outperform the 247 too. With the Depression squeezing every manufacturer, Douglas knew he would have to work extremely fast to have a chance at winning a contract" (Allen, 1988, p. 9).

This is a nice example of the market pull of product requirements by an application's customer matching with the technology push of a technically advanced manufacturer. The first technical idea was to get rid of the requirement for three motors. "Immediately Douglas's engineers decided to dispense with the trimotor idea. As the chief engineer, James H. Kindleberger, later said, 'Why build anything that even looks like a Fokker or Ford?' According to Kindelberger's assistant, Arthur Raymond, 'We all thought...that we had a chance to meet the requirement with two engines. The simplicity of the solution appealed to us. It would clean up the whole front of the airplane. You would not have the propeller in front of the pilot; you would not have the aerodynamic drag of the engine up there....You'd have less noise and vibration in the cabin, and no gas lines or fumes in the fuselage. You'd have a simpler, less costly design'" (Allen, 1988, p. 9).

These are engineers speaking—a simplification of design that met requirements with elegance and technical advantages. The design would incorporate all the best and most innovative ideas of the time. At Cal Tech, Raymond also was studying new methods for calculating airplane performance, and 2 days later, he finished calculations indicating that a two-engine plane could meet TWA's demands. The toughest requirement was that the 14,000-pound plane would have to be able to take off from a 4000-ft-high airport with just one engine. Within 10 days, Douglas's engineers had sketched a rough design for the plane, which they called the "Douglas Commercial No. 1" (DC-1). "The designers also agreed to give the plane a rounded, all-metal, stressed-skin semi-monocoque fuselage, backed by aluminum-alloy stiffeners that ran around its inside like barrel hoops, recently introduced cowlings in front of the engines to reduce drag, retractable landing gear, and wing flaps that had never been used on a large plane but held the key to being able to both fly fast and land slow" (Allen, 1988, p. 9).

TWA gave Douglas Aviation $125,000 to build a prototype. And hedging its bet, TWA also gave a different contract to General Aviation to build a traditional trimotor.

The Douglas engineers began designing the plane using a new wind tunnel at Cal Tech to refine the wing shape. They considered using either a Pratt and Whitney engine or a Wright engine, each with nine cylinders in an air-cooled radial configuration. They chose the Wright engine with 690 horsepower, which then had the highest power-to-weight ratio of any engine. They added a two-pitch propeller, which was a new invention to alter the angle of the blades either for maximum power at takeoff or for maximum fuel efficiency at cruise speed.

Then it was time to test the plane (Allen, 1988, p. 11):

> The DC-1 was wheeled out for its first test flight on July 1, 1933, less than ten months after TWA had signed the contract for it. The pilot was Douglas's vice-president for sales, Carl Cover. As Donald Douglas and hundreds of his employees watched from the grass beside the airstrip at the Santa Monica Airport, the just-completed plane accelerated down the runway, lifted into the air, rose—and then fell. One engine had gone out. Cover banked sharply, and as he descended, the engine came back to life. He started climbing again, and then both engines went dead....Cover again dropped the nose, and both engines came back. He then began gingerly coaxing the plane higher...until he finally had enough altitude to circle and make a safe landing.

We know that there are always problems in first designs. Fortunately, Cover survived, and the problem turned out to be simple. A new carburetor had been installed in the engine, but its floats had been put in upside down. The engineers found the trouble and fixed the carburetors. Two days later, the plane was tried again, and this time everything worked right. It flew both on two engines and on one.

Then Douglas was ready for the crucial test (Allen, 1988, p. 11):

> The plane was taken to TWA's highest airport in Winslow, Arizona, and loaded down with sandbags to give it maximum laden weight....Pilot Edmund Allen took the controls; on board as a representative of TWA was...Tommy Tomlinson. The cautious plan was to start with full power and cut one engine back to half-power just after the plane left the ground....But as the plane began its first lift-off, Tomlinson reached up and flipped not the throttle but the ignition switch on one engine....The plane sank at first—to within six inches of the ground....Allen shoved the throttle down all the way on the remaining engine, and the plane steadied and started to rise. It kept climbing.

This prototype plane, the DC-1, was the only one ever built. Before the plane went into production, the Douglas engineers were already seeing how to improve it. They stretched the fuselage a foot and a half and cut out a little of the cargo space and made it seat 14 passengers (exceeding TWA's requirement for 12 passenger space). They also put on more powerful engines, increasing the cruise speed to 196 mi/h. This new improved version they called the DC-2. "TWA put its first DC-2 in service on May 18, 1934. By the end of the month it had broken the New York-to-Chicago speed record four times....In August the DC-2 began transcontinental service between Newark and Los Angeles, with stops at Chicago, Kansas City, and Albuquerque....Within a year all the leading airlines except United, wedded to its 247, had ordered DC-2s" (Allen, 1988, p. 11).

The DC-2, and its enlarged successor the DC-3, technically obsoleted competing models. Later, they were technically obsoleted by jet-powered commercial aircraft. The DC-3 had a total product lifetime of 15 years from 1935 until 1950.

Product Life Cycles

Most products have a finite time on the market. A useful way of representing product lifetimes is by charting profits over time, as in Fig. 10.1. At first, the profits are negative as investment is made in R&D to design and produce the product. Then, when sales begin, the profits move in a positive direction (yet still remaining negative as additional investment in production is made). The break-even point occurs when sales exceed the costs of investment (year 6 in Fig. 10.1). Then annual profits increase as sales increase until either profitability per unit decreases as competitors enter with "me too" products and push the price down or sales decline (years 16 to 18 in Fig. 10.1). Finally, after the

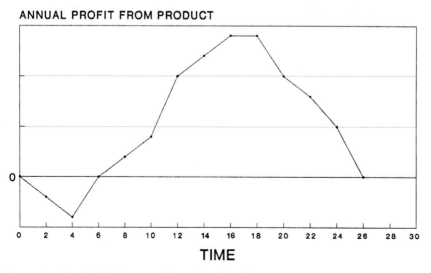

ANNUAL PROFIT FROM PRODUCT

TIME

Figure 10.1 Product lifetime—profits as function of time.

product is obsoleted by competition or by a new product model, profits continue to decline until the product is withdrawn from the market (year 26 in Fig. 10.1).

A product lifetime is the time between introduction of a new product into the marketplace and its withdrawal from the market.

There are five principal reasons why products have finite lifetimes:

1. Technical performance obsolescence
2. Technical feature obsolescence
3. Cost obsolescence
4. Safety obsolescence
5. Fashion changes

Performance or *feature obsolescence* in a product occurs when its performance and/or features are markedly less at the same price as those of a competing product. *Cost obsolescence* occurs in a product when the same performance can be obtained in a competing product at a lower price and the product cannot be produced at a cost to meet that lower price. *Safety obsolescence* occurs in a product when a competing product offers similar performance and price with improved safety of operation or when government regulations require safer features or operation in a product line. Finally, products in which technology, costs, and safety features are relatively stable can still become obsolete as a result of fashion changes (as in clothing). *Fashion obsolescence* occurs in a product in which product competition is undifferentiable in performance and price but is differentiable in lifestyle.

Recall from Chap. 3 that product life-cycle estimates are used in technology roadmaps for planning next-generation technology products in the strategic business units' product planning.

Technologies and Product Life Cycle

A product/technology matrix is useful for estimating product life cycles. Consider the form in Fig. 10.2. Along the vertical columns of the matrix one can list the different aspects of a product as a system:

- Boundary
- Control
- Components
- Connections

Aspects of product system	Tech1	Tech2	• • •	TechN
Boundary	R	S		M
Control	S	S		S
Components				
Connections				
Materials				
Energy				
Peripherals				
Fabrication				
Assembly	S	M		S

R = rapidly changing; M = moderately rapid changing; S = slowly changing or stable.

Figure 10.2   Product/technology matrix.

- Materials
- Energy
- Peripherals
- Fabrication and assembly processes

For the horizontal rows of the matrix one can list the technologies relevant to the product system and the substituting (or potentially substituting) technologies relevant to the product system.

On such a matrix one makes an entry in the boxes where a technology (or substituting technology) is relevant to the product system. Such an entry can be scaled in some manner, such as R for rapidly changing technologies, M for moderately rapid changing technologies, and S for slowly changing or stable technologies. Then, by reading down a column, one can identify which aspects of the product system may be affected by rapidly changing technology. And by reading across a row, one can identify rapidly changing technologies that affect one or more aspects of the product system. Finally, one can enter a final row to the matrix to identify which aspects of the product system are most likely to be affected by the pace of technological change and thereby provide the lower limit to a product life cycle.

ILLUSTRATION
Product Line Life Cycles

Tire chord product lifetimes.   Just as products have finite lifetimes, so too do entire product lines have finite lifetimes. As an illustration, a succession of four product line life cycles in tire chord materials is graphed

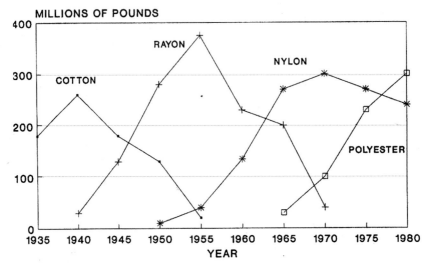

Figure 10.3 Cord materials for tires.

in Fig. 10.3. Tire chord reinforces the rubber or rubber-like material of the tire in order to contain the air pressure and sustain the strong forces acting on the tire.

At first, cotton was the material used in the early automobile tires. Just before 1940, a new artificially made plastic polymer, called *rayon,* was introduced for use as a tire chord. It had greater strength than cotton and eventually almost entirely replaced cotton as a tire chord material by 1955.

However, around 1947, an even stronger artificially made plastic polymer, *nylon,* was introduced as a tire chord material. It grew in sales volume until, by 1970, nylon had almost completely replaced rayon as a tire chord material.

Again, in about 1962, another new plastic polymer material, *polyester,* stronger than nylon, began replacing nylon so that, by 1976, the sales of polyester for tire chords had begun to exceed the sales of nylon for tire chord. Polyester was on its way to replacing nylon in tires.

Now, of course, neither cotton nor rayon nor nylon disappeared from the marketplace. For other customer applications, these materials had different performance characteristics. It was only in tire chord material that the combination of strength, weight, and price determined the market replacements of the sequence of product life cycles. Also note that the overall market for tire chord grew during this period as the market for the application—tires—increased.

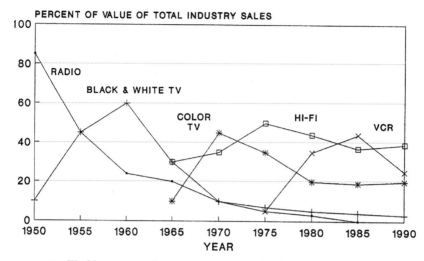

Figure 10.4 World consumer electronics product line lifetimes.

Consumer electronics product lifetimes. As another illustration of product lines that have affected the organization of industrial sectors, consider the various types of consumer electronics that transistors and ICs have fostered. Figure 10.4 depicts the various generic types of consumer electronics products as a percent of the total value of consumer electronics industrial sales (Valery, 1991).

In the figure, one can see that radio, the first consumer electronics product, declined from 100 percent of the industry's value prior to 1950 to a very small percentage by 1985. Of course, some radios are still made and sold, but as a percentage of industrial value, they had become very small. Moreover, as a technology in consumer electronics, the radio function was incorporated within high-fidelity (hi-fi) technology products (e.g., home hi-fi, home entertainment products, car hi-fi, Walkman-like cassette players, etc.).

One also can see that the technology of black-and-white television was substituted by color television. Monochrome television sets rose to a peak of 60 percent of industrial value in 1960 and declined toward 0 percent after 1990. Color television as an industrial technology rose rapidly after 1964 to a peak of industrial value of about 45 percent in 1971. Thereafter, color television continued as a mature technology in the industry at about 20 percent of industrial value in 1990. Color television sets were sold as replacement units or as multiple copies, with a consumer owning several sets. Hi-fi replacing radio peaked as an industrial technology at about 55 percent of industry sales in 1975, steadying at about 40 percent in 1990. The television video cassette recorder (VCR) began sales in 1975, peaking at just over 40 percent of sales in 1985.

What we see in this illustration is that an industry, such as consumer electronics, may have several generic technology products (e.g., radio, monochrome television, color television, hi-fi, VCR). Moreover, some of these technologies become obsolete and are replaced by a newer technology (e.g., monochrome television replaced by color and radio replaced by hi-fi). For technologies that are not replaced, there will be a growth of the sales of the products of that technology until their market saturates. Thereafter, sales will continue at a replacement rate and any growth of demographics in customers.

Product Line Life Cycles

Product lines are generic classes of products offered for similar applications. Technology affects both the performance and cost of product lines. Although products are obsoleted by technology, cost, safety, features, packaging, or fashion, product lines are obsoleted only by technology or safety. Changes in features, packaging, and fashion are all accommodated by changes within a product line.

The general form of a product line life cycle is a convex curve, as depicted in Fig. 10.5. The peak of the curve is a function of applications

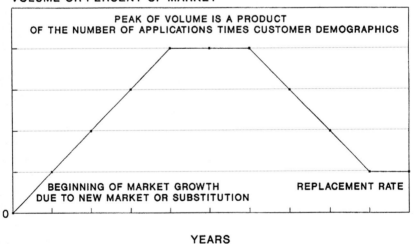

Figure 10.5 General form of product line life cycle.

and the demographics of customers for the applications. After a product line attains its maximum market, its sales volume is determined by replacement rate and any expansion in demographics or applications that require multiple ownership by customers.

Substituting next-generation product lines will have overlapping curves (as in the tire chord example in Fig. 10.3). The reason for this is that, initially, a next-generation product line will be priced higher than the existing product line. The higher price will be required because of the R&D costs of developing the product line, the initial capital cost of constructing production facilities, and the initially higher costs of materials and unit production.

Substitution will begin for the applications that demand the higher performance of the new-generation product line that justifies the price premium. After volume production and sales grow and competitive pressures reduce the price premium, then the new-generation product line begins to substitute more completely for the existing product line. The existing product line may lower prices to lengthen its lifetime or to form a permanent low-priced market niche for low-performance applications.

The preparation of product line lifetime forecasts requires judgments on the extrapolation of the price-performance charts and judgments on the minimum performance required for a class of applications.

If, for a new-generation product line, not all the performance parameters are superior to the older-generation product line, then there may occur a unique-performance market for continuing the old-generation product line as a market niche (as discussed in the previous chapter).

Industrial Technology Life Cycle

The emergence of new-technology product lines can create new industrial sectors. Conversely, the technical obsolescence of product lines can cause whole industrial sectors to die or be restructured. The concept of this level of impact has been termed the *industrial technology life cycle.* It was first articulated by William Abernathy and James Utterback (1978) and later elaborated by others, such as David Ford and Chris Ryan (1981).

Figure 10.6 shows a simplified form of a technology life cycle. Initially, there is a period of technology development before products create an applications market. Once a new product/process/service creates a market application, sales begin during an *application launch phase.*

MARKET VOLUME

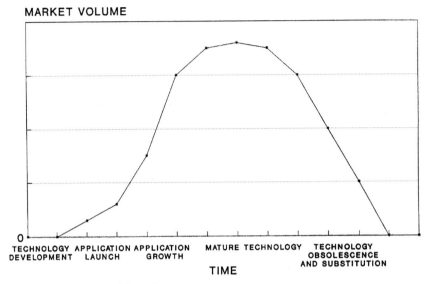

Figure 10.6 Technology life cycle.

An *application growth phase* follows next in which the market may grow to a substantial volume. When innovation in both product and process substantially slows, then the market may saturate when the customer population has fully purchased the product. Subsequent purchases are usually then only for replacement, and the market volume of sales levels off—this is called the *mature technology phase.*

In some technologies, a new replacement technology may be innovated as a product substitution, and then the existing technology becomes obsolete. Products based on the existing technology may no longer be purchased as customer preference is given to the higher-performance substituting technology-based products—this is called the *technology obsolescence phase.*

ILLUSTRATION
Automobile Technology Life Cycle

Figure 10.7 plots as a histogram the number of domestically produced automobiles sold in the United States over the period from 1899 to 1985 (plotted in odd-numbered years). If one fits a technology life-cycle curve over it (as show by the solid curve), one can see that the pattern of the technology life cycle fits roughly, but only roughly.

One also can see that there were four exceptions to the simple theoretical pattern of the technology life cycle:

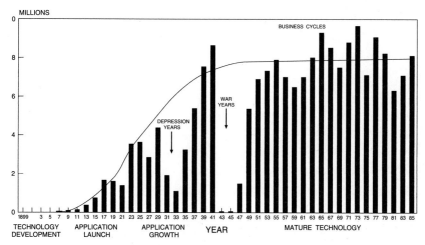

Figure 10.7 U.S. automobile sales—domestic production.

1. The steep decline in automobile sales due to the depression in the 1930s

2. The cessation of domestic automobile production during World War II (from 1942 to 1946) for conversion to weapons production

3. Business expansion-recession cycles on the tops of the pattern (from 1950 through 1985)

4. No technology obsolescence of the automobile, since no substituting basic technology for the internal combustion engine and petroleum-fueled automobile system has yet succeeded commercially

The technology life-cycle pattern ignores any interaction between business cycles and technology innovation. However, this interaction turns out to be important. Also, the mature technology phase of the American automobile industry from the 1950s through the 1970s is a curious story—not really fitting the pattern postulated in the technology life cycle. In fact, technology innovation slowed not due to natural limitations, but rather there were deliberate management decisions in the American automobile industry not to innovate (from 1960–1970).

By the 1990s, the basic technology of the automobile still had not become obsolete because the internal combustion engine was still the most efficient transformer of petroleum/gasoline energy to mechanical power and petroleum was still an available fuel. Competing technologies such as electric batteries or the electric fuel cell did not perform adequately or economically for powering an automobile. Technology obsolescence in an industry only occurs when there is an adequate substituting technology.

Despite these differences, however, one can still see

1. An applications launch phase as the new automobiles were marketed from 1903 to 1918

2. An industrial standard design when Henry Ford innovated the Model T in 1918

3. A linear growth phase as the market was expanded, with every family wishing for a family car

4. After World War II, an eventual saturation of the market with family multiple ownership of automobiles

5. A continuing mature technology market as long as the internal combustion engine automobile is not technically obsoleted

In the number of domestic automobile producers, over a hundred firms were started in the new industry. In 1923, after the design standard of the Ford Model T, most of these firms failed to produce the standardized automobile in the dramatically lowered price range, with only 9 of the more than 100 firms surviving. After that, fierce competition for market share further lowered their numbers so that by 1967 the surviving firms were then down to four—General Motors, Ford, Chrysler, and American Motors. In the 1980s, Chrysler purchased American Motors—leaving, by 1990, the last three American automobile manufacturers. While this was happening for domestic producers, foreign automobile manufacturers began capturing a large percentage of the American market after the 1970s.

Competitive Strategy in the Applications Launch Phase

The first application of a new technology is important. This is what makes forecasting the conditions of competition in the applications launch phase difficult. New technology will create new applications and new markets. This is easy to forecast. But which new applications and which new markets? This is what is difficult.

We do know that in the applications launch phase, the more radical the functional capability provided by a new technology, the more radical and different will be the applications created for the technology.

The best competitive strategy in the applications launch phase will be one that fosters a particular application and quickly moves the new product to additional applications.

ILLUSTRATION
Early Growth of Convex Computer Corporation

In the middle of the 1980s, new technology in the architecture of computers launched new companies, among which was the Convex Computer Corporation. By 1991, Convex was still growing and began challenging the 20-year established leader in the top end of the supercomputer market, Cray. Convex provides an excellent illustration of successful competitive strategy in the applications launch phase of a new technology.

By 1980, the performance of computers could be forecast as heading toward a *natural limit* of the then-current architecture of computers, the serial Von Neumann architecture. A new idea was to utilize parallel computation to avoid the natural limit on serial architecture.

In 1985–1987, over two dozen new firms were started to produce a "mini-supercomputer." Convex was one of these, and by 1991 it had emerged as the clear leader in the new industry. Beginning in 1985, Convex had grown from an initial sales that year of $10 million to a 1990 sales of $210 million.

In 1991, Convex launched its third model series of computers, the C3 series, which spanned the market spectrum from the minicomputer through the mainframe and into the lower end of the supercomputer market. In just 6 years Convex had emerged as a challenger to all the then giants of the computer industry sectors, Cray, IBM, and DEC, supercomputer, mainframe, and minicomputer leaders. This was possible because of the new technology of parallel computation. Why was it, then, that of the two dozen mini-supercomputer firms in 1985, Convex had pulled out ahead?

Convex had been founded in 1982 with two cofounders, one of whom was Robert J. Paluck. Paluck had been at Mostek before cofounding Convex. He had spent 11 years there, rising to vice-president of product development and marketing. Earlier, he had graduated from the University of Illinois in Urbana with a bachelor's degree in electrical engineering.

The second cofounder of Convex was Steven J. Wallach, who also had studied electrical engineering first as an undergraduate at the Polytechnic University in New York and second as a master's graduate at the University of Pennsylvania. He later obtained a second master's degree from Boston University in business. After obtaining his first master's degree in electrical engineering, he went to work for Data General in 1975 (one of the many new firms established to produce minicomputers in that technology discontinuity in the early 1970s).

In 1977, Data General's commercial fortune hung on a second-generation model line that was to be marketed as the Eclipse series of minicomputers. Steve Wallach was the manager of that design. Wallach knew how to design computers.

However, a funny thing happened to Wallach on the "way to the market," so to speak. He was trying to sell the Eclipse to customers, but many of them had already purchased a DEC Vax series computer (which was then the market leader in minicomputers). The Eclipse, as fine as it was, could not outperform a Vax. And what the Vax users really wanted was two or three times the Vax performance for the same price. Moreover, they also wanted it to run the same software they had already written for the Vax. At that time, however, these Vax customers had no other choice but to move to a much higher-priced DEC supercomputer, or to an IBM mainframe at multiples of the cost, or to the Cray at even greater multiples of cost to get greater performance. And they would have to rewrite all their software for any of them. It was a bind. The Vax customers had got minicomputer performance for a good price at DEC, but there was no path upward for improving performance.

What Wallach was learning as the 1970s drew to an end was that the real customer requirement for minicomputers was not for a new model of the minicomputer (like the Eclipse) but for a mini-supercomputer performing an order of magnitude faster than a Vax minicomputer but with Vax compatibility and a Vax price. This his own designed Eclipse series could not deliver. First, Data General had a different proprietary operating system from the DEC Vax. Second, Data General used similar technology to DEC and therefore had no real performance or cost advantage. Wallach saw a real market need that his company, Data General, could not and was not planning to provide.

Discouraged, Wallach left Data General and went to work for Rolm. But he had the market-pull problem of Vax customers always in mind and was looking for a way to satisfy it. Then, in 1981, he saw the technical solution—parallel computer architectures—the new computer science research frontier. "A parallel-processing computer, simply defined, is one that can perform operations using more than one processor simultaneously. You can generally divide parallel processing into three major areas of research: (parallel) algorithms and applications; (parallel) programming languages; and (parallel) architectures" (Obermeier, 1988, p. 275).

The key technical bottleneck in the way the serial Von Neumann architecture works is the way in which the central processing unit (CPU) addresses memory. When data are retrieved from memory, they are written to the CPU register, that register is incremented, and the new value of the register is put back into memory. During this data retrieval process, the CPU stands idle, waiting for the data. This has been called the *Von Neumann bottleneck,* and it "accounts for the sometimes slow and inefficient use of conventional serial-processor resources" (Obermeier, 1988, p. 275).

In a parallel-processing architecture, if several calculations can be performed simultaneously, then one can in theory dedicate a CPU to each of the calculation procedures occurring simultaneously. Then the delay due to memory access does not serially add up, as it does in the Von Neumann bottleneck. For computationally intense processes that use large numbers of simultaneous calculations (such as vector and matrices calculations), this speed up by parallel processing can be very much faster.

Dan Slotnick in 1966 had made the first operating parallel processor with 64 processors. Yet it needed to await the development of the microprocessor to make it commercially feasible. Thus the parallel-processing concept had been around, awaiting other technical developments—a frequent occurrence in technological innovation.

After the single-chip CPUs were introduced in the early 1980s, university researchers started experimenting with the building of parallel machines, and several companies were started to build parallel machines. The first commercial product to enter the market was a $7 million Denelcor machine in 1985. And by 1986, there were several machines on the market made by companies such as Bolt Beranek and Newman, Cray Research, DEC, Intel Allant, Encore, Thinking Machines, Sequent, and Convex.

Wallach understood that if he were to meet the frustrated Vax users' needs, he would have to provide a total solution—algorithms and applications, programming languages, architectures, and price. Convex would aim pricing at the top end of the minicomputer market but an order of magnitude below the Cray-dominated supercomputer market.

Just after Denelcor's announcement of its $7 million supercomputer machine, Convex announced a $300,000 C1 model. Alliant offered machines in the same price range ($200,000 to $1 million), as did Sequent and Encore. DEC formed a whole new company, which it terminated 3 years later. Convex prospered; Alliant, Sequent, and Encore survived. Those which went head to head against Cray failed. Wallach and Paluck were right about the price range. They were right in some other strategies:

1. They chose Convex's operating system to be Unix, an open system invented at AT&T and nonexclusively licensed.
2. They aimed at Cray and Vax users, ensuring that any program that ran either on a Cray or a Vax would run on a Convex.

3. They supported and encouraged the parallelization of Fortran as a language in which to write scientific applications.

The choice of Unix as an operating system encouraged applications developers to write for the Convex. Convex hired a research staff of program developers to work with customers on adapting any Cray or Vax software to run on the Convex. Convex's strategy was to provide an advantageous performance/price ratio and software compatibility.

For customers who already owned supercomputers (such as Crays), for one-hundredth the cost of a Cray, the Convex customer got one-tenth the performance—an order of magnitude improvement in the performance/price ratio. And one could run all the programs written for the Cray. Therefore, Cray customers could add Convex mini-supercomputers as kind of a departmental supercomputer to supplement the division's supercomputer. Convex called this a "departmental segment of the worldwide supercomputer market."

In addition, customers who already owned DEC Vax minicomputers could replace the Vax with a Convex and gain multiples of improvement in performance at the same price—again, a very advantageous performance/price ratio. And they could run all their application programs written for their Vaxes on the Convex.

This was Convex's unique competitive strategy among all the new mini-supercomputer manufacturers using the new parallel-processing technology. It was Cray/Vax compatibility, a Unix open operating system, and an advantageous performance/price ratio.

And it worked. By 1991, Convex dominated 73 percent of the departmental segment of the worldwide supercomputer market. And this was another competitive strategy that Convex boldly pioneered in 1987, which was very early to enter the international market for mini-supercomputers. Convex entered the European market just 2 years after their first entry into the American market. By 1990, 60 percent of the company's $200 million sales came from Europe and the world and only 40 percent from the United States.

The 1990 market distribution of Convex sales in supercomputer applications was in numerically intensive applications. In *computer-aided engineering,* designs for new products and processes were analyzed by customers using Convex supercomputers in the aircraft, automobile, and semiconductor industries. For *government and aerospace* customers, Convex supercomputers processed applications in laboratory research and signal and image processing. The *petroleum industry* customers used Convex supercomputers to generate, store, and manipulate seismic data for analyzing oil explorations and mapping petroleum and gas reservoirs. University and industrial researchers in the *advanced research* customer market used Convex supercomputers to explore new areas of scientific knowledge. *Chemistry* applications were in research and pharmaceutical development.

Competition and New Applications

In a new-technology industry, competitors fight to enter the market and for market share.

1. A clear performance/price advantage is necessary to enable product entry through product substitution.
2. When two competitors with new products begin substitution, the competitor who facilitates applications will gain the advantage.
3. Applications generation is encouraged most rapidly by standards or software compatibility that enables the immediate and easy transfer of the new technology to existing applications.
4. Applications generation is also encouraged by customer service that adapts or generates new applications of the technology.

Thus, in the applications launch phase, a competitive strategy to provide an order-of-magnitude increase in the performance/price ratio of existing technologies is *necessary*. In the applications launch phase of an industrial technology life cycle, the competitors that provide a better *applications focus* will have an advantage over other early competitors.

An applications focus on creating a new market, although slower to build, may over the long term prove beneficial and profitable by creating new market niches. Only after an innovative product is completed with the availability to the customer of the needed supplies and peripherals for an application system will the product succeed in the marketplace as a new application. An innovative product *without proprietary information in the design of the product or in its production* will eventually have no technology competitive advantage against competitors with "me too" products.

The pace of development of applications for a new-technology product is critical to the rate of market growth. Several techniques can facilitate the development of applications:

1. Research performed by the product manufacturers for user applications development
2. Performing applications-development contracts for customers
3. Assisting the formation of technical user groups and communicating with them and assisting them

4. Use of selected customers as test sites for first use of product prototypes

5. Provision of industrial standards or open architectures to facilitate
third-party development of applications software.

Innovation in the Technology Life Cycle after
a Design/Industrial Standard

In moving from the applications launch phase into the linear market
growth phase of an industrial technology life cycle, an important factor
is the lowering of price to facilitate market growth. This is often accomplished after the design of a product is relatively standardized by
bringing together the best features of early product designs and recent
innovations to design a product that well satisfies an application. This
has been called a *design standard* for a product line. All competing
products in the product line are then similarly designed.

The term *design standard* means that a product system is completed in terms of critical components and configuration sufficient for
adequate performance and cost for an application. Incremental innovations in both the product and the production process can still continue.

William Abernathy and James Utterback (1978) pointed out that
after a design standard (or an industrial standard), then innovation
emphasis in the technology may shift to process innovation—improving the production processes of the technology-based product. A
product design standard provides sufficient stability in the technology system of the product design for large-scale mass production to
begin.

*Production-process innovations facilitate growth of the market
during the applications growth phase by reducing the cost of the
product and improving product quality.*

Underlying this concept of technology life cycle for an industry is the
notion of a single dominant product type (standardized by a design
standard) and a single dominant production process. As sketched in
Fig. 10.8, the innovation rate in the product type would peak early in
the technology life cycle, followed by a rapid increase in the rate of process innovations in production. One sees an initial burst of innovations
in product peaking as the technology life cycle for the product enters a
linear phase. And one sees a later burst of innovations for production
processes, after a standard product design has been achieved. After the
technological rate of progress for production processes enters a linear

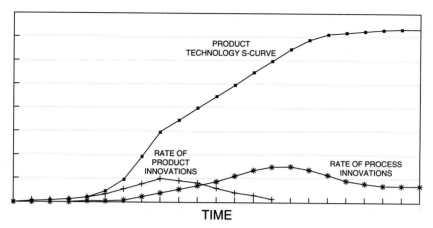

Figure 10.8 Comparing product and process innovation rates.

portion of growth, the technology is stable enough for the market growth portion.

After the technological progress in both product technology and production technology reaches its natural limit, all rates of technological progress slow and market growth continues until the market is saturated with a product in which functionality and performance are no longer changing. New performance or applications are no longer being created by market growth. Therefore, the eventual market size is determined by replacements.

Setting the design standard is critical to competitiveness. Firms whose products fit within the product design standard succeed in competition, while firms whose products do not fit the design standard must quickly redesign products to the standards or fail.

Competitive Conditions in the Mature Phase of an Industrial Technology Life Cycle

After an industrial product design standard is established, the business and consumer markets grow around the design standard as the technology continues to improve with successive improved design standard models. These markets then partition into luxury, middle, and economy markets with corresponding differences in performance and features in their product focus. Firms establishing reputations for quality prosper, and the struggle for survival in the economy markets becomes fierce, with price-cutting margins to the point where many firms fail.

Technology innovations will have shifted from a focus on product improvement to production improvement. Increasingly, the competitive conditions turn from performance (and products differentiated by performance) to similar products competing primarily in terms of price and quality.

In the mature technology phase, competitive strategy should focus on four principal factors: (1) cost/quality leadership in production, (2) market share and brand recognition, (3) product variation and market niching, and (4) fashion and demographic and lifestyle and environment/regulatory changes. Of these, improvements in product performance and functionality will likely play only a small role (due to the maturity of the product technology) in competitive strategy. Product differentiation will be limited mostly to market niching and fashion. Product improvement will, however, still continue, but mostly in the adding of features in luxury models, improving product dependability and safety and reducing product maintenance. Products also may have to be improved to meet increasingly higher standards of public safety and environmental quality. However, technological change will play the primary role in production technology through improving production quality and lowering production costs.

Comparing the maturity phase of a technology life cycle with its early applications phase, the major difference in competitive conditions is the later dominance of production cost/quality competitive factors over the early dominance of performance/applications competitive factors.

Complications in the Industrial Technology Life Cycle

The technology life-cycle pattern captures the most general aspect of the impact of technology change on markets. However, it is oversimplified, ignoring several factors. First, the attainment of a product standard (or industrial standard) may not occur neatly. Second, there may not be a single standard product design but several over time, with each defining a next-generation product technology and producing discontinuities in technical direction. Third, a new-technology market usually segments into types of customers and different industrial sectors. Fourth, a new functional technology is often composed of several technologies whose different rates of progress can structurally alter market growth and dynamics.

Some of the critical factors in the development of new markets include (1) development of applications, (2) improvement of performance, (3) lowering of costs, (4) development of standards, (5) development of safety, (6) development of training, (7) development of infrastructure, (8) market segmentation, and (9) excess production capacity in the industry.

The industrial technology life cycle shows a common pattern such that after a basic new technology is innovated, the market for it begins to grow as applications are found. Applications usually are found first for the military, then industrial, then business, then hobbyist, and finally, consumers as the performance is improved by technology and the cost is reduced.

Only after a design standard for the new-technology product is accomplished do the business and consumer markets really begin to grow to large volumes. As the technology continues to improve, the business and consumer markets usually partition into luxury, middle, and economy markets with corresponding differences in performance and features in their product focus. Technology innovations will shift after a design standard from a focus on product improvement to production improvement, and the rate of technology innovation in production increases.

Yet, during the course of growth, if technology discontinuities in product performance do occur, then new product design standards also will occur. These discontinuities also may partition the market into different industrial segments for high-end performance and low-end performance.

Finally, as technology progress in the product slows or stops, products by all producers will look alike in performance and be differentiated only by fashion or luxury features. In the so-called mature phase of the technology life cycle, products are then of a commodity type, i.e., functionally undifferentiated. Then innovation continues primarily in production processes, and competition is principally in price.

At this time, excess production capacity often occurs because the big market has attracted many producers and the market growth has encouraged increased production capacity. When the market growth slows and the market volume levels off or even declines, then many firms fail or exit the industry. Figure 10.9 presents a comparison of the growth of industrial capacity with market growth, and beneath that is the number of producing firms.

If a superior-performing new technology is then innovated, the existing product technology will be obsolete, and a process of product substitution in the market will begin. If, however, no substituting superior technology surfaces, the market for the technology will continue, with market volume in this technology maturity phase a function of product replacement and demographic growth, with business cycles conditions superimposed. If production improvements continue to lower the cost

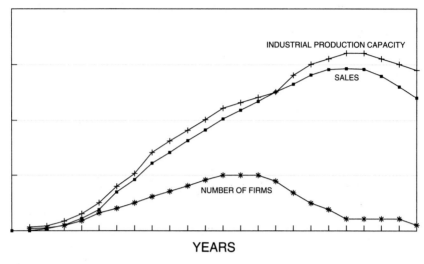

YEARS

Figure 10.9 Market dynamics between technology life cycle and production capacity.

of the product dramatically, the market may grow, even if market demographics do not change, as a result of multiple copies of products purchased by the market.

ILLUSTRATION
Economic Long Waves and
U.S. Technology Leadership

Let us now consider the longest periodic phenomenon associated with technology change—economic long waves, which occur when several technology life cycles together foster economic smoothing. For example, the American automobile, once a world competitive powerhouse, began to technologically slide behind international competition in the 1960s to the point of a major crisis in the 1980s. In the 1970s, America also lost an entire industry that played the major role in innovating, consumer electronics, to Japanese competition. Were there general economic/technological patterns behind these specific kinds of events?

Reviewing the troubled 1980s in the U.S. economy, Richard Nelson (1990, p. 117) took a historical perspective on U.S. economic/technological developments. "During the quarter century following the Second World War, the United States was the world's most productive economy, by virtually all measures....These differences held not just in the aggregate but in almost all industries....U.S. firms were in the forefront of developing the leading-edge technologies, their exports in these accounted for the lion's share of world markets, and their overseas branches often were dominant firms in their host countries....No longer. The U.S. technological lead has badly eroded in many industries and in some the U.S. now is a lagger."

Nelson saw the basis for this past dominance in the two factors of mass-production leadership and new high technology. "...There were two distinguishable components of the post-war U.S. technological dominance. One was a lead in mass-production industries, and this was of long standing. The other was in 'high tech,' and this was new. Each stemmed from different sources. And they eroded for conceptually separate, if institutionally connected, reasons" (Nelson, 1990, p. 117).

Figure 10.10 graphs the U.S. share of world exports and imports from 1960 to 1988 (U.S. Trade, 1989). There one sees the dramatic crossover in 1971 when the U.S. share of exports was exceeded by the U.S. share of imports. And Fig. 10.11 shows the trade balance in manufactured goods (U.S. Manufactures Trade) from 1980 to 1988. In 1982, the value of manufactured imports exceeded the value of manufactured exports—the U.S. began importing more manufactured goods than it exported.

Economic Long Waves Due to Technological Innovation

Underlying the preceding analysis of U.S. competitive problems is a theoretical idea of long cycles in industrial development. Technical change has had very long-term effects on national economies, occurring in cyclic patterns since the beginning of the industrial revolution.

The first industrial revolution began about 1765 in England—centered on the innovations of coke-based steelmaking, the steam engine, and powered textile machinery organized in factories. Later the innovations of railroads, steam ships, electrical lighting and power, chemical dyes and chemical-based plastics, telephone and radio, automobiles and airplanes, and many others all continued the technological basis of industrial revolution that spread around the world. North America began industrializing after about 1850. Japan began after 1865 and began a reconstruction after 1946. Thus one can view the last 200 years as both a series of basic technological innovations and a series of national industrializations.

The interactions of long-term economic patterns and periodic basic technological innovations was emphasized by N. D. Kondratieff,

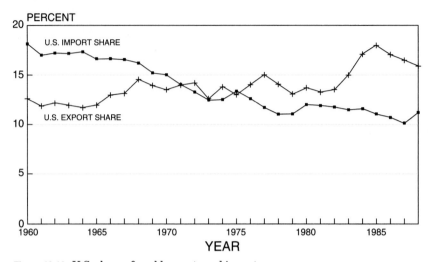

Figure 10.10 U.S. share of world exports and imports.

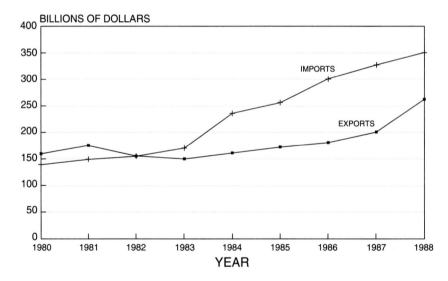

Figure 10.11 U.S. trade in manufactured goods.

who was an economist in Stalin's Soviet Union in the 1930s, where Marxist doctrine then dictated the eventual downfall of capitalism. Kondratieff looked at the historical price data of capitalist countries since the industrial revolution. Although a nominal communist, Kondratieff saw not the Marxist-predicted pattern of decline and fall of capitalistic countries. Instead, he saw a recurrent pattern of economic revitalization occurring in long-term economic cycles of about 50 years (with alternating economic advances and depressions). Kondratieff then explained the recurrent economic expansions as the capitalistic exploitation of new basic technological innovations. Of course, in the brutal, rigid, authoritarian, and doctrinaire regime in Stalin's Russia, this was an unwanted and totally unacceptable idea—that capitalism might periodically renew itself through technology. Stalin had Kondratieff arrested and shipped off to a concentration camp in Siberia—where Kondratieff perished.

Robert Ayres in summarizing the intellectual precedents of the idea of technology and long-term economic growth, pointing out that (prior to Kondratieff's observations on the impact of technological innovation on economic growth) Jevons, in 1884, had noted long cycles in prices and later Van Gelderen was the first to suggest that a period of rising prices was driven by rapid growth in leading industrial sectors. Schumpeter, Rostow, Mensch, and Forrester revived and extended the long-wave idea (Ayers, 1990).

The radical innovations that had direct economic impact on the early phases of the European industrial revolution can be clustered in several groups:

1770–1800: The beginning of the industrial revolution in Europe was based on the new technologies of steam power, coal-fired steel, and textile machinery.

1830–1850: Next, the acceleration of the European industrial revolution was based on the technologies of railroads and steamships and of the telegraph and coal-produced gas lighting.

1870–1895: Then, basic advances in steel making began the almost total substitution of steel for iron, and there were in innovations around the discovery and refining of petroleum, the invention of electrical power and lighting technology, the invention of the internal combustion engine, the invention of the telephone, and innovations in the chemical dyes industry.

1895–1930: In this period, the automobile and the airplane were invented, and the electron vacuum tube and radio were invented. The first chemically produced plastics were invented, and petroleum became the primary energy source (displacing coal) and a major material source.

There is not perfect agreement among students about the exact depiction of the economic long waves that were coordinated with these periods of major technology transformations. But most descriptions are similar to the following (Mandel, 1984):

1792–1825	Kondratieff first wave	Expansion	Iron, steam power, textile machinery
1825–1847		Contraction	
1847–1873	Kondratieff second wave	Expansion	Railroads, steamships, telegraph, coal gas
1873–1893		Contraction	
1893–1913	Kondratieff third wave	Expansion	Steelmaking, petroleum, electrical power, internal combustion engine, telephone, chemical dyes
1913–1939		Contraction	
1939–1974	Kondratieff fourth wave	Expansion	Chemical fibers, pharmaceuticals, television, computers, transistors, IC circuits

The Role of Science in Economic Development

Modern science basically began with the Newtonian synthesis of the mechanistic perspective around 1700 (although Copernicus, Kepler, and Galileo provided the important background for Newton). Modern chemical theory began late in the 1700s and advanced through the 1800s, providing an understanding for the technical developments of Neucommin's and Watt's steam engines and Otto's internal combustion engine and the chemical-based dye industry. The physics of electricity and magnetism were being discovered in the nineteenth century, providing a knowledge base for the inventions in the electrical power industry and in the electronics of communication of the telegraph, telephone, and radio.

The industrial revolution arose from technological innovation, which for the first time in history was based on scientific knowledge.

This does not mean that science invented technology. Technological invention is a different activity from science. For the first time in history, however, technological inventors were scientifically literate, rather than merely technically literate. This is a very important point. A technology-based industrial revolution in the history of the world had to await the social innovation of science.

Science provides a sophisticated and growing knowledge of nature in order for technological innovation to become frequent, recurrent, and continually occurring—thereby driving economic development.

The concept of technologically based economic long waves is that major economic growth can occur from a scientifically based radical technological innovation that creates significant new functionality for humanity. This can be depicted as follows (Betz, 1987, p. 85):

1. Discoveries in science create a phenomenal base for technological innovation.
2. Radically new and basic technological innovations create new products and applications that provide new functionality.
3. These products create rapidly growing new markets and new industries.
4. The new industries continue to innovate products and processes, expanding markets by adding applications, improving quality, and lowering prices and facilitating demand by providing employment.
5. As the technologies mature, many competitors enter internationally, eventually creating excess production capacity in all markets.
6. Excess capacity decreases profitability and increases business failures and unemployment.
7. Subsequent economic turmoil in financial markets may lead to depressions, locally and globally.
8. New science and new technology may provide the basis for a new economic expansion, awaiting innovation until financial and economic conditions improve.

It is important to remember that this pattern is historical and based on the origin of basic technologies. Thus the pattern can only repeat when *new* basic technologies are invented and innovated. Basic technologies are ultimately founded on nature and science. When (if ever) all relevant nature is discovered and science has understood and explained everything, the source for new basic technologies will have been mined out. Therefore, in principle, radical technological innovation may not be eternal; and in principle, Kondratieff long waves may not always recur.

Procedure: Anticipating Market Conditions

1. Anticipate where in a technology life cycle the core technology of a particular industry may be. Is there a generic product standard defining the industry? Is there a key production process in the industry? What are the key competitive factors currently in the industry? Can you expect any of these changes in order of priority as the phases in the technology life cycle occur?

2. For a radically new technology, identify the first applications for which it might substitute for existing products/processes/services? What are the required performance, features, and price for this substitution?

3. For an early technology life-cycle stage, has there emerged an industrial standard or a standard design? If not, what requirements should such a standard encompass?

4. In the industrial technology life cycle, what different product lines have emerged or are emerging? Are there potential substitutions either due to improved performance/price ratios among the product lines or to new substituting technologies? Anticipate the lifetimes of different product lines.

5. For products within such product lines, what are the most important factors in limiting product lifetimes: performance, safety, cost, quality, fashion, lifestyle, etc.? Estimate the expected lifetimes of the different products.

6. How are trends in global industrialization affecting your industry in terms of technologies, markets, labor costs, capital structures, etc.?

7. Are there new discoveries in science that might provide phenomena on which to base substituting technologies for the core technology of an industry?

Practical Implications

The lessons to be learned from this chapter are techniques for identifying and anticipating the impact of technological change on markets and competitive conditions.

1. There are several kinds of life cycles by means of which technological change affects markets:

   a. Product
   b. Product line
   c. Technology life cycle for an industrial core technology
   d. Science-based economic long waves

2. A product lifetime is the length of time between introduction into the marketplace of a product and its withdrawal from sales.
   a. Technological change can influence product lifetime though changes in product performance, features, safety, cost, or quality.
   b. A technology/product matrix is useful for estimating product life cycles.

3. A product-line lifetime is the length of time between introduction into the market of a product line and its replacement by a new product line.
   a. Technological change can influence product-line lifetimes through changes in the technology of the generic product system or generic production system.
   b. Next generations of technology systems obsolete current product lines, and substituting technologies can obsolete current product lines.

4. The technology life cycle is the time from the innovation of a new industry until its market saturation or replacement, with the following phases: applications launch, product design standard, applications growth period, and core technology maturity period.
   a. The best competitive strategy in the applications launch phase will be one that fosters applications adoption of the new technology product and tracks and quickly adapts applications to the new product and establishes the industrial standard.
   b. Comparing the maturity phase of a technology life cycle to its early applications phase, the major difference in competitive conditions is the later dominance of production cost/quality factors over the early dominance of performance/applications factors.

5. A pattern of a science-based, technologically driven economic long wave has characterized the industrialization of the world, based on the interrelations between science and technology and industry.

For Further Reflection

1. Identify three basic product technologies that respectively innovated in the periods 1850–1900, 1900–1950, and after 1950.
2. Have any of these technologies been technically obsoleted by substituting technologies?

3. Of the technologies that are still viable, what product lines evolved in the technology? What has been the product line lifetimes in these lines?

4. Describe the overall technology life cycle for these technologies. When and how was a product design standard first achieved? Have there been subsequent design standards? Was there one major production process in the technology or several? What were the key process innovations, and when did they occur?

5. The textile industry was one element of the original industrial revolution in England. Describe its origin. What contributions to the technology were made in different countries? Describe the current state of the U.S. textile industry in terms of product lines and imports and exports.

Forecasting the Rates of Technological Change

We have seen that the rates of change in technologies are important for planning new product development and are critical in business planning. Technological innovation provides distinct competitive advantages only in some commercial "window of opportunity." This chapter now turns to techniques for forecasting the rate of change in a developing technology.

The rate of technical change in a technology can be expressed as the rate of improvement in a generic performance parameter *for the technology.*

Generic performance parameters indicate that technical improvement will be useful to a wide range of applications. Progress in technical performance parameters will be accomplished through improvement in the logic or morphology of the technology.

There are basically three approaches to forecasting. The first is an *extrapolation* of past and current trends. The second is a *structural analysis* of underlying factors determining these trends. The third is an *exploration of possible changes* in the structural factors. One begins with the first approach, extrapolating trends. Then one can use the second approach to probe deeper into these trends by asking: What are the underlying structures that influence the trends? Having identified structural factors, one can then ask: What could alter these structures?

ILLUSTRATION
A Logic Tree for Forecasting
Techniques

Periodically, reviews of the literature on technology forecasting appear, such as in Landford (1972) or Godet (1983). In 1988, Jack Worlton

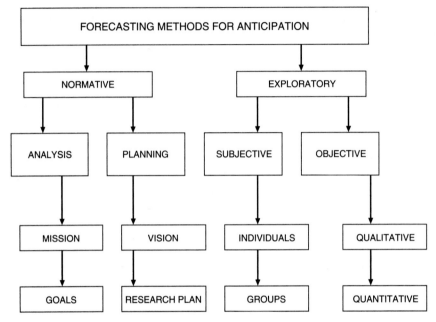

Figure 11.1 Worlton logic tree for forecasting approaches.

(p. 312) offered a logic tree as a guide to that literature, and one such logic tree is presented in Fig. 11.1.

By the term *exploratory,* Worlton meant "starting from the present and advancing step by step toward the future" (Worlton, 1988, p. 312). He should have used the term "extrapolation." Methods that have been proposed to do this have all been extrapolation of current trends into the future. The two most important methods have been (1) the *Delphi method,* for what Worlton called "subjective" and "group" explorations, and (2) the *technology S-curve,* for what Worlton called "objective" and "quantitative" explorations.

The logic line that Worlton called "subjective" and "individual" has not provided a useful technique, because there is no sound method for calibrating which individuals at any time in history are good predictors of the future. For example, Christopher Cerf and Victor Navasky (1984) collected some amusing examples of notoriously wrong predictions by some brilliant, successful, and famous folk:

- The scientist Lord Kelvin (after whom the absolute temperature scale is named) asserted in 1895 that "heavier-than-air flying machines are impossible." (The Wright brothers did not fly until 1904.)

- The famous scientist Albert Einstein in 1932 asserted that "there is not the slightest indication that [nuclear] energy will ever be attainable." (And at that time, extrapolating from what was then known, he was right; Hahn and Meitner discovered nuclear fission 4 years later in 1936.)

- The successful business man Tom Watson, whose son was to lead his firm, IBM, to become a giant in the computer business, said in 1943, "I think there is a world market for about five computers." (And at that time, knowing what he did about one of the world's first computers, the Mark I, which he had helped support Harvard's Howard Aiken to build, that was a pretty good estimate. He apparently then was not aware of Mauchley and Eckert's ENIAC at the University of Pennsylvania.)

- Even as late as 1977, just as the awkward new personal computers were being introduced into the market, the giant of the minicomputer industry who had founded Digital Equipment Corporation, Kenneth Olsen, said, "There is no reason for an individual to have a computer in their home." (Lacking the applications software that was to be later developed, such as spreadsheets and word processing, this was then not an unreasonable opinion.)

What Cerf and Navasky (with their naughty sense of humor) have really underscored is that until both a scientific basis has been discovered for a basic invention and a basic invention for a technology does occur, there is no reasonable basis upon which to predict technical progress. Moreover, even after a basic innovation is discovered or created, some experts may be extremely pessimistic about progress because they are familiar with the problems and difficulties of new technology. Thus one does require more objective and systematic procedures for technology forecasting than simply the subjective judgments of individuals (no matter how brilliant and successful).

By the term *normative,* Worlton meant "inventing some future and identifying the actions needed to bring that future into existence" (Worlton, 1988, p. 312). The approaches of *normative analysis* look at the underlying structures of current trends; they do not merely extrapolate trends. These approaches ask what alterations in structures can create new technology. The principal method here is a morphologic analysis of a technology as a system (which we will review in the next chapter).

Finally, on the logic branch of *normative planning,* this approach involves actual formulation of technology strategy and research programs to implement such strategy. We will focus on this topic in a later chapter.

In summary, from Worlton's logic tree we can see that technology forecasting is a companion to technology planning. One should try both to anticipate the future and to make the future that one desires happen. This is the essence of all action, particularly the essence of technology forecasting and planning. This chapter discusses techniques for extrapolation of technological trends, and later chapters examine techniques for research planning to advance technology.

Technology Performance Parameters

We recall that a technology is describable by a functional transformation that changes purposely defined input states to purposely defined output states. The measure of performance of this transformation is the efficiency of the transformation.

A technology performance parameter is the efficiency of the transforation as perceived by the user of the technology.

All technologies are systems whose operation is controlled by many different variables and specifications. The appropriate performance measure for a technology is one that directly relates technical progress to customer utility. The choice of the technology performance with which to characterize the utility of technical progress is therefore critical—and the choice may not be obvious. For example, in terms of inte-

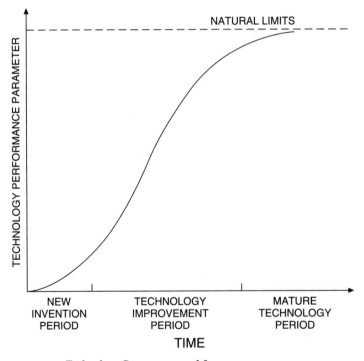

Figure 11.2 Technology S-curve, general form.

grated circuits, the choice of the central performance parameter has been the density of transistors on a chip. This was selected because increased numbers of transistors per chip allow more circuits per chip, which, in turn, allows increased functionality (or utility) to the chip's users. As another example, for commercial airplane manufacturers, two performance parameters have been important: cost per passenger mile and transit time.

The choice of a technology performance parameter to forecast technical progress for a technology system is a means of communicating the value of technical progress to the customer of the technology.

One can plot the rate of change of a performance parameter over time, and one will likely see a kind of S-shaped curve, such as in Fig. 11.2. This common historical pattern has been called the *technology S-curve*, and it has been used as a basis for extrapolative forecasts of technology change (Foster, 1982; Martino, 1983; Twiss, 1968.) Three general periods have been noted in the rate of progress of technology: (1) an early period of new invention, (2) a middle period of technology improvement, and (3) a late period of technology maturity. Since all technologies are based on natural phenomena, for any given technology S-curves there will be a natural limit to progress due to the underlying phenomena.

ILLUSTRATION
Technological Progress in Illumination

As an example of a technology S-curve, Fig. 11.3 plots the improvements over time in the efficiency (lumens per watt) of incandescent lamp technology. And one can see that the shape of the curve does look like a lazy kind of S. In the early years of lamp technology from 1889 to 1909, the rate of improvement in technical performance increased exponentially over time. From 1909 to 1920, however, further progress was linear over time. Then finally, the rate of technical progress turned over into an asymptotic region approaching a finite limit during the period 1920 to 1960.

The incandescent light bulb was innovated independently by the American, Thomas Edison, and the Englishman, Joseph Swann. Both used carbon as the filament in their bulbs. Thus the first point on Fig. 11.3 at 2 lumens per watt (lm/W) in 1889 was the efficiency of either Edison's or Swann's bulb.

Edison was more successful commercially and came to dominate the new electrical illumination industry. However, to sell his light bulb, Edison also had to innovate the supply of electricity for the light. He therefore innovated the first coal-fired power plant, which in 1882 was

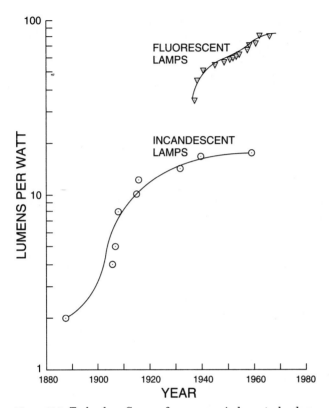

Figure 11.3 Technology S-curve for progress in lamp technology.

serving the Wall Street district of New York City. Later, the Edison Electric Light Company became the General Electric Company.

By 1900, the new General Electric Company (GE) was still producing the carbon-filament type of lamp that Edison had invented. However, GE was then facing new technological threats to its incandescent lamp (Wise, 1980). One threat was a "glower" lamp, invented by German chemist Walther Nernst, that used a ceramic filament. This produced a higher efficiency, 2 lm/W, because of the higher temperature that the ceramic filament could withstand. A second threat was the new mercury vapor lamp, developed by the American inventor Peter Cooper Hewitt. George Westinghouse, a major competitor to GE, supported Hewitt's work and had obtained the American rights to Nernst's lamp.

Employed at GE at the time was an outstanding engineer, Charles Steinmetz. He had visited Hewitt's laboratory and recognized the commercial threat to GE of Hewitt's mercury vapor lamp. At that time, Steinmetz suggested the urgency of GE performing basic research to meet such threats to ensure its long-term success.

Referring back to Fig. 11.3, one can see the basis for this threat, since the figure compares the progress in the technical performance of fluorescent and incandescent lamps. One sees that fluorescent lamps have an intrinsically higher efficiency than incandescent lamps. The reason is that fluorescent lamps are based on different physical phenomena than incandescent lamps. Fluorescent lamps utilize the passage of electrons through gases, in which electron-atomic collisions excite gas atoms to higher electronic states that relax by fluorescence, i.e., giving off photons. In contrast, incandescent lamps utilize the passage of electrons through solid materials, in which electron-atoms collisions yield energy as photons as a result of the electrons altering course.

The intrinsic difference in efficiency of conversion of electricity to light arises from the physical differences of these two processes. Here we see two different processes of nature, electrons traversing gases and electrons traversing solids. Both technologies, incandescence and fluorescence, provide light, but with intrinsically different efficiencies. Steinmetz recognized that the technical efficiency of Hewitt's lamp would always be higher than that of Edison's lamp because of their different physical processes. Accordingly, Steinmetz urged GE to do more fundamental research. On September 21, 1900, Steinmetz wrote a letter to a GE vice president, Edwin Rice (who was then head of GE's manufacturing and engineering functions), proposing a new "research and development center" for GE. Other key technical leaders for GE, specifically Albert Davis and Elihu Thomson, agreed with Steinmetz's proposal, and Rice obtained the approval of GE's president, Charles Coffin.

Rice and Davis then hired a young professor at MIT, Willis Whitney, to direct the new center. Steinmetz's proposal for the laboratory was for scientists to research the basis for new technologies: "'We all agreed it was to be a real scientific laboratory,' Davis later wrote" (Wise, 1980, p. 410).

Wise summarized the importance of this concept of using science to explore the basis for new technologies as follows: "He [Whitney] was not the first professional scientist to be employed in American industry, or even in General Electric. Nor was his laboratory the first established by that company or its predecessors....Whitney's effort marks a pioneering attempt by American industry to employ scientists in a new role—as 'industrial researchers'" (Wise, 1980, p. 410).

Technology S-Curves and Physical Phenomena Underlying Technologies

Underlying any technology is a physical phenomenon. The range of technical performance of the technology is determined by the nature of the underlying phenomenon. Technical progress in a technology based on one physical process will show the continuity of form along an S-curve shape. However, technical progress that occurs when a different generic physical process is used will jump from the original S-curve to a new and different S-curve.

Every technology S-curve is expressive of only one underlying physical phenomenon.

We see, therefore, two kinds of technical progress. The first is *incremental progress* along an S-curve as improvements are made in a technology utilizing one physical process. The second is *discontinuous progress* that jumps from an old S-curve to a new S-curve as improvements in a technology use a new and different physical process.

The Initial Exponential Portion of a Technology S-Curve

One can gain an understanding of why such a shape has appeared often historically in technological progress by examining the mathematics depicting S-shaped curves. One mathematical way to obtain exponential growth is through a phenomenon having, at any time, a growth rate proportional to the currently existing state.

First, we should remind ourselves of some calculus. Recall that the notation for the derivative of a continuous function $f(x)$ of a variable x is noted by $f'(x) = df(x)/dx$. Then recall that if the function $f(x) = \ln x$ (the "natural log" of x), then its derivative is

$$f'(x) = 1/x \qquad (11.1)$$

Also recall that if the form of $f(x)$ is a function $u(x)$ raised to an exponential, that is,

$$f(x) = e^{u(x)} \qquad (11.2)$$

then its derivative has the form

$$f'(x) = u'(x)e^{u(x)} \qquad (11.3)$$

Consider now a phenomenon where the quantity of the phenomenon in a state V grows from an initial state V_0 exponentially with time t:

$$V = V_0 e^{kt} \qquad (11.4)$$

where k = a constant rate of growth.

Now this is the mathematical form of the beginning of the S-curve, and we can show that this form is derivable from a particular kind of growth formula. First, using equation (11.3), differentiate this formula by time t:

$$\frac{dV}{dt} = kV_0 e^{kt} \tag{11.5}$$

Next, divide equation (11.5) by V on both sides:

$$\frac{dV/dt}{V} = \frac{kV_0 e^{kt}}{V_0 e^{kt}} \tag{11.6}$$

Simplifying this equation, we obtain:

$$\frac{dV/dt}{V} = k \tag{11.7}$$

And by multiplying both sides by V, we obtain

$$\frac{dV}{dt} = kV \tag{11.8}$$

What this equation says is that the rate of change of a state V at time t in exponential growth is proportional to the size of the state V at time t multiplied by a constant rate of growth factor k.

The conclusion here is that one way to generate exponential growth is to have a rate of growth from one state of the phenomenon to the next that is proportional to the number in the prior state. For example, Fig. 11.4 shows the increasing numbers of Ph.D.s in the United States after 1946. After World War II in the United States, both university faculty positions and university research support from the federal government increased very rapidly until 1965, providing financing for graduate education in engineering, mathematics, and the physical sciences. In the United States, the number of doctoral students produced at any time is proportional to the number of research faculty having external grant support (since grant support in the United States is the major source of funds for graduate student support during their thesis research). So the number of Ph.D.s in engineering, mathematics, and the physical sciences increased exponentially from 1946 to about 1955, since the growth rate was proportional to the existing state of research-supported faculty, which, in turn, was increasing proportionally to the numbers of graduate students produced. This pattern changed after 1965 when faculty positions were filled and the expansion of higher education slowed.

The technology S-curve requires that the new ideas for improving a new technology must be proportional to an existing set of ideas in the beginning of the new technology for an exponential growth to occur.

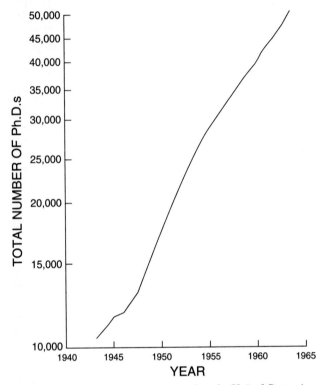

Figure 11.4 Number of Ph.D.s produced in the United States in engineering, mathematics, and the physical sciences.

Since the creation of new ideas is not necessarily proportional to the quantity of preexisting ideas, one must take the technology S-curve formula not as a model of technological progress but as a historical *analogy.*

It is true that for a brand-new technology, when ideas are new, the ideas stimulate new thinking and thus new ideas (but not in a quantitative relation). Thus this analogy states that there will be an explosion of new thinking which can roughly be described as exponential—as if it were quantitatively related—an analogy. When the ideas of a radically new technology are extremely exciting and intriguing, then many new people start to think also about how to improve it.

Asymptotic Portion of the S-Curve

To derive the mathematical form of the S-curve, one begins with the preceding differential equation (11.8) for exponential growth and adds another term that is negative and squared, that is,

$$- bV^2$$

so that a new differential equation is formulated:

$$\frac{dV}{dt} = kV - bV^2 \tag{11.9}$$

This means that the rate of growth (dV/dt) of a state V increases proportionally to the state (kV) and decreases proportionally to the square of the state $(-bV^2)$. When V is small, the positive term will dominate in the beginning to increase growth, but when V becomes large, the negative term will dominate, slowing growth. And this is the reason that the S shape results.

The derivation of the solution to this differential equation uses a technique called *partial fractions,* and results in the form

$$V = \frac{k/b}{1 + e^{-kt}} \tag{11.10}$$

At the origin when $t = 0$, $V = 1/2k/b$. Very late in time as t approaches infinity, V approaches k/b, so V increases from $1/2(k/b)$ to k/b as time lapses. The form of change in V is shown in Fig. 11.3, which is the S shape.

The ratio of the parameter k/b is called the *natural limit* of the performance parameter for the technology. In the example of the incandescent lamp (Fig. 11.3), the natural limit is nearly 20 lm/W, so here $k/b = 20$ lm/W.

In using a technology S-curve to forecast the rate of technological progress, the most important parameter to try to estimate is the natural limit (k/b).

A word of caution, however, about using the mathematics for technology S-curves is in order. The difference between a model of a phenomenon and an analogy of the phenomenon is that the model both describes and explains the phenomenon and an analogy may describe a phenomenon but does not explain it.

The mathematical form of the technology S-curve is not a model of the process of technological change but an analogous pattern frequently seen in the histories of technical progress.

The technology S-curve is not a model because it does not explain the rate of change of a technology's progress. Such an explanation must be derived from an understanding of the morphology of the technology and how progress altered that morphology. Accordingly, the technology

S-curve is merely a *descriptive analogy*. As an analogy, however, it can be used for an intelligent "guess" at the rate of change, i.e., anticipation and forecasting, but it cannot be used to "predict" how that change may occur.

ILLUSTRATION (*Concluded*)
GE's Response to the Lighting Threat

Let us return to the incandescent lamp illustration to see how technical change did occur. To defend GE's lamp business, Whitney's research strategy was to systematically explore different elements to find an improved substitute for the carbon filament in the incandescent lamp. He considered all feasible elements and assigned one of these to each of his scientists. As Whitney hired many scientists for the new center, one of them was a fresh Ph.D., William Coolidge (who had earlier just taken a job at MIT). Coolidge was an American who had gone to Germany to get his doctorate, graduating from the University of Leipzig in 1905 (obtaining doctorates from Germany was a frequent practice at the time for American scientists).

Coolidge was assigned the element of tungsten. Whitney explained to Coolidge the commercial importance of the research. Coolidge wrote to his parents: "I am fortunate now in being on the most important problem the lab has ever had....If we can get the metal tungsten in such shape that it can be drawn into wire, it means millions of dollars to the company" (Wolff, 1984, p. 84).

The promise of tungsten was that it had a high melting point (higher than carbon) and therefore might be usable to incandesce at a higher temperature, thus physically providing a higher lumen per watt efficiency.

Most of industrial research is developmental, some is applied, and a very little is basic research. Coolidge's problem with tungsten was of the nature of applied research. He knew the attractive feature of tungsten for the lamp application and the need for tungsten to be drawn into a wire for this application. But how to do this? The problem was the ductility of tungsten. It was too brittle to be drawn into a wire.

Coolidge tried many things, experimenting by forming mixtures of tungsten that had most of the properties of tungsten but with more ductility. In June of 1906, his first research break came. He observed that mercury was absorbed into hot tungsten, forming an amalgam. Next, he found that cadmium and bismuth also could be absorbed into hot tungsten, again forming an amalgam, and that this mixture was sufficiently ductile to be squeezed through a die to make a wire. In March of 1907, Coolidge discovered that this tungsten amalgam could be bent without cracking when heated to 400°F. He was on the right track!

By the fall of the next year, Coolidge thought that he had the right mixture, and he then decided to try to find a production process for making tungsten amalgam wire. He visited factories in New England that made wire and needles, seeing a swaging technique for wire making. Swaging is a process of gradually reducing thickness by pulling the wire through smaller and smaller dies. In May of 1909, he purchased a commercial swaging machine for his research and modified it to pull tungsten into wire. It made wire. But when the wire was heated within a lamp, it broke!

Problems! No research is ever without problems. No technical project is ever without problems. Technical progress consists of understanding, invention, and problem solving.

Fortunately, Coolidge understood a similar problem. He knew that in making ice cream, glycerine was sometimes added to prevent the formation of ice crystals as the milk freezes. He also understood that crystallization of the amalgam as it solidified was the source of the problem of breakage when the wire was heated. Therefore, he tried adding other substances to the amalgam to prevent crystallization. One substance he tried was thorium oxide, and it worked. He had invented and solved the problem. On September 27, 1910, Coolidge demonstrated a successful tungsten filament lamp to his supervisor, Whitney.

GE immediately innovated the new product made with Coolidge's process. Within a year, GE had discarded all earlier lamp filament equipment and was producing and selling only tungsten filament lamp bulbs. The research effort had cost GE a total of $100,000 and had taken 5 years of effort. By 1920, nine years later, two-thirds of GE's profit came from the new lamps.

One sees in Fig. 11.3 that after Coolidge's invention of the tungsten amalgam filament in 1910, technical progress in incandescent lamp efficiency leveled off asymptotically. From nature, there has been no better element for incandescent lamp filaments than tungsten. Incandescent lamp technology had reached a natural limit for the underlying physical process of heating from electric current through metals.

Characteristics of Technology S-Curves

Let us summarize the concepts that we have been illustrating.

Definition

> A technology S-curve is an analogy of the frequent patterns of incremental progress in a technology's principal performance parameter over time.

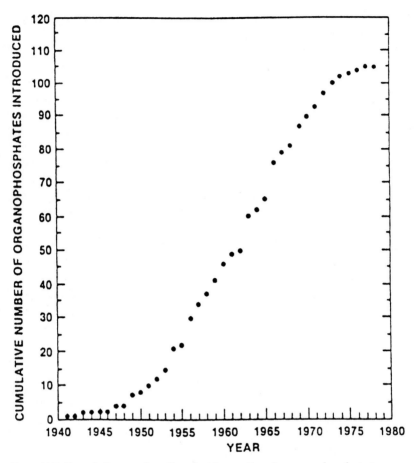

Figure 11.5 Cumulative number of product innovations in organophosphate insecticides.

Properties

A technology S-curve is dependent on an underlying generic physical process.

The natural limit of progress on a technology S-curve arises from the intrinsic factors in the underlying physical process.

Discontinuities in technical progress for a technology occur when alternate physical processes can be used in inventions for the technology.

Such discontinuities express themselves as different and disconnected S-curves.

ILLUSTRATION
American Cyanamid Company:
Organophosphate and
Pyrethroid Insecticides

Robert Becker, director of the Agricultural Research Division of American Cyanamid Company, and Laurine Speltz, staff assistant, decided to evaluate the S-curve technique and chose to conduct a retrospective study of a class of insecticides, organophosphates. First they had to chose a measure of technological progress, and they decided to use the cumulative number of organophosphates introduced by industry: "based on the assumption that each introduction, i.e., a new chemical moiety, represented some improvement in terms of cost, spectrum, market niche, or the like, over those products already in the marketplace" (Becker and Speltz, 1983, p. 31).

Figure 11.5 displays their results, forming an S-curve. In 1974, American Cyanamid discontinued any more discovery research on organophosphates. "The analysis suggests that research should have been discontinued in about 1968. In fact, we might better have spent the associated man-years on a new technology" (Becker and Speltz, 1983, p. 31).

Encouraged by this example, Becker and Speltz next tried the technique on one of their current research efforts, another class of insecticides, the pyrethroid insecticides. "At the time this [pyrethroid] analysis was made in 1982, American Cyanamid had established a commercial position in the pyrethroid market. A pyrethroid, 'PANACTO' cypothrin, had been introduced in 1979 and a second, 'PAYOFF' flucythrinate, in 1982" (Becker and Speltz, 1986, p. 21).

Figure 11.6 displays their results. Since this class of insecticides was then relatively new, Becker and Speltz had only points from 1975 through 1983 to graph, which showed clearly that the technology was yet in an early exponential growth rate phase. "The S-curve for the pyrethroids...its shape is not obvious" (Becker and Speltz, 1986, p. 21).

Therefore, the important analytic problem was to establish the level of natural limits to the technology. Becker and Speltz next identified five properties they thought key to significant technical improvement: (1) spectrum of activity, (2) increased soil activity, (3) systemic activity, (4) reduced fish toxicity, and (5) reduced cost.

The *spectrum of activity* is the range of parasites against which the insecticide is effective. A major weakness of the class of pyrethroid insecticides is marginal performance against mites. "It was believed that this weakness could lead to the discovery and development of several niche products, but the size of the market opportunity of such a product was questionable" (Becker and Speltz, 1986, p. 23).

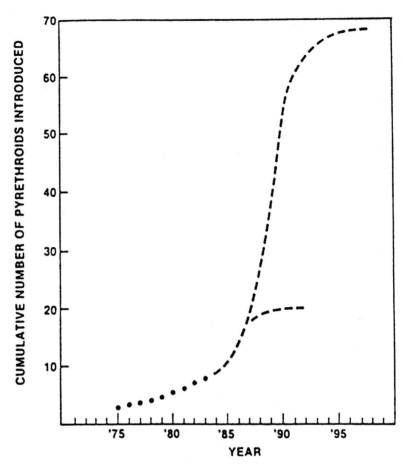

Figure 11.6 Forecasted cumulative number of pyrethroid insecticide innovations.

The property of *increased soil activity* indicated if the nature of the soil impaired insecticide performance. Becker and Speltz knew that pyrethroids were deactivated by soil binding. Their own research gave them little hope that there was much chance of overcoming such deactivation. The property of *systemic activity* indicated if the molecular structure offered opportunities for technical improvement. Becker and Speltz also thought that owing to the hydrophobic nature of the pyrethroids, there also was little hope that systemic activity could be much improved. The property of *reduced fish toxicity* indicated an important safety feature of the insecticide in its impact on the environment. However, here Becker and Speltz felt that improvement would not have much impact on commercial sales. And on the final criteria of *reduced cost,* they also thought that not much improvement could be made over the existing pyrethroids.

Accordingly, their forecast was that the lower of the two possible curves in Fig. 11.6 would prove to be the eventual outcome of the rate of growth of the pyrethroid insecticide technology. "It is recognized that this analysis is only a forecast, and that it has the uncertainty and risks associated with any forecast. Additionally, while we believe the development of a pyrethroid which truly represents a technical advance is unlikely, we do not believe that the development of a `me too' pyrethroid which could be commercially important is precluded. Furthermore, every R&D manager must deal with the difficult decision to change the direction of a research program in the face of uncertainty about the outcome, and any decision to change a research program invariably involves some type of forecast" (Becker and Speltz, 1986, p. 23).

Physical Processes and Technological Discontinuities

When progress in innovation has occurred through a substitution of a new underlying physical phenomenon for the technology, then a chart of the performance parameter's rate of change will "jump" onto a new S-curve. Technological discontinuities also have been described as progress in the form of "next generations of technology."

Figure 11.7 sketches how several next generations of technology can themselves be enveloped by an overall S-curve (the example used is the previous one for lamps). Yet the envelope of a series of discontinuous S-curves is not simply a kind of the "sum" of S-curves. Each S-curve represents a particular phenomenal base to a technology. When phenomenal bases change, then S-curves change, and it is the phenomenal bases that provide the natural limits to a technology. Thus using an "envelope" as a simple S-curve can hide the most important idea that an S-curve can really present—which is the notion of a natural limit to a given phenomenally based technology.

The discontinuous-technology S-curves for changes in the phenomenal bases of technology are some of the most important information that technology forecasting can provide for business and competitive strategies. As we shall see in a later chapter, historically, most technologically based failures of businesses have occurred just at such technology discontinuities. Conversely, most successful new high-tech businesses have been started at just such technology discontinuities.

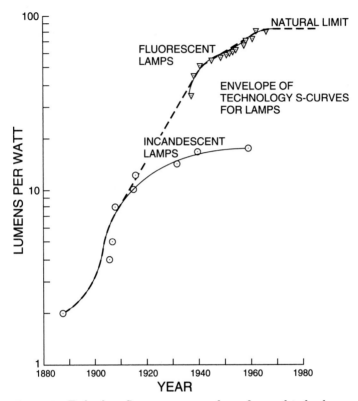

Figure 11.7 Technology S-curve as an envelope of several technology S-curves each representing a different physical basis.

ILLUSTRATION
Next Generations of Integrated
Circuit Technology

A historical example of major jumps in technical progress can be seen in the technical progress in semiconductor chips. This has proceeded in generations of technologies, which represent not a simple S-curve but jumps from one S-curve to another. Figure 11.8 shows the density of transistors achievable on an IC chip over time. The graph is plotted on a semilog scale, and one can see that *this is not the form of the technology S-curve.*

Recall that the semilog scale converts an exponential function to a linear function as the logarithm of the exponential. Recall also the exponential growth equation:

$$V = V_0 e^{kt} \tag{11.4}$$

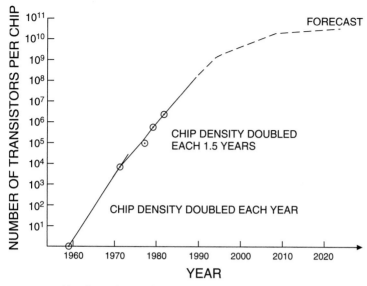

Figure 11.8 Envelope of technology S-curves for several generations of technology in semiconductor chip density plotted on semilog scale.

Taking the natural logarithm of both sides gives

$$\ln V_0 = \ln V + kt \qquad (11.11)$$

Graphing this equation of $\ln V$ against t yields a straight line with slope k and vertical intersect at $\ln V_0$, as shown in Fig. 11.8. Although the rates of progress from 1959 to 1972 and from 1972 to 1984 were both exponential, the rates of exponential increase differed in the two periods. Initially, from the invention of the chip in 1960 until 1972, the growth rate was such that, on average, the density of transistors on a chip were *doubled each year*. However, in the second period from 1972 to 1984, while the growth was still exponential, the rate was such as to achieve an average *doubling each 1.5 years*.

In the year in which this forecast was made (1983), researchers then estimated that the natural limits to chip density would eventually restrict the number of transistors per chip to the range from 10^9 to 10^{10}, or a natural limit on the order of 1 billion to 10 billion transistors per chip.

This estimate was presented by Dr. James Meindl (then director of Stanford University's Center for Integrated Systems) at a research meeting composed of expert researchers in chip technology (the Third Annual International Electron Devices Meeting, which was held in Washington, D.C., in December of 1983). "Projection of the future course of semiconductors has become a liturgical requirement of integrated circuit meetings since Gordon Moore of Intel Corporation...in the mid-1970s...observed that the number of transistors on a chip had been roughly doubling each year since Texas Instruments and Fairchild Semiconductor independently developed the integrated circuit in 1959" (Robinson, 1984, p. 267).

At that time, Meindl also expressed a consensus of experts that anticipated an eventual limit to the growth. "Meindl's message...is that further moderation of the growth curve is in store....The reason for the anticipated decline in rate is that engineers are approaching a number of theoretical and practical limits on the minimum size of transistors....Meindl calls this future era [of natural limits] ULSI, for ultralarge-scale integration, as opposed to the current VLSI or very-large-scale integration epoch" (Robinson, 1984, p. 267).

Here is the clue to how one interprets this example in terms of the usual technology S-curve. Note that an expected linear portion of a technology S-curve will have been historically absent. The reason is that progress in chip density has occurred discontinuously. Rather than a steady incremental improvement occurring, chip manufacturers leap-frogged technologies as epochs or eras. These leap-frogging, discontinuous jumps advanced technology in generations of technology: MSI, LSI, VLSI, ULSI.

From the innovation of chips in 1960 until the mid-1960s, improving skills and tools and processes steadily increased chip density. By then ideas had progressed to new tools and processes that a new generation of industrial processes were installed to jump the density from the order of tens of transistors per chip to hundreds of transistors per chip. Then, around 1965, chip manufacturers installed new equipment which enabled them to design and manufacture chips at middle-scale integration (MSI), with hundreds of transistors per chip.

Progress was then steady, and by the beginnings of the 1970s, five years later, new kinds of equipment and refined processes were available to again jump ahead into manufacturing transistors at large-scale integration (LSI), with thousands of transistors per chip. Now it was in this epoch that Japanese firms began producing chips, providing competition to American chip manufacturers.

After about another 5 years, from the mid-1970s to the beginning of the 1980s, progress in chip technology steadily increased. And again, new ideas for tools and processes and design procedures provided another opportunity for a jump, a leap, from chips with thousands of transistors to chips with millions of transistors. In 1981, another new generation of technology for chip manufacturers began as very-large-scale integration (VLSI), with millions of transistors per chip.

This was the history behind Meindl's projections at that meeting in 1983, as the experts then looked ahead to one more generation of chip technology, ultra-large-scale integration (ULSI), with billions of transistors per chip. Were they then, however, anticipating this as the last generation? Only if no new physics were explored; but then quantum well devices were begun.

In summary, the linear portion of any technology S-curve for chips has been absent. It has been due to the fact that technical progress in processing chips to achieve increasing densities has occurred through the *substitution of new physical processes.*

Fitting Technology S-Curves

We may now summarize the technique of fitting technology S-curves to technical data on the rate of progress of a technology in the following manner:

1. Identify a key technical performance parameter.
2. Collect existing historical data on technical performance since the date of innovation of the technology, and plot them on a time graph.
3. Identify intrinsic factors in the underlying physical processes that will ultimately limit technical progress for the technology.
4. Estimate the magnitude of the natural limit on the performance parameter, and plot this asymptote on the graph of the historical data.
5. Estimate the time of two inflection points between the historical data and the asymptotic natural limit (first inflection from exponential to linear rate of progress and second inflection from linear to asymptotic region).
6. Expert forecasting of the exact times of inflection will likely be more unreliable than their anticipation of the research issues required to be addressed for inflection points to be reached.

The natural limit of progress on a technology S-curve arises from the intrinsic factors in the underlying generic physical process. Expert researchers base their predictions on an understanding of the physical processes underlying technology and on imagination about the research activities that may enlighten and lead to technology invention. Discontinuities in technical progress for a technology occur when alternate manipulations of nature are invented for the technology. Such discontinuities express themselves as different and disconnected S-curves. Since the mathematical form of the technology S-curve is not a model of the process of technological change but an analogous pattern frequently seen in the histories of technical progress, it can only be used to "guess" at the rate of change but not to "predict" how that change may occur.

ILLUSTRATION
Extrapolating U.S. Exports

The rawest sort of extrapolation is fitting near-term event data to recent event data when the underlying form of the curve, the generic pattern of the class of events, is unknown. This can be done by arbitrarily fitting straight lines to series that appear to be monotonically increasing or decreasing. If the series appears to be periodic, one can arbitrarily fit sinusoidal curves. If the series appears to have no underlying pattern, one can extrapolate using the average of the last three points. One also can use even more sophisticated running-average methods, such as the Box-Jenkins methods.

However, no matter how clever one is in fitting curves or averaging points, one is still left with the basic weakness that little is known about the real underlying forms or structures of the events one is trying to forecast. For example, E. Mahmoud, J. Motwani, and G. Rice (1990, pp. 375 and 380) compared forecasts for U.S. exports using time series and econometric models and came up with the following conclusion: "Exports have usually been forecast using econometric methods. Nevertheless, some studies have shown that time series methods can also predict exports....the methods studied have been sophisticated ones such as Box-Jenkins....Research has indicated the simpler forecasting techniques can be more accurate than Box-Jenkins techniques....Our findings suggest that time series methods can provide as accurate if not more accurate forecasts than an econometric approach."

What this example illustrates is that forecasts that presume a fixed underlying structure for events (e.g., econometric models) or forecasts that use no underlying structure or form (e.g., time series methods) are about equally accurate and equally inaccurate.

Any forecast will be much improved by an understanding of the underlying forms and structures than by refinements of curve-fitting techniques that ignore or treat underlying forms as fixed.

Methodology of Forecasting

We should now pause and discuss some of the methodologic issues about forecasting in general. Any forecast is an attempt to anticipate the course of future events. This attempt to anticipate, however, can proceed with different levels of knowledge about the nature of the events being forecast and can result in deepening levels of forecasts: (1) extrapolation, (2) generic patterns, (3) structural factors, and (4) planning agenda.

When a forecaster has almost no knowledge about the events except historical data on past occurrence, he or she can do little more than *extrapolate* the direction of future events from past events. Extrapolation forecasting consists of fitting a trend line to historical data.

When a forecaster has some knowledge about the general pattern of a class of events but little knowledge about the specific exemplar of that class at hand, he or she can use the *generic pattern to fit the extrapolation* of the specific exemplar case. The technology S-curve of this chapter is just such a generic pattern for the class of innovations. Fitting a generic pattern to an extrapolation results in more knowledge than mere extrapolation because one knows beforehand the form of the curve to be extrapolated.

In addition to knowing the generic pattern of an event, knowing something about the kinds of factors that influence the direction and pace of the events provides a basis for even better anticipation. Extrapolations from past data always assume that the *structure* of future events is similar to the structure of past events. Changes in structural factors will render extrapolation meaningless and create the most fundamental errors of forecasting.

The deepest level of forecasting requires an understanding not only of the generic pattern of the class of events to be anticipated but also an understanding of the structure of the events, and then it proceeds to intervene in the future by *planning* to bring about a desired future event. A research agenda provides an anticipatory document to bring about a technological future.

ILLUSTRATION
Experts Forecasting
Quarterly Hog Prices

What about the use of experts to make forecasts? Experts certainly know a little more about structure and are sensitive to factors that alter structures. But experts do not necessarily have quantitative models of structures. Consequently, some experts will be accurate sometimes and sometimes not.

An illustration of experts performing short-term forecasts is seen in the quarterly hog price forecasts made from 1976 through 1984 and reported by the U.S. Department of Agriculture. Two groups of experts, faculty in the agricultural economics departments of the University of Missouri and Purdue University, provided forecasts with a quarter of the year lead time (e.g.; 1976-01, 1976-02, 1976-03, 1976-04).

David Bessler and Peter Chamberlain (1987) compiled the quarterly forecasts from the Missouri and Purdue experts and compared them with actual data. Their results are shown in Table 11.1. In the table, one can see that about two-thirds of the time both the Missouri and Purdue experts anticipated the direction of change in hog prices correctly for the next quarter and about one-third of the time they misforecasted the direction of the price changes (let alone missing the magnitudes).

The trouble with using experts to forecast is that there is no way to

TABLE 11.1   Comparison of Two Forecasts to Actual Hog Prices by Quarters from 1976 to 1984 (dollars per hundredweight of hog)

Year	Actual price	Missouri forecast	Purdue forecast
1976-01	47.99	47.00	47.00
1976-02	49.19	47.00	48.50
1976-03	43.88	47.00	45.00
1976-04	34.25	36.00	35.00
1977-01	39.08	34.50	35.00
1977-02	40.87	35.50	32.50
1977-03	43.85	42.50	44.00
1977-04	41.38	37.50	37.50
1978-01	47.44	39.50	36.00
1978-02	47.84	48.50	42.00
1978-03	48.52	46.50	51.00
1978-04	50.05	48.50	45.00
1979-01	51.98	48.50	51.00
1979-02	43.04	45.00	48.00
1979-03	38.52	40.00	43.00
1979-04	36.39	35.00	32.50
1980-01	36.74	38.50	32.50
1980-02	31.18	35.50	37.00
1980-03	46.23	37.50	42.00
1980-04	46.44	43.50	45.00
1981-01	41.13	45.00	50.00
1981-02	43.62	42.00	44.00
1981-03	50.42	52.00	52.50
1981-04	43.63	49.50	48.00
1982-01	48.17	44.50	48.50
1982-02	56.46	53.00	51.00
1982-03	61.99	58.00	60.00
1982-04	55.12	59.50	61.00
1983-01	55.00	57.50	58.50
1983-02	46.74	53.50	52.00
1983-03	46.90	46.50	45.00
1983-04	42.18	41.50	40.50
1984-01	47.68	48.50	50.50
1984-02	48.91	49.50	52.00
1984-03	51.21	56.50	55.00
1984-04	47.54	45.50	46.50

anticipate an event or to calibrate the reliability and accuracy of any given expert.

The accuracy of experts in forecasting can only be judged in hindsight, and even accurate past performance is no guarantee of accurate future performance.

Use of Experts in Technology Forecasting

Within R&D infrastructures there exist experts (researchers) in science and engineering who attempt to create knowledge for inventing and improving technologies. One technique some managers use to forecast technology has been to gather experts and ask them to predict the future.

One of the first efforts at this was a study commissioned in the 1930s by the National Research Council and carried out by the National Resources Committee, chaired by a sociologist William Osburn (Osburn, 1937). Ayres commented on this report: "This ambitious study used technological trend curves almost exclusively to illuminate the past rather than to forecast the future" (Ayres, 1989, p. 52).

After World War II, the military began supporting formal methods of forecasting technology. Theo von Karman chaired a group to produce a technology forecast for the U.S. Air Force. The Rand Corporation was then created to continue and extend this work. "It [Rand] carried out many such forecasts and pioneered in developing technological forecasting methodologies such as the 'Delphi' method" (Ayres, 1989, p. 52).

Experts can be brought face to face to produce a report, such as the von Karman report mentioned above. However, N. C. Dulkey and Q. Helmer (1963), in developing what they called the "Delphi method," also saw a usefulness sometimes in querying experts by questionnaire and recirculating questions until a consensus among experts was obtained.

Joseph Martino has succinctly summarized the Delphi technique: "Delphi has three characteristics that distinguish it from conventional face-to-face group interaction: (1) anonymity, (2) iteration with controlled feedback, and (3) statistical group response" (Martino, 1983, p. 16). During a Delphi technique, the members of the group do not meet and therefore may not know who else is included in the group. Group interaction is carried out by means of questionnaires that are circulated to members of the group. Each group member is informed only of the current status of the collective opinion. Reiterations are continued until some consensus is reached.

Used for forecasting, such Delphi groups have attempted to predict what technologies will occur and when. However, technology-predicting forecasts by experts have often turned out to be notoriously wrong, particularly about dates.

Gene Rowe, George Wright, and Fergus Bolger (1991) reviewed the studies performed on the efficiency of Delphi methods versus any method of consulting experts. They concluded that the technique is probably no more useful than any technique for gaining group consensus among experts. "We conclude that inadequacies in the nature of feedback typically supplied in applications of Delphi tend to ensure that any small gains in the resolution of 'process loss' are offset by the removal of any opportunity for group 'process gain'" (Rowe, Wright, and Bolger, 1991, p. 235).

By *process loss,* they meant factors in group decision activities, including such factors as the fact that in some circumstances the group's collective judgment may be inferior to the judgment of the group's best member or premature closing of group judgment results in poor decisions or individual jockeying for power may skew results.

The Delphi technique focuses on reducing group process interaction and not on eliciting structural information.

Any refinement of techniques for forecasting that does not focus on eliciting information about the underlying structures of forecasted events is not likely to improve the forecast.

In 1981, Martino, Haines, and Keller studied about 2000 technological forecasts and found that about 700 of them used methods of extrapolation, leading indicators, causal models, and stochastic forms (Martino et al., 1981). All forms (Delphi consensus, extrapolation, leading indicators, etc.) for technology forecasting depend on expert opinion. Therefore, it is important to understand what opinions experts are really experts about.

Are experts expert in predicting the future? Probably not. What research experts really know are (1) a technological need, (2) the scientific theory underlying the technology, and (3) imaginative research approaches to attempt to improve a technology.

Therefore, the proper use of experts in technological anticipation and creation is to

1. Identify future technological needs

2. Identify areas that require improved understanding in order to advance technology

3. Propose research directions to study these areas

4. Use detailed information about the research directions to forecast the direction and pace of technological progress

5. Use the research directions to structure research programs

ILLUSTRATION
Changing Structures in the
U.S. Energy Scenario

William Peirce (1985) reviewed a 1962 forecast of natural gas production and consumption just before the energy crises of the 1970s occurred and the subsequent actual production and demand. Figure 11.9
shows the forecast made by the U.S. Federal Power Commission in
February of 1962. The commission forecasted a monotonic growth in
demand for natural gas from 1966 through 1990. The commission reviewed known gas reservoir capacities and projected a slight rise and
then a decline in the overall supply of domestic natural gas. To this the
commission added a projection for imported gas. Then the commission
depicted what it called an "unsatisfied demand gap" from 1970
through 1990.

In fact, actual gas consumption declined slightly after a peak in 1971.
Peirce pointed out that the forecast turned out to be wrong principally because it ignored the effects of price on supply and demand. "...The dominant 'official' forecast met with substantial opposition. Economists within
the Federal Energy Administration and the successor Department of
Energy, as well as outside, raised questions about the absence of price in
the discussion of supply and demand" (Peirce, 1985, p. 437).

It should be noted that the forecasters were aware of price as a factor, but gas producers had long been operating within a price-regulated
structure and could not predict how that structure could be altered
(Peirce, 1985, p. 435):

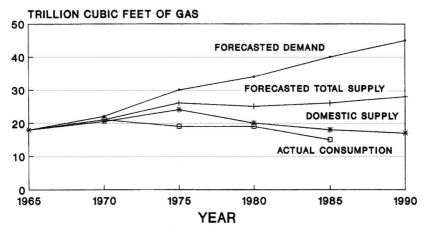

Figure 11.9 U.S. Federal Power Commission forecast of natural gas production and consumption (1962).

The "forecast demand" and the production data [in the forecast] were close to those of the American Gas Association....National gas policy during the 1970s was essentially one of pessimism about resources, restriction of uses and allocation of supplies by government directive, and attempt to prevent economic rents from being transferred from consumers to producers. Not only were reserves decreasing year by year, the conventional wisdom held that there was very little gas left to discover and that demand was inelastic. Under those conditions, there was little point in allowing prices to rise to market clearing levels. Rather, the wellhead price would be kept low and the limited amount of natural gas would be reserved primarily for household use.

However, price as the factor did change dramatically after the foreign oil producers in OPEC began functioning as an effective petroleum cartel in that decade. Political organization altered the structure of the pricing situation.

In terms of gas, two structural factors could be affected by higher prices—domestic exploration for natural gas and energy conservation practices by customers. "Even among those who saw little hope of increasing reserves, raising prices had some appeal. As the decade of the 1970s wore on, more conservationists began to advocate higher prices as a device to encourage investments in saving energy. Furthermore, decontrolling prices would spur the search for conventional gas" (Peirce, 1985, p. 437).

Policy options then took two courses, subsidizing synthetic coal-gasification projects and eventually allowing a rise in gas prices. The rise in gas prices then undercut the rationale for coal-gasification demonstration projects in the near future. When prices sharply increased, consumption then decreased.

The two structural factors, economics and nature, still play a role in forecasting gas supply (Peirce, 1985, p. 439):

Appraisals of the resource base for natural gas have become very controversial during the past decade. If one looks only at "conventional" sources...the potential recoverable gas is guessed by the Potential Gas Committee of the American Association of Petroleum Geologists to be around 1000 TCF. With production running around 20 TECF per year, that indicates about 50 years of potential resources....What these estimates leave out, however, are the less conventional possibilities: gas deeper than 30,000 ft, geopressured aquifers, coal seams, Devonian shales, tight gas sands, or other sources that cannot be economically produced with current technology.

What we see in this illustration is that the perspectives of gas experts were influenced by contemporary views on politics, economics, science, and technology, and these perspectives bounded and focused the forecasts that were made for gas supply and demand. As these external factors altered, the assumptions on which gas forecasts were made also altered. And with these the forecasts altered.

Forecasts and Underlying Structures

In any forecast, four general classes of structural features will underpin events:

1. Structures of current technological capabilities
2. Structures of economic activities and markets
3. Structures of nature and natural potential
4. Structures of demographics and cultures

Whenever underlying structures alter, forecasts based principally on extrapolation will be in error.

Practical Implications

The lessons to be learned in this chapter are both the techniques for extrapolating a rate of a technology and the kinds of factors that can invalidate such extrapolations.

1. Forecasts of technological change extrapolate the rate of improvement of a key performance parameter of the technology, which measures how efficiently the technology at that time performs its function.
2. A technology S-curve is an analogy of the commonly found pattern of incremental progress in a technology's principal performance parameter over time.
   a. Each S-curve is dependent on an underlying generic physical process.
   b. Discontinuities in technical progress for a technology occur when alternate physical processes can be used in inventions for the technology, and such discontinuities express themselves as different and disconnected S-curves.
   c. The natural limit of progress on a technology S-curve arises from the intrinsic factors in the underlying physical process.
3. The technique of fitting technology S-curves to technical data requires
   a. Identifying a key technical performance parameter

    b. Collecting existing historical data on technical performance since the date of innovation of the technology and plotting them on a time graph

    c. Identifying intrinsic factors in the underlying physical processes that will ultimately limit technical progress for the technology

    d. Estimating the magnitude of the natural limit on the performance parameter and plotting this asymptote on the graph of the historical data

    e. Estimating the time of two inflection points between the historical data and the asymptotic natural limit (first inflection from exponential to linear rate of progress and second inflection from linear to asymptotic region).

4. Whenever underlying structures alter, forecasts based principally on extrapolation will be in error.

    a. In any forecast, four general classes of structural features will underpin events: (1) structures of current technological capabilities, (2) structures of economic activities and markets, (3) structures of nature and natural potential, and (4) structures of demographics and cultures.

5. Attempts to anticipate change can proceed with different levels of knowledge about the nature of the events being forecast and can result in deepening levels of forecasts.

    a. Time-series extrapolation forecasts

    b. Generic historical pattern forecasts (e.g., technology S-curve)

    c. Identification and analysis of structural factors underlying extrapolation forecasts

    d. Planning agenda for actions to alter extrapolated forecasts

6. The proper use of experts in technological anticipation and creation is to

    a. Identify future technological needs

    b. Identify technology areas that require improved understanding in order to advance technology

    c. Propose research directions to study these areas

    d. Use detailed information about the research directions to forecast the direction and pace of technological progress

    e. Use the research directions to structure research programs

For Further Reflection

1. Identify several technologies and their key technical performance parameters. Why are these particular performance parameters used?

2. Plot the historical technology S-curves for the technologies. Which have shown one or more technology discontinuities? What were the physical sources for these discontinuities?

3. For one of the technologies, forecast its future progress. What are the likely sources of that progress? What kinds of technical experts should be consulted to identify the technical issues for that progress?

Forecasting Directions of Technological Change

Merely forecasting a rate of technological change is not sufficient for technology strategy; one also must forecast the direction of change. Wherein will the changes occur in the technology? The way to do this is to describe the technology as a system. The concept of a *technology system* provides general principles by means of which a technology can be varied and improved.

Progress in a technical system will occur by improving points within the system that limit the overall system performance—technical bottlenecks. Anticipating these bottlenecks allows one to know where progress must be made if the overall system is to be improved.

Recall from Chap. 5 that any technology is a mapping of a functional logic to a physical structure.

The analysis of a technology as a system will require two descriptions, a logic scheme and a corresponding morphology.

As illustrations of technology systems, this chapter will compare the technical evolution of two very different technologies: the automobile and the computer. Together they can illustrate the range of technology systems having very complex morphologies and simple logics to having simple morphologies and complex logics.

ILLUSTRATION
The Automobile as a
Technology System

The functional transformation of the automobile is intended to provide land transportation for moving passengers and goods from one location on land to another, so the basic functional scheme requires powered movement over land. In logical order, an automobile must be fueled, entered, started, directed, and stopped. The technology system of the automobile must provide the subfunctions of fueling, starting and stopping, control and direction, and comfort and safety.

To accomplish this, the physical structure of the automobile has several subsystems which together configure the automobile system. There is a body subsystem, a fuel subsystem, a starting and ignition subsystem, an engine, transmission, and drive train subsystem, a wheels and suspension subsystem, a breaking subsystem, and a steering and clutching subsystem. In addition, there are also subsystems for passenger comfort, that is, seating, windows, and climate control. And there are subsystems for safety, such as seatbelts, padding, and airbags.

We can see that the automobile system is simple in logic but complex in morphology. Historically, the automobile was a systems-type innovation begun late in the nineteenth century. It occurred from an inventive synthesis of ideas from carriage, bicycle, and engine technologies. Horse-drawn carriages had been made in the eighteenth and nineteenth centuries with iron hubs, wooden spokes, and iron-banded wheels. They also may have used steel spring suspensions on carriage or seat and lighter-weight, wood-laminated boxes. The combination of iron and steel materials with wood contributed to its refinement.

The bicycle was invented just after the middle of the nineteenth century. Its invention took advantage of new high-strength and lightweight steels. Some bicycle manufacturers created a new systems innovation by adding the newly invented gasoline engine onto basically a four-wheeled bicycle—and the automobile was invented. "Alan Nevins observes that the [automobile] industry was born from the consumer's desire for a light personal transportation vehicle, a desire stimulated by the bicycle boom of the 1890s....Men with experience in the bicycle industry were the first to see the possibilities of the automobile as a means of personal transportation. Their technological orientation led them to improve the automobile's performance through lightweight designs, high-strength materials, and low-friction ball bearings rather than increased motor power" (Abernathy, 1978, p. 12).

The first morphologic issue in early automobile design centered on the choice of motive power, the engine. There were three early contenders, steam-, gasoline-, and electric-powered engines. "Before 1900 both steam and electric cars were more successful and reliable....In 1900 a gasoline-powered car defeated electric and steam cars for the first time in a free-for-all race at Washington Park racetrack in Chicago; and, after this, improvements and market acceptance came rapidly" (Abernathy, 1978, p. 12).

The competitive issue then was performance. The gasoline-powered car could go faster than the steam-powered car and both faster and further than the electric-powered car. So the first morphologic choice in the automobile technology system was the gasoline engine. (It is interesting to note that battery technology, even a century later, had not progressed sufficiently for the electric car to replace the gasoline internal combustion engine car.)

Other configurational choices were made next. In 1903, Buick relocated the engine from the rear to the front of the car (as in most current designs). The bicycle-like drive chain was replaced by a direct drive shaft, connecting the front-placed engine to rear wheels. In 1904, Packard patented the four-position transmission with positions in the shape of an H, which subsequently became the standard for manual transmissions (with later alternative choices of four or five speeds). Ford redesigned the earlier one-piece block-and-head engines in two separate pieces for ease of casting and machining. By 1907, therefore, the automobile system began to look more like modern designs than like the early carriage/bicycles with rear-mounted engines.

Recall that a product design standard for the automobile occurred with Henry Ford's introduction of the famous Model T. Ford's focus then was on the rural market (since around 1900, half of Americans still lived on the farm). He aimed to build a practical, high-quality automobile priced at $450. His strategy was price. The rural application required an inexpensive, reliable, and durable car that also had a high clearance for dirt roads and easy maintainability by mechanically minded farmers. Ford needed both to innovate and to use the best design features of other automobiles then in existence.

The key to his innovation lay in the weight and strength of the chassis of the automobile structure. Materials costs in the early automobile were a very large part of total cost. In 1908, when Henry Ford introduced the Model T, the average per-pound price of an automobile hovered around $4000. If Ford could reduce the weight of the Model T by at least half of then competing designs, that technology would produce an enormous competitive advantage for his grand strategy of a "car for the people."

Ford's innovation for decreasing the weight of the automobile was to use high-strength steel for the chassis, made with the element vanadium as an alloy. Henry Ford learned of this new steel when he attended an automobile race. In one of the unfortunate accidents that day, a racing car imported from France was wrecked. "In 1905 [Ford] saw a French automobile wrecked in a smash-up. Looking over the wreck, he picked up a valve stem, very light and tough....it proved to be a French steel with vanadium alloy. Ford found that none [in the United States] could duplicate the metal. [Ford]...found a small steel company in

Canton, Ohio [and]...offered to guarantee them against loss. The first heat was a failure....the second time the steel came through. Until then [he]...had been forced to be satisfied with steel running between 60,000 and 70,000 pounds tensile strength. With vanadium steel, the strength went to 170,000 pounds" (Nevins and Hill, 1953, p. 349).

Making the chassis of this steel meant that Ford could reduce the weight of the chassis by nearly a third and get the same strength. It was a technological breakthrough that allowed Ford to imagine an innovative new product design. "Charlie, [Ford said to Charles Sorensen, who helped him design the Model T] this means entirely new design requirements and we can get a better, lighter and cheaper car as a result of it" (Sorensen, 1956, p. 98).

Recall that design, like invention, is also an embedding of functional ideas into the physical configuration of an object (with invention being the first time a particular mapping is created and design being the repeated times it is done thereafter). Ford used the new vanadium steel to fabricate the chassis of the automobile, which reduced the overall weight of the Model T to about half that of existing automobiles. In addition, Ford also innovated the design by mounting the motor to this chassis with a three-point suspension. The prior practice had been to bolt the engine directly to the frame, and often even the cylinder blocks of those engines had been twisted in half by the enormous strain that occurred when the automobile bounced over a hole or rut.

Ford also designed the Model T to be the "best of breed." He used the best ideas in other contemporary automobiles. For example, he replaced the traditional dry-cell batteries for starting the car with a magnet-powered ignition (one cranked the Model T to start it). He also designed his car with a high road clearance for rural roads. The Model T became a design standard for automobile technology that lasted long after its time. "For eighteen years the design of the Model T chassis was not significantly changed. During this period the industry's production of passenger cars increased nearly sixtyfold, from 64,500 cars annually to 3,700,000....Ford maintained about a 50 percent market share through 1924" (Abernathy, 1978, p. 18).

After introduction of the Model T, the improvements in automobile product technology were incremental for a long time. For example, in 1921, Hudson marketed a car with an enclosed steel panel body that had steel panel over a wooden body frame. This morphologic change made automobiles more durable to weather. Later, Studebaker shaped the steel panel for stiffness and eliminated the wooden body frame. Chrysler streamlined the steel body shape for greater airflow efficiency. Just before World War II, General Motors innovated the automatic transmission.

After the war, the American automobile industry entered a period of stable technology. Management focus from 1950 turned to refining existing technology, automating production techniques, styling, and reducing numbers of parts. Only in the 1980s did technology in the automobile system begin to incrementally change again with the introduction of electronics into the control systems of the automobile.

Systems

Let us next review the general concept of systems and see how it applies to technologies. It is in the perception of technology as a system that we can describe and envision incremental advances in a technology.

There are two general kinds of systems, open systems and closed systems (Betz, 1974). An *open system* receives inputs from its environment, processes these, and exports outputs to its environment. In a *closed system,* the system processes are wholly enclosed, neither receiving inputs from nor exporting outputs to its environment.

An example of an open physical system can be found in the energy processes on the surface of the earth; that is, the earth receives energy from the sun and reflects some back into the environment, absorbing the remainder to fuel the internal processes of earth. The subsystems of the planetary surface energy system are inanimate physical processes (air and water) and animate physical processes (plant and animal). Examples of physical systems on earth that are closed systems are the planetary chemical element cycles, such as the carbon and calcium cycles. Here the subsystems are plant and animal processes of respiration, birth, growth, and death.

All systems have a boundary and an environment. Within the boundary, the system will contain parts and connections between parts. There will always be at least two subsystems within the system, a transformation subsystem and a control subsystem. A system is defined by its dominant process or processes. The boundaries of an open system are the points at which inputs are received from and outputs exported to its environment. The functionally dominant process is the transformation that converts inputs to outputs.

All technology systems are functionally defined open systems, accepting inputs and transforming inputs into outputs.

The boundaries of a technology system are the points of the physical structure that receive inputs from and which export outputs into the environment.

The kind of transformation of generic inputs into generic outputs defines the functional capability of the technology system.

For example, the automobile as a technology system displays two dominant processes, energy conversion and motion. For energy conversion, there are four boundary points to the automobile:

Inputs   The fuel tank inlet and the air filtration input

Outputs   Exhaust and radiator

For automobile motion, there are five boundary points:

Input   Four tire contact points with the ground

Outputs   Space immediately in front of the automobile

The generic technology system of the automobile can thus be defined as a physical device that inputs energy sources from the environment and converts this energy to mechanical power to transform the physical location of the device.

All functional open systems, such as a technology system, must be described with two levels mapped to each other:

1. *Logic schematic*—logical scheme of the functional transformation

2. *Physical morphology*—constructed physical structure whose processes map in a one-to-one manner with the logic scheme

The logic schematic is a step-by-step laying out in order of the discrete transformations required to take a generic input to a generic output of the technology's functional transformation.

A logic schematic for a technology system is represented as a topologic graph of the parallel and sequential steps of unit transformations.

The physical morphology of a technology system is the assembly and connection of the physical structures and processes whose physical operation can be interpreted in a one-to-one manner with the logical steps of the schematic.

The morphology of a technological device also may be analyzed as an open system defined by a boundary and containing components, connections between the components, and a control subsystem.

Recall that the foundation of any technology is the invention of the mapping of a logical procedure in a one-to-one manner with the sequences of a physical process. The technology system, having both a functional schematic and a physical morphology, articulates this same idea.

ILLUSTRATION
The Computer as a
Technology System

In comparison with the functional transformation of the automobile as translocation, the functional transformation of a computer is data computation—numerically, geometrically, and logically transforming inputted information into outputted information.

In Chap. 5 we reviewed the story of the invention of the stored-program computer with the Von Neumann architecture. This architecture generically consists of input and output devices, a central processing device, and a memory device—all connected together on a data bus. The architecture provides the basic configuration for a computer as a system, since it allows programs and data to be inputted from the environment, stored in memory, and transformed by the central processing device and the computational results to be stored back in memory and then read out to the output device. Thus a computational transformation would have been performed on input data, with the logical steps of the transformation defined by the stored program, and the computation returned to the environment—a functionally defined open system—with the function being computation.

This is accomplished physically by containing input devices, central processing units, data buses, active memory, and storage memory in boxes and connecting them with appropriate cables. Within each box are printed circuit boards, IC chips, connections, some mechanical devices for cooling, and disk storage devices.

Historically there were several critical system element advances after the use of electronic tubes in the first computer. One was the invention of the ferromagnetic core memory; another was the substitution of transistors for vacuum tubes; and a third was the fabrication of transistors and circuits on semiconducting IC chips. The ferrite core memory first made possible a large main memory, but it was eventually replaced by the substituting technology of semiconductor integrated circuits (dynamic random memory chips). Transistors eventually replaced electronic tubes in the computer's logic circuits, and individual transistors were later replaced by semiconductor integrated circuits. Progress in the new technology system of the computer thus occurred first in the component elements of the computer subsystems, logic and memory.

When Mauchly and Eckert and later Von Neumann and Goldstein built their first vacuum tube stored-program computers, their major problem was the use of vacuum tubes for memory. For example, two tubes were required for one memory bit and with, for example, a 12-bit word length, each word stored in memory required 24 tubes. To have a memory capacity of, say, a thousand words (a very small memory) would require 24,000 tubes for memory alone. Thus, while vacuum tubes might be suitable for the first logic circuits in the general-purpose computer, they were not suitable for memory circuits.

This first technical bottleneck in the computer system was addressed by Eckert and Mauchly in their EDVAC. They invented a memory unit consisting of an electroacoustic delay line in a longitudinal tube filled with mercury. Their idea was to store a string of bits of 1s and 0s morphologically as successive acoustic pulses in a long mercury tube. An electrical transducer at the start end of the tube would pulse acoustic waves into the tube corresponding to a 1 bit or no pulse corresponding to a 0 bit. As the 1-bit pulses and the 0-bit spaces between them traveled the length of the tube, the word they represented would be stored temporarily. When the string of pulses began to reach the end of the mercury tube, the acoustic pulse would be retransduced back into electric pulses for reading by the computer system, or if they did not need to be read, they would then be reinserted electrically back into the front end of the tube as a renewed string of pulses and spaces morphologically expressing the stored word. This storage could repeat in cycle again until the word was "read" and a new "word" was temporarily stored in the mercury tube.

This was one of Eckert and Mauchly's inventions resulting from the ENIAC, and the conflict with the University of Pennsylvania over the rights to the invention was one reason they left the university. They obtained a $75,000 contract from the Census Bureau to develop the computer and used this contract to start their new company, the Eckert-Mauchly Computer Company. Subsequently, they received a $300,000 contract from the Census Bureau to build what they would call the Universal Automatic Computer (UNIVAC). However, this money was not sufficient for the development costs, and Eckert and Mauchly were forced to solve their financial problems by selling out to Remington Rand (a typewriter manufacturer) on February 1, 1950. The mercury storage tube was used in that first UNIVAC built for the Census Bureau and delivered from Remington Rand on March 31, 1951. For permanent storage, the UNIVAC read data and programs from and to magnetic tape and/or punch cards.

Meanwhile, another idea for memory was being developed by Jan Rajchman at RCA in a device called a "selectron tube." When Von Neumann was building his computer at the Institute for Advanced Study in Princeton, he had entered into a joint R&D contract with the newly established RCA Research Laboratory in Princeton, with the latter providing the memory device. The selectron tube was a cathode-ray tube, similar to a television tube, in which areas on the phosphorous-coated face were bit-mapped, as two voltages, positive and negative. Reading out a bit interpreted a positively charged spot as a 1 bit and a negatively charged spot as a 0 bit. About the same time, F. C. Williams and T. Kilburn at the University of Manchester also were working on this same kind of device. Their version was used in IBM's first all electronic commercial computer, the IBM 701 (Pugh, 1984).

Yet both the mercury storage tube and the cathode-ray tube were poor memory devices because they operated much more slowly than the electronic circuits. Memory remained the bottleneck.

However, while Eckart and Mauchly were working on their computer, there was another computer project at MIT, the Whirlwind Project, directed by Jay Forrester. Historically, it was as important as the Eckart and Mauchly project. Forrester's computer made the next breakthrough required for progress in the technology system of the computer—the ferrite core main memory. The Whirlwind Project also influenced IBM and accelerated its entry into computers, positioning IBM for early dominance in the mainframe computer industry.

Jay Forrester graduated in engineering from the University of Nebraska in 1939. He went to MIT as a graduate student in electrical engineering, obtaining a doctorate. This occurred during World War II, and Forrester worked in military research at the Servomechanisms Laboratory at MIT. During the war, in 1944, Forrester participated in studies for an aircraft analyzer that led a project called the Aircraft Stability and Control Analyzer Project. "Jay Forrester was described by people in the project as brilliant as well as 'cool and distant and personally remote in a way that kept him in control without ever diminishing our loyalty and pride in the project.' He insisted on finding and hiring the best people according to his own definition, people with originality and genius who were not bound by the traditional approach" (Pugh, 1984, p. 63).

In August of 1945, the Servomechanisms Laboratory received a feasibility study contract from the Naval Office of Research and Invention for the aircraft analyzer. Two months later, in October, Forrester attended a conference on advanced computation techniques. He wanted to learn about the ENIAC, which Mauchly and Eckert had built. He was interested in using digital electronic computation in his Aircraft

Stability and Control Analyzer Project. When he saw the digital circuit technologies developed for ENIAC, he thought he could use digital techniques. "His decision to shift from analog to digital techniques was crucial to the future of his project, and it may also have been the first time a major control function was undertaken by digital techniques" (Pugh, 1984, p. 64).

With this in mind, Forrester decided to redirect the analyzer project. In January of 1946, Forrester went back to the Navy with a new project proposal to (1) design a digital computer and (2) adapt the computer to the aircraft analyzer. The crux of the problem was the main memory subsystem for the computer. The need for response in real time for control in the aircraft simulator made both the mercury delay line and the rotating magnetic drum technology too slow for this application....Memory was then the technical bottleneck in the system.

International events resulted in a reorientation of Forrester's project. (Pugh, 1984, p. 68):

> Then in the fall of 1949 it was learned that the Soviet Union had detonated an atomic bomb, making government leaders increasingly concerned about the potential threat of Russian bombers. The Air Defense System Engineering Committee was created in January 1950 under the chairmanship of George E. Vally of MIT to study the problem and make recommendations....within a week Vally was considering the possibility of using the Whirlwind computer for air defense. Not only was the development of the computer well along, but project Whirlwind had the largest collection of digital computer engineers in the country....The aircraft analyzer project was stopped, and the technical effort redirected to this new objective [a national defense system].

The Whirlwind Project was redirected by this new application. Yet Forrester had not solved the technical bottleneck in the computer system, i.e., memory. In April of 1949, he happened to see an advertisement for a new material called "Deltamax." He noted that it had the kind of ferromagnetic properties he had envisioned in his novel idea for a three-dimensional memory array.

Deltamax was made of 50 percent nickel and 50 percent iron that had been rolled very thinly; this concept had been developed during the war by Germans. After the war, U.S. naval scientists brought samples of the material back to the United States as well as one of the special machines required to make it (a Sendzimir mill). These scientists encouraged an American firm, Arnold Engineering (a subsidiary of Allegheny-Ludlum), to make the material as a kind of tape. Its important property was that it had a sharp threshold for magnetization reversal when an external magnetic field was applied to it.

Thus another new invention was conceived of as function mapped onto morphology. The magnetic direction of a loop of the metal could be mapped to represent the numerical unit of a 1 or a 0. "Forrester documented his ideas meticulously in his engineering notebook, as was his custom. His first entry, dated June 13, 1949, showed a nearly rectangular magnetic hysteresis loop and adjacent to it an XY array for addressing magnetic toroids" (Pugh, 1984, p. 69).

What Forrester invented was an array of small magnetic toroids strung in three dimensions with wires and through which a simultaneous set of pulses on three x, y, and z wires could select a particular toroid to magnetize in one direction or the opposite direction. One direction of the magnetized state could then be translated as the storage of a logic bit 1, and the other direction could be translated as the storage of a logic bit 0. Assembled in three-dimensional arrays, the tiny toroids could function as a main memory of a computer, since changes in direction of magnetization could be accomplished fairly rapidly. In fact, the ferrite-core array memory made the first computers really practical.

However, like many inventions motivated by need, more than a single inventor may come up with the same idea at nearly the same time. Thus it was with the ferrite-core array memory. A. N. Wang had just completed his Ph.D. studies in May of 1948 and joined the Harvard Computation Laboratory. This laboratory was headed by Howard Aiken, who had built the Mark I mixed tube/relay computer with IBM and Navy sponsorship. Aiken suggested to Wang that he think about ways of using magnetic media for information storage. About a month later, in June of 1948, Wang had an idea for a two-dimensional memory-shift register using magnetic cores.

Wang applied for a patent for his device and then described it at a technical symposium in September of 1949. A faculty member at the University of Illinois, Ralph Meagher, attended and listened to Wang's talk. When he returned to Illinois, he suggested to a graduate student, Mike Haynes, that magnetic-core memories might be a good topic for his Ph.D. dissertation. Haynes began research on stringing cores in two-dimensional arrays for a memory and described his research on these arrays in his Ph.D. thesis.

Then Haynes went to work for IBM. There, along with Gordon Whitney, Edward Rabenda, and Erich Block, the ferrite-core memory found its first commercial application in February of 1953 as an 80 by 12 ferrite-core card input memory buffer in the IBM 407 printer and the IBM 714 and 759 card reader and control unit. Since Wang held the patent on the technology, IBM paid him $300,000 for the rights to his invention. With this money, Wang set up his own company, which in the late 1970s was to make a big commercial success by innovating the first commercially successful computerized word processor (Betz, 1987).

However, Wang, Haynes, and Forrester had not been the only people thinking about magnetic core memories. At RCA, Jan Rajchman, who had worked on the selectron memory storage tube for Von Neumann's computer, also had the same bright idea. Rajchman had been thinking of better storage devices. He had once read about some square-loop magnetic sheets developed by Germans, and it was now obvious to him that the magnetic loop could be used as a memory. In September of 1950, Rajchman filed a patent. Forrester filed for a patent on May 11, 1951. As a result of these independent inventions of the ferrite-core memory, there occurred a legal suit about the Forrester and Rajchman patents, and a patent interference was declared on September 25, 1956.

The Whirlwind computer built at MIT was the first computer to use ferrite-core arrays for main memory. "In the September 15, 1952, *Quarterly Progress Report* of MIT's Digital Computer Laboratory, Bill Papian reported that three 16×16 magnetic core arrays were in operation" (Pugh, 1984, p. 70).

Progress in Technology Systems

The general way of identifying potential or actual opportunities for technological advances lies in either progress in the logic schematic or progress in the morphology. Technology advance may occur by (1) extending the logic schematic, (2) alternating physical morphologies for a given schematic, or (3) improving performance of a given morphology for a given schematic by improving parts of the system.

Technical progress may occur by adding features and corresponding functional subsystems to the logic. Technical progress may occur in alternate physical processes, structures, or configurations for mapping to the logic. Technical progress may occur in improving parts of the system, its subsystems, components, connections, materials, energy sources, motivative devices, or control devices. Thus, viewed as a system, technical progress can occur from changes in any aspect of the system: (1) critical system elements, (2) components of the system, (3) connections of components within the system, (4) control subsystems of the system, (5) material bases within the system, (6) power bases of the system, or (7) system boundary.

A technology system cannot be innovated until all the critical elements for the system components, connections, control, materials, and power already exist technologically. For example, the steam engine was a necessary critical element before the railroad could be innovated. The gasoline engine was a critical system element before the airplane could be innovated. The vacuum electronic tube was a necessary element before radio and television could be innovated. The vacuum tube and the ferrite-core memory were necessary elements for the general-purpose computer. Critical elements are the building blocks for a system's components, connections, control, material, or power bases.

Technical progress in a system may occur from further progress in the components of the system. For example, over the early 30-year history of the computer technology, rapid and continual progress in the central processing units, in memory units, and in storage units drove the technological progress of the computer.

Technical progress in a system also may occur in the connections of the system. For example, in the telephone communications industry, fiberoptic transmission lines began to revolutionize the industry in the 1970s. For computers, in the 1980s, the development of local area network connections and protocols and the fiberoptic transmission systems also contributed to the continuing revolution of the industry.

Technical progress in a system also may occur in the control systems of a technology. For the automobile industry in the 1980s, technical progress was primarily occurring into developing electronic control systems. In computers, major technical progress had occurred in developing assembly languages, higher-level cognitive languages, compiler systems, interactive compiling systems, artificial intelligence languages, etc.

The material bases in the framework of a system, in the devices of a system, in the connections of a system, or in the control of a system are also sources of technical change and progress. An example of technical progress in the framework of the airplane was the beginning in the 1980s of the substitution of organic composite materials for metals in the construction of airplanes. An example of advances in materials in a device was the substitution of silicon for germanium in new transistors in the 1950s, and in the 1990s the attempts to substitute gallium arsenide for silicon in the semiconductor chips.

Finally, energy and power sources may provide sources of advance in a technology system. The classic examples are the substitution of steam power for wind power for ships in the nineteenth century and in the twentieth century the substitution of nuclear power for diesel power in the submarine.

ILLUSTRATION
Systems Control in Automobiles
and Computers

All technology systems require a control subsystem. For example, in the automobile, the control subsystem consists of the steering, acceleration, braking, clutching, and instrumentation sensors and controls. Controlling actions consist of a planned route, ignition, starting, steering, and stopping. To accomplish location transformation, the automobile has two separate and parallel control subsystems that need to be coordinated at all times of operation: the energy transformation process and the power and steering process.

As the car is accelerated or decelerated, held at constant velocity, or stopped, the energy transformation must be controlled in exact coordination as fuel and air are injected into the engine and ignited and the torque from the engine's flywheel is transmitted through appropriate gearing to the drive wheels.

The control of a computer also requires the coordination of two subsystems, a system level control and an application program level control. However, whereas the automobile energy and power subsystems are operated in parallel, the system program and the application programs operate hierarchically. The systems control turns on the computer and checks that the devices are ready to operate. It also receives an application control program and stores it in memory. Next, it accepts data input and executes the program instructions on the inputted data. It temporally stores results and transfers these results to long-term storage and/or output devices.

Control systems also can be fixed or variable in decision capability. For example, an important difference between the control systems of the automobile and of a computer is in the fixed nature or variability of the control. The morphology of the automobile system is fixed, and only one functional transformation can occur, transportation. The morphology of the computer is variable as different application programs can reconfigure the functional operations in the central processing unit.

Recall that the most basic logical operations in the computational transformations of the computer are the primary logical and arithmetic operations and that each of these is morphologically mapped to different electronic circuits. The electronic circuits in a modern computer are constructed so as to be differently configured for each operation by a third level of system control called the *microinstruction set* of the microprocessor unit.

The heart of the computer systems level control is the internal clock that sequences the operations of the computer. Recall that the basic operation of the computer is to represent a computation as a sequence of program instructions which step by step convert input data into computed results. The internal clock controls the sequencing of steps, such as (1) load the current instruction into the program register, (2) configure the ALU for the operation indicated in the instruction, (3) fetch the data to be operated on from memory to temporary data registers, (4) perform operations on the data, (5) store the results in memory, (6) advance the clock, (7) fetch the next instruction into the program register, and (8) cycle again until all the instructions in the application program have been executed.

System Control

All technology systems require control of the transforming function of the technology.

> *System control requires at least three parallel subsystems of control: (1) input of material resources for fueling the transformation activities, (2) control of the transformation activities, and (3) control of the purposes of the transformation.*

And any or all of the parallel control subsystems also may contain a hierarchy of subsystems. For example, in the automobile, the gasoline and air subsystems provide energy for the process, the power subsystem transforms energy to locomotion, and the driver and steering subsystems direct the motion of the automobile. In the computer example, the electrical power subsystem provides energy for computer functioning, and the three hierarchical levels of control (system, application, microinstruction) control the computational execution. In addition, the programmer who has written the application program (and perhaps also an on-line user who interacts in real time with the system) provides the directing purpose of the computational activities.

In these examples we see a technology system in which partial control lies in the purposive direction of an application programmer or an interactive user (e.g., automobile driver). For this reason, there is also a human component in any technology. To include this human operator component, we also must see that all technologies are embedded within a sociotechnical system.

The design of the control system for any technology system must therefore also be encompassed within a larger design of the sociotechnical system. It is here that the market perspective and the technology perspective collide head to head. Any technology system that does not perform well in the sociotechnology system will be a commercial failure.

Morphologic Analysis
of a Technology System

The term *morphology* literally means the "science of form," since *morphe* is the Greek word for "form." However, we will use the term in a specialized sense to mean the *form of the physical structures and processes of phenomena used in a technology system.* F. Zwicky (1948) introduced this use of the term in a technique to systematically list alternative physical forms of a technology system.

Zwicky proposed that one can systematically explore sources for technical advance in a system by logically constructing all possible combinations of envisioned alternatives in any of these aspects of the system. First, a clear statement of the function of the technology must be made. Next, a general logic scheme of the functional transformation must be designed. Then a list must be made of all the physical processes mappable into the logical steps.

Morphologic alternatives can then be searched for by listing all conceivable variations on the logical architecture and all possible physical substitutions in the logical steps. Then an abstraction must be made of the physical form of the technology system in terms of components, connections, materials, control, and power. One then makes a list of all conceivable alternative forms of each of these in terms of the physical processes. Then one makes a list of all possible combinations of these alternative constructs—each offering an alternate morphology for the system.

Zwicky used the example of alternative structures for jet engines, identifying each major abstracted element:

1. Intrinsic or extrinsic chemically active mass

2. Internal or external thrust generation

3. Intrinsic, extrinsic, or zero-thrust augmentation

4. Positive or negative jets

5. Nature of the conversion of the chemical energy into mechanical energy

6. Vacuum, air, water, earth

7. Translatory, rotatory, oscillatory, or no motion

8. Gaseous, liquid, or solid state of propellant

9. Continuous or intermittent operation

10. Self-igniting or non-self-igniting propellants (Jatsch, 1967, p. 176)

Next, Zwicky enumerated all possible combinations of these features and came up with 36,864 alternative combinations. Now many of these alternatives are not technically interesting, and these Zwicky eliminated. Actually, when Zwicky performed this evaluation in 1943, he did use many fewer parameters and reduced his alternatives to be examined to only 576 combinations. Zwicky's procedure in its full combinatorial form is cumbersome.

Generally, most technological research is actually focused as a limited form of morphologic analysis, considering alternatives for physical features and performance in a technology system.

Most industrial research laboratories use a modified version in organizing their research areas. They describe their principal product lines and production processes as technology systems and then organize programs of research to improve different aspects of these systems.

Some kind of morphologic analysis provides the basis for all technology planning of the physical aspects of a system.

What constitutes the boundary of a technology system is an important judgment, since it focuses the system. The view of the boundaries of a computer system may stop at the principal device of the system or may include peripheral accessories, application accessories, maintenance accessories, infrastructure support, or networking into other systems.

To summarize the procedure for conducting a morphologic analysis of a technology system, one must start with the problems and technical limitations of the current technology system. One can then redefine the desired functional and performance capabilities of the technology (using current and potential embodiments of the technology in product systems and in applications systems.)

Next, one must abstract the principal parameters of the desired functional transformation. The different logical combinations of these parameters (which are mutually physically consistent and ordered by hierarchy of physical influence) provide a logic tree of alternative morphologies for the technology system. To examine alternatives,

1. One should begin with an existing boundary and architecture that provides the functional transformation defining the technology.

2. Next, one uses the applications system forecast to classify the types of applications currently being performed by the technology system and also to envision some new applications that could be performed with improved performance, added features, and/or lowered cost of the technology system.

3. One can then identify the important performance parameters and their desirable ranges, desirable features, and target cost to improve existing applications and to create new applications.

4. These criteria then provide constraints and desired ranges under which alternative morphologies of a technology system can be bound.

5. Next, examine critical elements of components and connections whose performance would limit the attainment of the desirable performances, features, and cost.

6. The performance of critical elements that do not permit desired system performance can then be identified as technical bottlenecks.

7. Next, one can examine alternative base materials and base power sources for alternative morphologies, again selecting only those which would fulfill desirable system performance.

8. One also can examine alternative architectural configurations and alternative types of control that might attain the desired higher performances.

9. Combining and recombining various permutations of critical elements, base materials and power, and alternative architectures and control systems provides a systematic way of enlarging the realm of possible technology developments.

ILLUSTRATION
Morphologic Analysis
of Molding Technology

Dominique Foray and Arnulf Grubler (1990) provided an illustration of a morphologic analysis in the area of molding technology. They emphasized that inclusiveness of approach, i.e., the inclusion of alternatives, was essential to the usefulness of the method. "A morphological analysis starts with building a morphological space for a particular set of technologies or products, in order to understand and thus not to 'miss' a technological route of possible future development" (Foray and Grubler, 1990, p. 537).

This first step is to carefully define the technology system in terms of its *functional capability* (system transformation). "First, the problem to be solved [for the functional capability desired] must be stated with great precision: in our case the problem consists of realizing ferrous metal products by a casting process (molding technology)" (Foray and Grubler, 1990, p. 537).

The second step is the abstraction of the key parameters of the functional capability (system transformation). Foray and Grubler considered four parameters:

P1   The nature of the pattern

P2   The nature of the mold cavity

P3   The stabilization force

P4   The bonding method

Next, Foray and Grubler (1990, p. 537) listed the principal alternatives for each parameter:

P1 (pattern)   Permanent (P11) or lost (P12)

> The molding methods can be classified according to the nature of the pattern....the pattern is used for a large number of castings....the pattern is used only once.

P2 (mold cavity)   Hollow (P21) or full (P22)

> The molding methods can be classified according to the fact that the mold cavity is hollow (the pattern is extracted before the casting) or full (foam polystyrene as expandable pattern is gasified by the molten metal during the casting), and by taking the place of the expandable pattern, the molten metal fills in the full cavity as it would fill in a hollow cavity.

P3 (stabilization force)   Chemical (P31) or physical (P32)

> ...The kinds of bonding systems used for stabilization of individual granules of molding material....

P4 (bonding method)   Simple (P41) or complex (P42)

> ...Both chemical and physical bonding methods can be described as simple (mechanical and inorganic chemical binder) or complex (magnetic field, vacuum and organic chemical binder).

Although all combinations of alternatives should be considered for possible progress, Foray and Grubler emphasized that one gets to more interesting technical alternatives faster by utilizing logical relations between system parameters. "...Each parameter corresponds to a given level of aggregation (or integration). The four levels can be ordered hierarchically, according to the relation between the parameters. For example, the stabilization force (P3) influences the bonding method (P2), or the nature of the pattern (P1) influences the nature of the model cavity (P4)" (Foray and Grubler, 1990, p. 537).

They next created combinations of alternatives ordered according to a logical hierarchy beginning with P1 (pattern), progressing to P2 (mold cavity), then to P3 (stabilization force), and finally, to P4 (bonding method). Figure 12.1 summarizes their ordering of combinations.

What this morphology logic tree shows is the combinations of technical parameters that can define different technical processes (a process system), 16 in all for the 4 parameters. After Foray and Grubler eliminated impossible solutions from these 16, what they had left were all possible processes in the actual technological state of the industry. An example of one of these combinations is as follows: "The GP process involves investing an injection molded foamed-polystyrene pattern in a

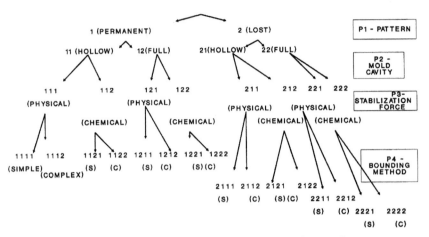

Figure 12.1 Morphologic ordering of ferrous casting technologies—four parameters. These four parameters allow 16 possible processes.

free flowing magnetizable modeling material. Immediately prior to pouring, the molding material is rigidized by a powerful magnetic field. During casting, the polystyrene pattern volatizes in the face of the incoming metal stream which occupies the void left by the falsified pattern. Shortly after the casting has solidified, the magnetic flux is switched off and the flask containing the casting is taken to the knockout station" (Foray and Grubler, 1990, p. 540).

Functional Analysis of a Technology System

Complementary to a morphologic analysis, a functional context analysis of a technology system also should be made from the application systems perspective.

1. Identify the major generic applications of product/production/service systems that embody the technology system.

2. From the applications perspective, identify desirable improvements in performance, features, and safety that would be recognized by the users (final customers) of the product/production/service systems.

3. Determine the prices that would expand applications or open up new applications.

4. Determine the factors that facilitate brand loyalty when product/production/service systems are replaced in an application.

Technology System and Product System

It is important to distinguish between a technology system and a product system. The system configuration of the generic technology system indicates only the principal logical steps necessary for the functional transformation of the technology. The system configuration of a product realizing the technology adds secondary logical steps and processes needed to fill in the details and fully connect the principal logic steps and add other secondary technology systems to complete the product.

The differences between a technology system and a product system are the details and fully connected steps of the logical transformation and completing secondary technology systems.

This means that to know generally how a technology works is not sufficient to know how to design a product incorporating that technology. The differences in knowing how to design a product provide a source of technical differences between competitors. The other source of differences between competitors' products lies in materials and production processes and production quality. Thus two product systems of exactly the same technology can differ tremendously.

Technology innovations (1) to improve the details, connectivity, and completeness of a product design and (2) to improve the quality of materials and production are important to competitiveness.

ILLUSTRATION
Technology System and
Product System: The PDP Computer
Product System

We next turn to an analysis of a product system that embodies a principal technological transformation. As an illustration of a specific product system of the generic technology of stored-program computation, let us review the design of one of the first minicomputers, innovated around 1970, the Digital Equipment Corporation's PDP-8.

As noted previously in the progress of the computer system, basic devices had to be improved in both the logic circuits and the memory circuits. The invention of the semiconductor integrated circuit (in 1959) provided the part morphologic basis for both logic and memory circuits in computers. By the late 1960s, technological progress in IC circuits made possible a new product line below the then-existing product line of mainframe computers. One of the first of these, the PDP-8, was innovated by a new company, Digital Equipment Corporation (DEC), founded by Ben Olsen.

After having worked on the whirlwind computer project at MIT under Jay Forrester, Ben Olsen founded DEC to produce transistorized circuit boards for computer applications. By the late 1960s, Olsen saw that he could make a relatively simple transistorized computer that, although low powered in comparison with the expensive mainframe computers of the time, could be afforded by a single scientist rather than only by a whole research laboratory. This was the commercial focus of the PDP-8, and over 100,000 of these units were sold to scientists. It launched the minicomputer industry as a market segment beneath the higher-priced mainframe industry. Also, only scientists bought these units because they were very difficult to program.

The configuration of the PDP-8 used a Von Neumann architecture. Its central processing unit was an arithmetic-logic unit (ALU) that performed both arithmetic (such as addition, multiplication, integration, etc.) and logical operations (such as and, or, if-then). It had a memory unit (MEM) that stored 4096 words, with each word of 12-bit length. It had a data bus (internal electrical connections to move data around inside the computer from input to memory, from memory to ALU, from ALU back to memory, and from memory to output). It had an input/output and control panel (I/O) with switches and a lighted display register for inputting, displaying output, and overall control. In the simplest form, data were entered manually by the graduate student (or technician). Most scientists also added to the PDP-8 other devices for automatic data entry and output (such as punched tape drives and readers).

These units constituted the central components of the generic technology system of the stored-program computer. In the PDP-8, however, in addition to these principal components, the architecture also contained several other devices. The first addition was an accumulator (AC) register that stored the intermediate computational results from the ALU as it stepped through a computation. The PDP-8 also had a storage register, called LINK, to temporarily store arithmetic overflow bits in the calculations (such as add the units and carry the overflow to the tens, etc.). It also had a memory address (MA) to indicate where in memory to find the data (or instruction) to read and where to store the computed data result from the ALU. There also was a memory buffer (MB), where read data from memory was temporarily stored to be ready at the exact time the ALU needed them. The ALU ran on a very strict and rapid sequence, according to the stored-program instructions. There also was a program counter (PC) to specify the location in memory of the next instruction that was to be next executed by the

ALU. In the PDP-8, there also was an instruction register (IR) into which the current instruction the ALU was executing was read from memory.

All these other devices (LINK, MA, MB, PC, IR) were necessary in order to carry out the general logical steps of using a stored program to guide computation on a data set. The difference between the principal units in the PDP-8 (arithmetic-logic unit, memory, input/output) and the secondary units (arithmetic overflow, memory address, memory buffer, program counter, instruction register) is an illustration of one of the important differences between a *generic technology* and an actual *product* that implements the technology—additional features necessary to make the technology perform adequately for an application. In addition to a product adding features to a generic technology, a product also may arrange a hierarchy of technology subsystems.

Recall that the concept of the stored-process computer used the idea that a binary number system could be mapped directly to the two physical states of a flip-flop electronic circuit. This circuit was one in which one tube was constantly conducting electricity (on) until a signal might arrive to turn it off. Then it remained off (nonconduction) until another signal might arrive to turn it on again. The logical concept of the binary unit 1 could be interpreted as the on (or conducting) state, and the binary unit 0 could be interpreted as the off (or not conducting) state of the electric circuit. In this way the functional concept of numbers (written on the binary base) could be mapped directly into the physical states (morphology) of a set of electronic flip-flop circuits. The number of these circuits together would express how large a number could be represented. This is what is meant by *word length* in the digital computer.

The PDP-8 was designed with a 12-bit word length, which required 12 transistorized flip-flop circuits to be constructed wherever a word was to be expressed morphologically in the computer. Recall also that these binary numbers not only must encode data but also must encode the instructions that perform the computational operations on the data. So a word length of 12 bits had an important constraint of limiting the complexity of the instructions. One of the morphologic points of progress in computer technology has been how to get a long word length the computer can use. For example, in microcomputers (computers on a chip), the word length began at 8 bits in 1975, increased to 16 bits in 1980, and increased to 32 bits in 1988. Each increase in word length led to faster processing speeds and faster memory access.

In addition to mapping the logical concept of numbers (binary) into the morphology of the electronic states of the transistorized circuits, the logical concepts of logic operations and arithmetic operations also had to be mapped into the morphology of electronic circuits and their connection. The arithmetic-logic unit (ALU) of the PDP-8 was designed not only to represent binary numbers but also to perform arithmetic operations and logic operations.

The PDP-8 ALU was constructed of sets of circuits that could be connected together to perform the basic logical operations, and these could be combined to perform basic arithmetic operations. And the ALU could then be programmed to perform any mathematical operations, such as addition, multiplication, division, integration, etc.

We see in this illustration that the logical functions of computation were morphologically mapped into the PDP-8 as a hierarchy of higher-level mathematical operations constructed from lower-level logical operations. The design of the principal technology system of the PDP-8 product used a hierarchy of subsystems.

Complex Technology Systems

A complex technological system may be constructed of parallel subsystems or of a hierarchy of subsystems or of both parallel and hierarchical subsystems. *Parallel subsystems* are component systems of a system that operate simultaneously but independently in the system's transformation function. *Hierarchical subsystems* are functionally lower-level systems whose operations determine the operation of the system at a synthetic upper level.

As we saw earlier, the architecture of the computer system is principally constructed of parallel subsystems: central processing unit (ALU), memory unit, registers, data bus, input/output units, etc. In addition, the computer system is a good example of a system constructed of hierarchical subsystems: beginning with transistorized bit word lengths and then scaling up to basic logic operations, basic arithmetic operations, and programmed mathematical and logical operations.

The automobile system is also principally constructed of several parallel subsystems: chassis, body, fuel subsystem, engine subsystem, engine-cooling subsystem, engine-exhaust subsystem, drive subsystem, suspension subsystem, braking subsystem, steering and control subsystem, etc.

A technology system may be simple (such as hammer with striking head, claw, and handle and fastener) or complex (such as the automobile or computer). System complexity is handed by constructing the system with parallel and hierarchical subsystems. The subsystems of complex systems must be configured, connected, and controlled.

Product Systems, Production Systems, and Service Systems

Let us now generalize on the embodiment of technology systems in the different kinds of economically relevant products, services, and production. Technology systems can assume three different manifestations: product systems, production systems, and service systems.

A product system *is a completed and connected transformational technology system used by a customer.*

A production system *is a completed and connected set of transformational technology systems used in producing a product.*

A service system *is a completed and connected set of transformational technology systems used in communicating and transacting operations within and between producer/customer/supplier networks.*

Technology systems can thus be classified as product systems, production systems, and service systems. A product system is a device usable in an application. A production system is a set of equipment and procedures for fabricating products. A service system is a set of devices, techniques, and procedures for communicating, transacting, and facilitating services.

A product system embodies in physical morphology a principal transformation of a generic technology. A production system also embodies in physical morphology a sequence of unit-process transformations that construct and reshape materials as parts and components of product systems. A service system is a set of devices controlled by procedures and algorithms to facilitate information, decision, communication, and transaction.

In general, any firm will utilize product technologies in the design of its products, production technologies in the production of its products, and service technologies to provide internal services within the firm and external services to customers and from suppliers.

Applications Systems

Product, production, and service system technologies can be used together (or individually) in the applications systems of a customer.

An application *is a generic class of productive or procedural activities defined by a purpose.*

Productive applications are activities focused on the outcome, or product, of the activities. *Procedural applications* are activities focused on the process during the application and not on the output. For example, a unit manufacturing process such as milling, molding, forging, or die casting is a productive application, since the focus is on the product produced by the unit process. In contrast, an accounting system would be an example of a procedural application, since the focus is not on the product of the accounts but on the business control procedure that accounting assists. Another example of a procedural application is any recreational activity (e.g., golfing, sailing, entertainment, etc.) where the focus is on the recreational experience.

The relationships between the technology system, product system, applications system, and business system can be envisioned in a Venn diagram, as in Fig. 12.2. The product system connects the technology system to the business system and also connects the applications system to both the technology and business systems.

Forecasting Applications Systems Development

Technology systems become embedded in both product/production/service systems and applications systems, for the applications systems will incorporate combinations of product, process, or service systems as subsystems and as components and connections in the applications system. To understand the requirements for technical progress as viewed from the applications system, one should ask the following kinds of questions:

1. What level of performances in a combination of product/production/service systems is minimally acceptable for an application, and what increments in performances would be clearly noticeable in the application?

2. What features of the product/production/service systems are used in the applications or would be used, and how are they used?

3. What peripheral devices to product/production/service systems are essential or very helpful for an application?

4. What aspects of the combination of product/production/service systems cause glitches and breakdowns and require frequent maintenance in an application?

5. What aspects of the product/production/service systems of the environment within which the systems are applied create safety or pollution risks?

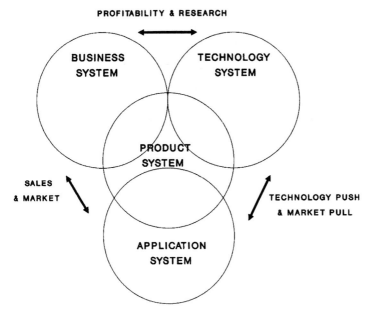

Figure 12.2 Product system as a linking system.

6. How does the current cost of the product/production/service systems or of peripherals limit the number of applications or duration of applications?

7. What factors in the applications system determine the product/production/service systems replacement rates?

8. What factors in the product/production/service systems inhibit or facilitate brand loyalty in replacement purchases?

Looking at a technology system from the perspective of an applications system anticipates the market requirements for technical change. The most frequent reason that new high-tech products fail commercially has been their *incompleteness* for use in an application system.

Practical Implications

The lesson to be learned from this chapter is how the concept of a technology system provides the basis for analyzing the technical bottlenecks in a technology and planning the directions of technological progress in the system.

1. The concept of a technology system provides general principles by means of which a technology can be varied and improved.

    a. Technical progress in a technical system can be made by improving points, technical bottlenecks, within the system that limit the overall system performance.

    b. Technology systems can be classified as product systems, production systems, and service systems.

    c. The differences between a generic technology system and the realized technology product system are the "details and completed and fully connected steps" of the logical transformation.

    d. Much of the technical differences between the technological capabilities of two different competitors lies in the details of the completeness and connectivity of the product design.

2. The analysis of a technology as a system requires two representations, a logic and a morphology, mapped in one-to-one correspondence.

    a. A logic scheme for a technology system is represented as a topologic graph of the parallel and sequential steps of unit transformations.

    b. The morphology of a technological device is also analyzed as an open system defined by a boundary and containing components, connections between the components, and a control subsystem.

    c. For control, technology systems require three subsystems:

        (1) To input material resources for fueling the transformation activities

        (2) To control the transformation activities

        (3) To direct the purposes of the transformation

    d. Technology advance may occur in

        (1) Extending the logic schematic

        (2) Alternate physical morphologies for a given schematic

        (3) Improvement of performance of a given morphology for a given schematic by improving parts of the system

3. A morphologic analysis of a system can systematically and thoroughly explore sources for technical advance in the physical aspect of a system by logically constructing all possible combinations of envisioned alternatives in any aspect of the system.

4. Complementary to a morphologic analysis, a functional context analysis of a technology system should be made from the applications systems perspective.

    a. Looking at a technology system from the perspective of an applications system anticipates the market requirements for technical change.

For Further Reflection

1. Identify a product system technology, a production process system technology, and a service technology.

2. Describe each as a generic technology system.

3. For the product system technology, identify a commercial product embodying the technology. Do the products of different makers of this product differ technically? If so, how?

4. Choose one of the preceding technology systems, and perform a rough morphologic analysis on that system. Can you identify some obvious morphologic alternatives of the system that have been commercialized?

5. Identify an applications system that uses one of the preceding technology systems. What technical features in the applications system are highly valued by customers?

13

Forecasting
Technology Discontinuities

Forecasting and planning technological change for incremental progress in the technology systems that a firm uses are relatively straightforward, as we have seen in the preceding chapters. The firm can identify the appropriate technologies and then extrapolate the rates of technological changes with technology S-curves and then can plan the directions of technological change with a morphologic analysis of the technology systems.

However, if a discontinuity in technological progress in a technology system occurs—a time when progress advances extraordinarily rapidly in a multiple of a performance parameter—then forecasting and planning technological change create special problems and opportunities for firms. This happens because (while incremental technological change reinforces the current competitive position and product lines of a firm) discontinuous technological change obsoletes product lines, placing existing firms at risk.

Recall that there are several types of innovation in terms of size and dimension:

1. Radical innovation

2. Incremental innovation

3. Systems innovation

4. Next-generation technology innovation

Also recall that the technical progress in a technology S-curve describes a history of *incremental* progress in a technology. Large leaps in technical progress "jumps" technology from one S-curve to a new one. Such leaps in progress occur as a result of the invention of different phenomenal bases for the technology and require then different practices in product design or production. Therefore, these are called *disconti-nuities* in technical practice.

A next-generation technology results from such a major leap in the performance of an existing technology system. It may occur as a result of an accumulation of incremental innovations of different parts of the system, a major improvement in a critical component of the system, or changes in the boundaries or architecture of the system.

Large, discontinuous leaps in technology have different economic impacts than the smooth, incremental progress in technology. When technological change is continuous, research planning can fit within the firm and its R&D organization. When technological change is discontinuous, the firm must depend on and be responsive to research outside the firm and innovation that does not fit into the current organization of the firm. Technology discontinuities create innovative situations in which one must literally "bet the company."

Incremental technological progress reinforces an industrial structure, while discontinuous technological progress alters industrial structures.

ILLUSTRATION
Technology Discontinuities and Progress in the Chemical Industry

As an illustration comparing incremental and radical innovations, Basil Achilladelis, Albert Schwarzkopf, and Martin Cines (1990) traced the history of radical innovations in the chemical industry. They found that radical innovations have been frequent in the chemical industry and that radical innovations were more often successful commercially than incremental innovations.

Achilladelis and associates identified the innovations introduced by the chemical industry from 1930 to 1980 in the two product areas of pesticides and organic chemical intermediates. They identified 634 pesticide innovations and 821 organic intermediate innovations. Next, they asked specialists to rank the innovations in terms of two criteria: originality and market success. They found that the more original (the more radical) the innovation, the more likely it was to be a commercial success. "The relation is strongest in organic chemical intermediates, where 66 percent of radical innovations are market successes and 84 percent of incremental innovations are market failures. In pesticides, 37 percent of radical innovations are market successes and 83 percent of incremental innovations are market failures" (Achilladelis et al., 1990, p. 3).

Radical product innovations provide more differentiation in competition than incremental innovations and therefore are more likely to be commercially successful.

Incremental product innovations do not provide as much differentiation, and therefore, any particular incremental product innovation is a look-alike product that is not as likely to be successful. This still does not negate the importance of incremental innovations, however.

The impact of radical innovation is to stimulate a new market, whereas incremental innovations enable continued dominance in an existing market.

For example, Achilladelis and associates plotted the numbers of patents for pesticide innovations from 1930 to 1980, as shown in Fig. 13.1. One can see that radical innovations were intermittent, spread from 1930 through 1976. The total of innovations (radical plus incremental) was consistent and growing from 1930 through 1970 (peaking in numbers toward the end of the 1960s). Moreover, the sales of products from these patented innovations began growing after 1950. One can see that there was a period of 20 years of radical and incremental innovations before the launching of a rapidly

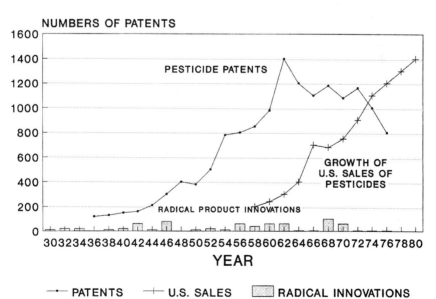

Figure 13.1 Pesticide patents and relative levels of U.S. sales.

growing market.

As products emerge from radical innovations and markets begin to grow, capital investments in plants for increased production capacities grow. Figure 13.2 presents a plot of the number of process innovations for organic chemical intermediates, the sales growth, and capital investment in production capacity. One can see two peaks in capital investment, one in the mid-1960s and the other in the mid-1970s. Capital investments are judgments based on both market growth and profitability. Profitability depends on the numbers of competitors. Fig. 13.3 plots the rate of competitive entry into the organic chemical intermediates industry and their cumulative number. One can see a declining rate of entry after 1965.

Recall that early in new markets companies see opportunities as products are introduced, and new companies enter until capacity exceeds demand and prices drop. Then profitability declines, and new companies cease entering. Achilladelis and associates (1990, p. 23) saw these four stages in the chemical industry over the period from 1930 to 1980:

> The first [stage] is characterized by few but important innovations, low patenting activity and low demand....For most chemical sectors this stage covers the period 1930–1950. The second [stage] is one of rapid growth innovations, patents and sales. It is the period when companies rush to take

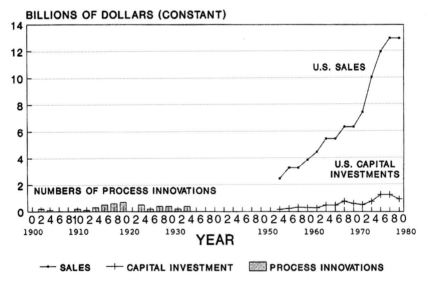

Figure 13.2 Organic chemical intermediates—U.S. capital investments and sales.

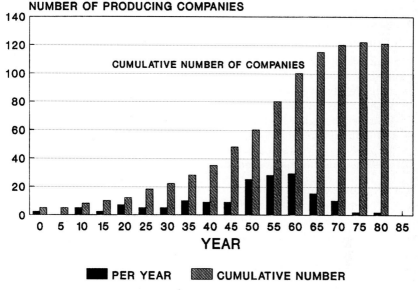

Figure 13.3 Number of companies entering organic intermediates market.

advantage of a proven technology and a fast-growing market. For the chemical sectors, this stage extends from 1950 to the mid-1960s....The third [stage] is one of maturity, when the innovations, patent and sales curves flatten out. This stage for the chemical sectors extends roughly from the mid-1960s to the early 1970s. The fourth [stage] is that of decline. It is characterized by a dropping curve of innovations and a rather flat sales curve which follows the GNP or other indicators of economic growth. For the chemical sectors this stage covers the period 1974–1982.

Recall that these stages are characteristic of a technology life cycle. The difference, however, is that the chemical industry was driven by several technologies, and therefore, its technology life cycle is a composite of several underlying technology life cycles (pesticides, organic chemical intermediates, plastics, etc.). As new basic technologies surfaced as a result of radical innovations, the markets for the chemical firms continued to grow.

At the end of a period of radical innovations in an industry, markets mature, excess production capacity occurs, and fierce competition for maintaining market shares together reduce profitability for all participants in the industry.

Achilladelis and associates (1990, pp. 24–25) also identified the important role that scientific research had in providing the base for the radical innovations:

> Scientific advance was undoubtedly an important driving force for radical innovations; organic chemistry had a particularly creative period since the 1860s. Polymer composition and polymerization mechanics were substantially elucidated by Sudinger, Carothers and Mark [industrial scientists] in the 1920s and early 1930s. Physical chemistry advanced in strides. Advances in engineering, particularly for reactions under high pressures and temperatures and for continuous catalytic processes, were made in the previous 20 years so that the design of pilot and manufacturing plants could be undertaken with confidence.

This is an example of a science-based Kondratieff economic long wave. The major source of radical innovation was new science. As fundamental discoveries and understanding in science are used for new technologies, radical innovations are made. Radical innovation slows when an area of science relevant to a technology system becomes well understood and scientific progress slows.

This does not mean, however, that technology cannot progress until science progresses. The interaction between scientific progress and technological progress is complex. A scientific discovery may be used for a radical technological innovation, but then incremental progress for that technology may push science to deepen and broaden its understanding of nature. Achilladelis and associates commented on this interaction of science and technology in the chemical industry. "The question of whether scientific advance leads innovation or vice versa has been the subject of intense argument in the history of science and innovation literature. Our evidence for the chemical industry indicates that they occur simultaneously. *The spark that set the motion was in most cases a radical innovation which was introduced when the scientific knowledge on which it relied was only partially available and market demand was neither gauged nor firmly established*" (Achilladelis et al., 1990, p. 25).

This incompleteness of scientific knowledge is a major opportunity for planning new technology.

> *Pursuing further scientific understanding of the phenomenon on which a radical innovation is based provides a fertile area for improving the technology.*

This also provides one of the major mechanisms for planning technological advance—cooperation between academic and industrial research. While academia and industry have different goals, the interactions between science and technology provide a bond. "Academic and industrial research are driven by different forces: the advancement of science and profit....But when a radical innovation...whose scientific background has not been fully established becomes also a market success, then the forces driving both academic and industrial research are activated simultaneously and attract both research-intensive companies and gifted academic researchers" (Achilladelis et al., 1990, p. 25).

The modern world of technological innovation involves ties between industrial and academic researchers. Whole new industries in addition to the chemical industry, such as computers and biotechnology, have arisen from the university and industrial ties.

Radical innovations played a strong role in setting the directions of technical competence in individual chemical firms. "...Radical innovations and market successes tend to concentrate at the earlier stages of a technological trajectory and...there is a remarkable coincidence of originality and market success of individual innovations, so that companies which enter a sector early have more chances of succeeding technically and commercially" (Achilladelis et al., 1990, p. 28).

Achilladelis and associates (1990, p. 28) called the further development of product/processes from radical innovation the setting of *corporate technological traditions:*

> ...radical innovations which become market successes are the steam engines that drive corporate research. They provide the most successful way by which a company enters a new sector because technological leadership and profit making are simultaneously achieved. The initial success is then perpetuated by the generation of diverse and powerful corporate forces....The cumulative effect...leads to the establishment of corporate technological traditions which acquire their own momentum, so that over periods from 15 to 30 or 40 years a disproportionate number of innovations results from the same technology.

Technology, Science, and Engineering

Although throughout this book I have been using the term *technology* as if it were a commonly understood term, now it will be helpful to more precisely fix its definition with respect to science. The reason this is now important is that in the history of modern economic development, new scientific phenomena have generally provided the basis for radical technological innovations. We will need mutually consistent definitions also for the terms *science* and *engineering*. Such terminology will enable us to describe accurately the relationships between universities and industr y in the advancement of technology—particularly during discontinuities.

The historical derivation of the term *technical* comes from the Greek word *technikos,* meaning "of art, skillful, practical." The portion of the word *ology* indicates a "knowledge of" or a "systematic treatment of." Thus the derivation of the term *technology* is literally "knowledge of the skillful and practical."

This is certainly the sense in which we use the term. However, for our purposes, this definition is too general to imply how one may predict knowledge of the useful before it exists. For this reason, let us use a slightly different definition of technology.

We will define technology *as the knowledge of the manipulation of nature for human purposes.*

Note that we have retained the notions both of knowledge and practicality (human purposes) but have added the notion of the manipulation of nature. This points out that all practical skills of technique ultimately derive from alterations of nature. Technology depends on a base in the natural world but extends the natural world through manipulation. It is for this reason that the discovery of natural phenomena has often resulted in a vision of its technological application.

Consider next the derivation of the term *science* from the Latin *scientia,* meaning "knowledge." However, the modern concept of scientific research has come to indicate a specific approach toward knowledge—one oriented toward nature that results in discovery and in explanations of nature.

For this reason, we will define science *as the discovery and explanation of nature.*

The link between science and technology is engineering, and we will adopt the following definition as consistent with the preceding definitions.

Engineering *is the understanding and manipulation of nature for human purposes.*

In these definitions one should first note that science is not focused by human purposes, as are engineering and technology.

Science is relevant to human purpose but not focused by it; rather, science is focused by the ubiquity and generality of existence of nature.

This is the reason in R&D management why basic research (or pure science, as it is sometimes called) is difficult to manage in a corporate environment. Corporations need research focused by purpose, whereas scientists are oriented by their profession to disciplinary classes of ubiquitously existent phenomena.

Although research activities in science focus primarily on discovery and understanding, sometimes inventions do occur in scientific research (notably instrument, technique or materials inventions). When inventions do occur in basic science, they are usually the basis for radically new technologies (such as the laser or the transistor).

Next, one should note that while technology is the knowledge of how to manipulate nature for human purposes, it does not necessarily imply a deep understanding of nature that allows the manipulation.

Engineering differs from technology by focusing on combining the technical knowledge of how to manipulate nature with scientific explanations of nature as to how and why the manipulation works.

Although technologically focused research concentrates on inventions and their development toward products, considerable levels of research effort in the engineering disciplines (in addition to the scientific disciplines) are also focused on advancing the understanding of nature. Such research also has been called *engineering science.* Engineering research for technological purposes draws on and uses the natural phenomena upon which technologies are based (the nature that is manipulated).

The better engineers understand nature, the more and clearer ideas they may have for refining manipulation.

For this reason, engineering research is organized not only on specific technologies but also on phenomena underlying technological sectors, such as mechanical engineering, electrical engineering, chemical engineering, and civil engineering.

The importance of understanding the relationships among science, technology, and engineering is that research for radical innovation, both basic new technologies and next generations of technologies, requires *advancing the underlying scientific base.* University-based science has often provided the basic invention for a new technology, but industrial research provides the innovation of the basic new technology. The research spectrum connecting science to technology has been one of the bases for university and industrial research cooperation.

Historically, the companies that participated in the early stages of a radically new technology have often been the firms that built strong and large organizations to exploit the new products and new markets of the new technology.

ILLUSTRATION
Radical Innovation and IBM's Entry into Computers: The SAGE Project

Let us now look in detail at how a university-based radical innovation provided an opportunity for a firm (IBM) to enter a new product line that obsoleted its established product line. After World War II, IBM substituted the electronic computer for its own punch-card-sorting product line. This was a technological discontinuity that IBM anticipated and exploited to raise itself to industrial dominance in a new industry. The basis of IBM's entry into the new computer technology was a partnership with MIT to produce the first military computers for the U.S. Air Force.

Recall that, historically, there were two major electronic computer projects that directly fostered commercial computers: (1) the Mauchly and Eckert project, EDVAC, at the University of Pennsylvania and sponsored by the U.S. Army, and (2) the Forrester project, Whirlwind, at the Massachusetts Institute of Technology and sponsored by the U.S. Air Force.

The Mauchly and Eckert EDVAC project had commercial transfer into the first commercial computer UNIVAC, and Mauchly and Eckert left their university to start a new company (which was subsequently purchased by Remington Rand). Recall that the Forrester Whirlwind project had commercial transfer into IBM's first computers, since IBM contracted to manufacture MIT's Whirlwind computer for the Air Force. Also recall that Forrester's project borrowed Mauchly and Eckert's solution for logic circuitry but that Forrester went on to break the memory problem by inventing the ferrite-core memory.

It was the ferrite-core memory that gave an important early competitive advantage to IBM in the new computer industry of the 1950s. In June of 1952, John McPherson of IBM participated in a committee meeting of a professional society and there talked to Norm Taylor of MIT. "'One of the best payoffs that belonging to a professional society could produce,' John McPherson recalls, 'I should have gotten a finder's fee.'...Norm Taylor of MIT advised him that the MIT Digital Computer Laboratory was looking for a commercial concern to manufacture the proposed air defense system. Was IBM interested?" (Pugh, 1984, p. 93).

IBM was interested. McPherson returned to IBM headquarters and discussed the project with IBM executives. It was the kind of opportunity that Tom Watson had been looking for in order to rebuild IBM's military products division and to improve electronics technology capabilities. IBM told Forrester of their interest. Forrester and his group were reviewing several companies as potential manufacturers of the Air Force computer. They visited Remington Rand, Raytheon, and IBM. Forrester chose IBM. "In the IBM organization we [the MIT group] observed a much higher degree of purposefulness, integration and esprit de corps....Also, of considerable interest to us was the evidence of much closer ties between research, factory and the field maintenance in IBM" (Pugh, 1984, p. 94).

With the deal between MIT and IBM concluded, IBM rented office space at a necktie factory in Poughkeepsie, New York, and got to work. The Whirlwind II project was renamed SAGE. By the summer of 1953, there were 203 technical and 26 administrative people working on the IBM part of the project.

The system was to have many digital computers at different sites around the country and in continuous communication with each other. They were to share data and calculate the paths of all aircraft over the country in order to identify any hostile aircraft. The computers were to use electronic vacuum tube logic and ferrite-core memory.

At this time of the innovation in the new computer industry, all the pioneering research had been performed at universities: Aiken at Harvard, Mauchly and Eckert at the University of Pennsylvania, and Forrester at MIT. Moreover, their research was primarily funded by the federal government: Aiken by the Navy (with IBM assistance), Mauchly and Eckert by the Army, and Forrester by the Air Force. Next, the transfer of technology into commercial applications then occurred both through the formation of new firms and through existing firms entering the new industry. Mauchly and Eckert formed a new company financed by a government contract from the Bureau of Census, and IBM produced the SAGE computers financed by the Air Force.

The pattern of government support of basic and applied research performed by universities and transferred into the defense industry as applied and development work became a common pattern in the United States after World War II.

When IBM built the SAGE computers for the Air Force, it provided IBM with a commercially important technology advantage. "The decision to join MIT on Project SAGE was probably the most important decision [IBM] management made during this period [1950s]. It gave IBM an inside track to the most advanced computer technologies in the world and propelled the company into a leading position in ferrite-core memories" (Pugh, 1984, p. 262).

IBM quickly gained a world lead in the technology of producing ferrite-core memories. "The memory technologies developed for SAGE were modified and refined for commercial products and then pushed to the limit in Project Stretch. By the time decisions for System/360 were made, IBM was already the world leader in ferrite-core memory technologies. It had produced some of the fastest, largest, and most reliable memories in industry, and its accomplishments in reducing manufacturing costs were unmatched" (Pugh, 1984, p. 263).

Introducing Radical Innovation into the Product-Development Cycle

Recall from Chap. 1 that there are two different processes for radical innovation and incremental innovation—the linear and cyclic processes. The problems of managing both processes in a firm arise from the firm's organization of its businesses. In the preceding illustration, IBM's production of the SAGE computers began its entry in the radical innovation process of first producing computers. Afterwards, later computer models were produced in a cyclic process during IBM's product-development processes. The problem that arose next was the creation of different product lines (low-performing and lower-priced computers and high-performing and higher-priced computers were organized into different divisions that created their own product-development cycles).

Recall that the linear innovation process for radical innovation involves a series of knowledge transformations from the following areas:

- Discovery and understanding
- Scientific feasibility
- Invention
- Technical feasibility prototype
- Functional application prototype
- Engineering prototype
- Manufacturing prototype
- Product production

Also recall that the cyclic innovation process for incremental innovation in the product-development process involves a cyclic set of activities:

- Anticipating technical change
- Creating technical change

- Implementing technical change in product/process/service development

- Introducing high-technology products into the marketplace

We have also seen (in the illustration concerning the chemical industry) that radical and incremental innovations occur at different rates, with radical innovation being relatively unpredictable. Thus, in an established industry organizing research primarily for incremental innovation, radical innovation is always a surprise. And therein arises the problem, for management aims at reducing the uncertainty in operations to predictability, and therefore, surprise is always unsettling to management planning.

Also recall that in the short term, the strongly competitive firm is one that has brief product/process-development cycles for continuously improving product quality and lowering production cost. And thus in the short term the speed and correctness of the product-development and manufacturing cycle are the critical competitive factors in a firm.

Over the long term, however, these product-development procedures may be interrupted by radical innovation. Therefore, recall that in Chap. 1 we emphasized that the difficult problem of strategic technology management centers on a profound but important conflict—trying to manage both radical and incremental innovation processes that have mutual incompatibilities.

Finally, recall that the concept of next-generation technologies is an important means of reconciling the management of both short-term incremental innovation and long-term radical innovation in the same firm. Furthermore, properly managed, the linear innovation process can appropriately work with the cyclic innovation process by feeding into it the technical feasibility of a radical innovation. The connection between the two innovation processes is the *functional prototype* of a next generation of a technology system.

> *The step of functional prototyping in the linear innovation process establishes the timing for introduction of technology discontinuities into the cyclic innovation process (product-development cycle).*

In any modern technology, some knowledge must be acquired from outside the firm through some kind of technology transfer. However, the mere transfer of technical knowledge is seldom sufficient for competitive advantage. Ordinarily, a firm also should create and add some proprietary knowledge of their own—either in product design or in manufacture. However, at a technology discontinuity that radical innovation produces, the firm must again depend heavily on technical knowledge produced outside the firm.

ILLUSTRATION (*Continued*)
IBM's First Product Line Crisis: The
System/360 Computers

Now we will see that the first technology discontinuity IBM faced in its new computer business was in the transition from electron vacuum tube computers to transistor-based computers. Initially, IBM had good commercial success with its early computers that used the ferrite-core memory technology. "In the fall of 1954, a small but increasing portion of IBM's revenue was coming from electronic stored-program computers. The IBM 704 and 705 computers with ferrite-core memories had just been announced. The information IBM sales representatives obtained from customers for electronic computers was providing a much better understanding of customer requirements and of design deficiencies in the systems" (Pugh, 1984, p. 160).

Recall that the first applications of a new technology are important to further refine the focus of the technology and to define the required performance and specifications for the new technology. The result then is to define a new model for the product aimed toward the increasing requirements for improved performance. Therefore, IBM engineers were thinking about how to build a next generation of faster computers. "In a meeting of IBM executives in January 1955, it was noted that all of their customers for larger computers desired machines which are as fast as possible, which have large amounts of random access memory, and which can be made available in 1957 or earlier" (Pugh, 1984, p. 161).

Very early in computer applications, the technology performance parameters (of speed and memory size) emerged as the most important for applications. IBM's problem then was how to finance further research for a next generation of high-performance computers. In the meeting mentioned above, the decision was made for IBM to seek government funds to assist in developing a next-generation computer. Next generations of a technology are expensive and risky. The government funding of the SAGE computers had helped IBM develop and manufacture ferrite-core memory technology in the first generation of its computers. IBM executives saw benefit in government-sponsored research for its next generation of computers.

IBM learned that Edward Teller (then the associate director of the University of California Lawrence Radiation Laboratory in Livermore, California) was ready to sponsor a high-performance computer. The Livermore Laboratory was a university-administered and government-sponsored research laboratory for hydrogen bomb research. In 1955, both IBM and Sperry Rand submitted proposals to Teller for the project, but Sperry Rand won the contract.

IBM continued to refine its planned next-generation computer, which the company called *Stretch*. The Stretch computer was to be wholly transistorized, replacing the electronic vacuum tubes of the first generation of computers. Undaunted, IBM next approached the National Security Agency (NSA) with its proposal. In January of 1956, IBM received a contract for memory development and a computer design effort totaling $1,350,000. Then IBM won a second contract from the Atomic Energy Commission's Los Alamos Laboratory (to deliver a Stretch computer to the AEC within 42 months for $4.3 million). As the Stretch project continued, IBM received another contract from the Air Force to prove a transistorized version of its 709 computer.

However, with its radically new technology of transistors, the next-generation Stretch project ran into technical difficulties (Pugh, 1984, p. 183):

> The Stretch effort had been long and costly. The system delivery date had slipped by almost a year from its originally contracted date. Personnel on Stretch exceeded 300 by 1959, and the total program cost over $25 million, almost twice the cost estimated in February 1956. Government development contracts plus the purchase price for the system covered little more than half of this....Although Stretch was nearly completed, one serious problem remained. Stretch had been tested on several customer problems and found to take about twice as long as projected.

As often happens with risky next-generation technology, Stretch was twice over budget and only delivering half the targeted speed performance. "The credibility of the company was far more important to Watson than profit or loss on any one sale....Watson believed IBM had a moral commitment to supply this performance or to adjust the price" (Pugh, 1984, p. 184).

IBM reduced the price and took the loss. Still there were commercial benefits from the Stretch technology. "The use of Stretch technology in the IBM 7090 computer also provided IBM with the first fully transistorized large-scale computer to be placed in production" (Pugh, 1984, p. 185).

Thus government-sponsored research was not always profitable for IBM, but it kept the company at the cutting edge of technology in the early days of the industry. IBM's ability to translate the research into commercial technology provided IBM with economic and competitive advantages. However, IBM had already become large enough that it was beginning to have trouble coordinating innovation within its divisions and research laboratories. Then, on the commercial side, the IBM 7090 computer had been introduced in 1959, and thereafter IBM was selling several computer models from an entry-level machine to advanced models.

Yet none of these models were compatible. When a customer desired to move up from entry-level performance, they had to rewrite software applications. By 1960, among the different computers produced by IBM, the incompatibility and diversity of the products had become a severe problem. IBM was organized as two divisions: the General Products Division (GPD), for the lower-priced machines, and the Data Systems Division (DSD), for the higher-priced machines. These divisions designed their own computer lines independently. And in addition, two different research laboratories also were designing incompatible computers—the IBM Poughkeepsie Laboratory and the Endicott Laboratory.

For example, the Endicott Laboratory was then applying electronics to conventional electromechanical accounting machines. Charles Branscomb at the Endicott Laboratory had designed a low-cost, spartan computing machine that was sold as the IBM 1401. This product had became the mainstay of the General Products Division. It had used the transistor circuitry and packaging developed in project Stretch. By 1962, 2000 of the 1401s had been sold (and eventually 10,000 were sold). The Endicott Laboratory also designed a transistorized successor to the first-generation IBM 705, which was announced in September of 1958 as the IBM 7070 (but not scheduled for production until 1959).

In June of 1959, IBM was reorganized into three product divisions: (1) Data Systems Division (DSD), for high-end computers, (2) General Products Division (GPD), for low-end computers, and (3) Data Processing Division, for sales and service. Responding to this reorganization, back in the Poughkeepsie Laboratory, a frantic effort began to regain control of the design of the high-end computers. "Because the 7070 was a large [computer] system, the reorganization placed responsibility for its development in the Data Systems Division in Poughkeepsie—in the hands of its original opponents. Steve Dunwell thought he could stop the 7070 project before it went any further, and told his group to design a replacement that was twice the speed and half the cost...which they called the 70AB" (Pugh, 1984, p. 189).

In this illustration, we see the rivalry and conflict that exists over technological strategies among divisions and research units in a large firm.

The two research laboratories, Endicott and Poughkeepsie, were rivals with different designers and different judgments on best design. At the time of the reorganization, T. Vincent Learson had been made group executive responsible for the three divisions. When Learson learned of Dunwell's decision, Learson ordered that the 7070 be continued, with maximum effort.

Yet this still left the Poughkeepsie Laboratory without a product line that it had designed. "Thwarted in that effort, the 70AB group began to target their machine for the lower performance region just above the 1401....Once again the 70AB project was blocked....The Poughkeepsie engineers...began to consider expanding the 70AB into a family of computers, called the 8000 series, to serve across the marketplace" (Pugh, 1984, p. 189).

In January of 1961, DSD formally presented its plans for the new 8000 series to top management in an auditorium: "...One person...was extremely distressed by what he saw...and stood out like a dark cloud in the back of the auditorium. And that person was T. Vincent Learson, the corporate executive in charge of both DSD and GPD and later the chairman of IBM" (Strohman, 1990, p. 35).

Learson, the new group executive responsible for all three IBM divisions, was disappointed in the planned performance of the 8000 series. Learson noted that the performance of the 8000 model was not making its goals. Although some improvements had been planned in circuit technology, much older technology was being used in order to get the machine out fast. Learson also saw great trouble in the lack of mutual compatibility of the 8000 series computers. He saw no plans whatsoever to make machines within the 8000 series compatible with one another.

Next-generation product planning in a large firm is a complex issue because of the differing technical and market views held by different divisions and research laboratories of the firm.

Learson also knew that the other division, GPD, also was continuing to design incompatible computers. Seven incompatible families of IBM computers had been created in the last decade. As a result, Learson saw that IBM was spending most of its development resources propagating a wide variety of central processors and that little development effort was devoted to such areas as programming and peripheral devices. Thus, in 1960, IBM had three different organizations developing similar but different products: the Data Systems Division with its 8000 series, World Trade with its SCAMP series, and the General Products Division with its 1400 and 1600 series. After the DSD presentation, Learson decided to force IBM to formulate a technology strategy. Learson moved Robert Evans (a Ph.D. in electrical engineering) into the Data Systems Division—just below the vice president of the division—but just above the division's chief computer designer, Frederick Brooks (who was responsible for the 8000 series plans). Then Learson instructed Evans to review the 8000 series plans. "His [Learson's] words were essentially, 'If it's right, build it; and if it's not right, do what's right,' Evans says" (Strohman, 1990, p. 35).

Evans decided it was not right. He wanted a product plan for the whole business, not just the upper line of the DSD for which the 8000 series had been planned. This brought Evans and Brooks (who was responsible for the 8000 series) into direct conflict. As Brooks (who lost the battle) recalled the conflict, "Bob [Evans] and I [Brooks] fought bitterly with two separate armies against and for the 8000 series. He [Evans] was arguing that we ought not to do a new product plan for the upper half of the business, but for the total business. I was arguing that was a putoff, and it would mean delaying at least two years. The battle...went to the Corporate Management Committee twice. We won the first time, and they won the second time—and Bob was right" (Strohman, 1990, p. 35).

Evans then proceeded to get IBM to plan for a wholly new and completely compatible product line from the low end to the top end. "Evans, winning the battle to kill the 8000 but realizing he had many wounded egos to heal, took all his adversaries, including Brooks 'Away to the, Gideon Putnam, a hotel in Saratoga Springs, to look at our belly button and decide how we were going to organize and what we were going to do.'...[There Bob Evans] made a dramatic overture: 'I so respected Brooks for his intellectual ability, I surprised him by offering him the senior job on what became System/360'—the job of systems manager. 'He surprised me by taking it'" (Strohman, 1990, p. 37).

The disagreement was an honest one about how important to the company was a better technology versus the time to do it. Technical disagreements and disagreements about the relative commercial benefits occur in any large, successful organization.

Decisions on next-generation product plans affect the careers of the different individuals involved, and therefore, different technology strategies can have different impassioned champions in a large firm.

Evans and Brooks, having bitterly fought the battle, still respected each other and worked together afterwards. Then Evans' next job was to get the General Products Division aboard for his total compatibility strategy. "Hanstra was with GPD, which was producing one of IBM's most popular and profitable machines, a small processor known as the 1401. The 1401 was threatened by a lower-priced competitor, the Honeywell 200, which came with a software tool that permitted it to run 1401 programs....Without GPD's support, system/360 would never see the light" (Strohman, 1990, p. 37).

Evans needed the help of Donald T. Spaulding, who led Learson's staff group. "...In October 1961, Spaulding proposed an international top-secret task force to plan every detail of the New Product Line (System /360) and make necessary compromises along the way. And in a stroke of political genius, he named as chairman of the group none other than John Hanstra. To no one's surprise, the vice-chairman was Bob Evans" (Strohman, 1990, p. 37).

This task force of top technical experts represented all the company's manufacturing and marketing divisions. "Throughout November and December 1961 the thirteen [of the committee] met daily in strictest secrecy at the Sheraton New Englander Motel, in Greenwich, Connecticut. Their lengthy sessions were punctuated by heated technical arguments. When they concluded, on December 28, they had hammered out the framework for the System/360 product line, a detailed technical document dubbed the SPREAD report" (Strohman, 1990, p. 37).

We note that it took a top-level executive to lead a strategic technology vision—compatibility across the product lines. It also required a management commitment, worked out through internal political battles and eventual consensus, to translate that technical vision into a concrete technology plan.

The resulting technical plan was called the *SPREAD report,* and it focused on seven basic technical points:

1. The central processing units were to handle both scientific and business applications with equal ease.

2. All products were to accept the same peripheral devices (such as disks and printers).

3. In order to reduce technical risk to get the products speedily to market, the processors were to be constructed in microminiaturization in transistors (although integrated circuits had been considered).

4. Uniform programming would be developed for the whole line.

5. A single high-level language would be designed for both business and scientific applications.

6. Processors would be able to address up to 2 billion characters.

7. The basic unit of information would be the 8-bit byte.

Next, there were problems in product development. They were solved, however, and the first 360 computer was shipped in April of 1965. "The financial records speak for the smashing ultimate success of IBM's gamble. In the six years from 1966 through 1971, IBM's gross income more than doubled, from $3.6 billion to $8.3 billion" (Strohman, 1990, p. 40).

Technology and Strategic Reorientation

In the preceding illustration, a previously successful company with several products ran into a technology discontinuity. Moreover, it was a technology discontinuity that obsoleted not only the company's several product life cycles but also its whole product line life cycle. It is difficult for a company to formulate a new technology strategy when it hits a technology discontinuity.

The principal difficulties in formulating a technology plan for a technology discontinuity are (1) the technical uncertainties of a new technical vision and (2) the differing perspectives among the different product-group managers and the technical staff about that technical vision.

In order for a company to develop a radically new product line technology strategy, the whole company must fight out different product plans.

To formulate a technology plan for a technology discontinuity, it is necessary for a high-level executive to force the issue and to organize the effort necessary to formulate and implement a technology strategy for the whole company.

Technology discontinuities are a common problem for firms that were successful initially in a radically new technology. What happens to a firm that was successful in the early stages of a radical innovation such that its corporate technological traditions were set for 15 to 40 years and then that technology is displaced by a new technology? A technology discontinuity is a time of crisis for such a firm.

The reason for the crisis is that technological discontinuities force not only changes in technology strategy but also changes in business strategy. This is why they are so difficult.

Technology discontinuities are a major source of organizational instability.

Many students of organizations have argued that instability is a periodic experience for any organization. For example, Michael Tushman and Philip Anderson (1986) argued that organizations experience periods of relative stability interrupted by sharp strategic reorientations. They and others (e.g., Norman, 1977; Miller and Friesen, 1984) have seen organizational change as a kind of evolution stimulated by and responding to change in business and economic structures.

In addition, the importance of effective top-level leadership in making a strategic reorientation has long been recognized. Michael Tushman and Elaine Romanelli (1985, p. 298) have nicely summarized the connection between executive strategy, environmental (structural) change, and organizational decline:

At least part of the reason for substantial organizational decline in the face of environmental change lies with the executive team. A set of executives who have been historically successful may become complacent with existing systems and/or be less vigilant to environmental changes. Or, even if an executive team registers external threat, they may not have the energy and/or competence to effectively deal with fundamentally different competitive conditions. The importance of an effective executive team is accentuated in industries where the rate of change in underlying technologies is substantial.

A major source of changes in the business environment and within the economic structure are technological discontinuities. For example, Tushman and colleagues (1985) looked at the minicomputer industry, focusing on 59 firms started between 1967 and 1971. They compared firm records of success and failure over a subsequent 14-year period. In their analysis of the reasons some firms survived and many failed, they argued that one must understand how the *conduct* of the firm in the *context* of the changing economic structure affected the *performance* of the firm.

The conduct of firms consists of the strategic, tactical, and organizational activities guided by the executive team [chief executive officer (CEO) and other principal executive officers]. Conduct must alter as the context of the firm changes when alterations in the economic structure affect competition. Changes can arise from (1) technological changes, (2) market changes, (3) resource changes, (4) regulation changes, and (5) competitive changes. Performance then depends on the executive team anticipating a correct strategy and organization for the future structure of the economic system. Such strategy and organization correctly anticipate technological opportunities for new products, market changes for new needs and applications, resource changes that affect the availability and cost of materials and energy, and changes in government regulations affecting safety, monopolies, taxes, etc.

Over the periods in which industries faced changes in economic structural factors, Tushman and associates found that those firms whose executive teams made no changes in strategy and organization had products that failed after the structural changes occurred. Those firms whose executives constantly made changes in strategy, organization, and product (whether or not structural changes were occurring) also failed.

The firms that survive and prosper through a technology discontinuity are those whose executive teams (1) correctly anticipate the discontinuity and prepare for it with appropriate technology strategy, product planning, and reorganization and (2) after making the appropriate changes hold a steady course to produce with quality and lowered costs.

ILLUSTRATION (*Concluded*)
IBM's Second Product Line Crisis:
The Obsolescence of the
Mainframe Computer

After IBM's successful System/360 series of mainframe computers in the late 1960s, technical progress continued in the computer industry—minicomputers, personal computers, engineering workstations, and mini-supercomputers. However, IBM did not lead technically in the commercialization any of these new generations of technologies.

Although at the different times involved the research capabilities of new technologies actually existed in IBM's many research laboratories, IBM management no longer incorporated radically innovative research in a timely fashion into business strategies in order to dominate the new markets (as they had previously done in the mainframe market). For example, as we saw in the illustration involving the DEC PDP8 computer in Chap. 12, around 1970, the technical progress in semiconducting IC chips provided the opportunity for a new product line of minicomputers. These were pioneered by such new companies such as DEC, Control Data, Prime, etc. IBM entered the minicomputer market late in the decade and captured only a small portion of the market (around 14 percent), in contrast with its dominance by then of the mainframe market (around 70 percent).

As a second example, we also saw that around 1980, a new generation in technical progress in semiconducting IC chips (the microprocessing chip) provided the opportunity for another new product line: personal computers. In this market, IBM did enter relatively early, but it lost its market lead because it was not a technological leader in this product line. Instead, recall that Apple became the technology leader by borrowing advanced research ideas from Xerox. By 1990, IBM had allowed its market lead to slip back to around 20 percent of the market, being neither a technical leader nor a cost leader in personal computers.

As a third example, by the middle of the 1980s, a new product line of high-performance personal computers for engineers, the engineering workstation, emerged. Again, new companies pioneered the market such as Apollo and Sun, using processors that ran a reduced instruction set chip (RISC). IBM entered later, gaining a modest market share, but again not dominating as a technical leader nor a cost leader.

As a fourth example, also occurring in the middle of the 1980s, another new product line, the mini-supercomputer, emerged by using microprocessors in parallel architectures. Again, many IBM researchers were in the research vanguard of parallel processing. But IBM management chose not to produce mini-supercomputers, perhaps because they competed directly against their own minicomputer and mainframe product lines (and as we saw, a new firm, Convex, emerged as the leader in this product line).

Now as a result of all these failures to lead aggressively in next generations of computer technology, by 1991, IBM's once dominance of the computer industry had passed to other firms as the product lines of the industry proliferated. As the personal computers and workstations and mini-supercomputers were linked together in distributed computing networks, not only was the minicomputer product line obsolete but also the mainframe product line (IBM's mainstay) was threatened with obsolescence by 1991. Thus, only 20 years after the great success of the System/360 project, IBM was again in technical and commercial trouble.

In the fall of 1990, for example, a popular business journal, *Business Week,* did a special report on the changing conception of the computer system. And in that article, the author pictured the chairman of NCR Corporation as smiling very broadly (then NCR was a long-time and second-running competitor to IBM). "'We're on the brink of a major crossover between micros and mainframes,' says Charles E. Exley Jr., CEO and chairman of NCR Corp. 'The microprocessor revolution is becoming an irresistible force. Ultimately, every business will want to move from their mainframes to microprocessor-based systems'" (Verity, 1990, p. 117).

Exley was convinced that the structure of the computer system was now the network, in which the mainframe was becoming obsolete. "So convinced is Exley that he has made the proposition the basis for NCR's entire strategy. Within a few years, every computer in the NCR line...will be based on one or more of Intel microchips. Gone will be NCR's proprietary mainframe designs...." (Verity 1990, p. 117).

Other firms, such as AT&T, agreed with Exley's technology strategy, in which he envisioned the mainframe computer as becoming obsolete in a product line type of life cycle. In fact, AT&T bought NCR. "American Telephone & Telegraph Company's hard-fought battle to become a major presence in the computer world was realized yesterday when NCR Corp., the nation's fifth-largest computer maker, agreed to a $7.4 billion merger of the two companies. 'The company that will emerge will be uniquely equipped to meet what customers will need in the future—global computer networks as easy to use and as accessible as the telephone network is today,' said Robert E. Allen, chairman of AT&T in a joint statement with Charles E. Exley, Jr., the chairman of NCR" (Skrzycki, 1991, p. C1).

As another example of the turmoil in IBM's business condition at that time, in June of 1991, *Business Week* ran another special feature on the computer industry. This time, however, the authors pictured IBM's chairman, John F. Ackers, as barely smiling. "'The fact that we're [IBM] losing share makes me goddamn mad....Everyone [in IBM] is too comfortable at a time when the business is in crisis.' What makes [this] message positively riveting is that it comes from no less than John F. Ackers, chairman of IBM and chief of the enterprise the world has viewed as the brightest paragon of American business" (Byrne, Depke, and Verity, 1991, p. 17).

IBM in crisis again! The company's problems centered on its being hit with technology discontinuities that were simultaneously playing havoc with three different life cycles for IBM products:

- Product life cycles
- Product line life cycles
- Technology life cycles.

First, IBM was struggling to maintain its *product life cycles*. "IBM...is now hurting in a key market, Japan. While its Japanese rivals deployed thousands of so-called systems engineers to help clients at the low-margin business of developing software, IBM held back....But now, IBM is paying the price as the Japanese exploit their knowledge of customers' needs to win new mainframe sales. IBM Japan...posted a 1.1 percent sales gain last year....Fujitsu and NEC saw sales surge more than 12 percent....Hitachi's rose 10 percent" (Byrne, Depke, and Verity, 1991, p. 30).

At the same time, the *product line life cycles* were obsoleting large main frames in the computer network. The mainframe computer as a product line had been designed to be multifunctional, providing computation, communications, data storage, data base server, distributed access, and other functions. Other product lines were providing some of these functions more efficiently and cheaper when hooked together in a computer network:

- The PCs and workstations were providing distributed processing.
- File servers were providing storage and retrieval.
- Mini-supercomputers were providing computationally intense service.

This was the vision in the previous example (about Exley at NCR Corporation) that the multifunctional mainframe was becoming technologically obsolescent within a sophisticated computer network. An illustration of this had occurred in 1990, when one of IBM's traditional Fortune 500 customers decided to get rid of their IBM mainframe. "In 1987, Joseph Freitas had a radical idea....As director of investment banking systems for Merrill Lynch & Co., Freitas wanted to build a 13-city computer network based entirely on microprocessors....Freitas started phasing out the IBM mainframe that had served as the heart of his information system since 1983. Until this month [November 1990] when Merrill gave the machine back to IBM, the firm had been paying $1 million per year in maintenance and leasing fees for it" (Schwarz, 1990, p. 122).

The reason that Merrill Lynch no longer needed the IBM mainframe in its computer network was that it had been primarily serving as a central library to the network. "For several years, Merrill had been taking more and more work off the mainframe. By last year, it was doing little processing and was acting more as a central library that served information to 1100 personal computers. Freitas realized that a computer designed to be a data base server could do that for far less. He brought a $500,000 system, based on multiple Intel microprocessors, from Sequent Computer Systems, Inc." (Schwarz, 1990, p. 122).

Merrill Lynch processing needs were almost entirely distributed to the personal computers in the network. The mainframe no longer did processing but only file serving. A $1 million-per-year lease and maintenance cost for a file server was too expensive. For only a half a million dollars one could buy a good file server outright, and the system could be improved with better software. In the new system, reports could be entered anywhere on any PC and then were available to all the other PCs on the network. "For example, a banker can ask for current and projected price-earning ratios for five companies. The system then calls across the network for the latest earnings reports, gets price quotes from another source and earnings projections from a third. This is done by bankers who don't know where each data base is. Merrill believes its new network and software gives it an edge over many of its competitors" (Schwarz, 1990, p. 122).

What then was a mainframe doing in a network? In future computer systems, the answer (as Freitas had the vision to see) was nothing! The mainframe was as obsolete as a dinosaur as a time-sharing system for data transfers. PCs and workstations could prove to be cheaper and more efficient for distributed computation and local storage. File servers were more efficient and cheaper for central data storage and distribution. And if one needed really intensive computation, the new mini-supercomputer parallel-processing machines did even this faster and cheaper than a mainframe.

The mainframe then still did powerful jobs and represented enormous potential. Only when the cost/performance ratios were clearly distinctive to the customer was it then substituted by networks of computers. In the Merrill Lynch example of a primarily transactional distributed network, the cost advantage in using specialized file servers was clearly distinctive to Merrill Lynch.

Therefore, in addition to IBM struggling with trying to maintain market share in the mainframe computer product, the product line itself was becoming obsolete for many applications. And at the same time, the technology life cycle of the computer industry was entering a linear growth phase with excess competition and production capacity. Tom Watson, Jr., had lead IBM first into computers after World War II and through the next generation into transistorization of computers. However, executive leadership in IBM after Watson seemed no longer capable of managing the integration of radical and incremental innovation and commercially exploiting next generations of technology in a timely and aggressive manner.

The percentage of the world computer market changed for the different product lines during the 1980s. In 1983, mainframe sales declined below 40 percent of total computer sales, continuing to fall below 30 percent by 1988. Minicomputer sales held just above 30 percent until 1985, after which they began a rapid decline toward zero as an obsolete product line. Meanwhile, personal computers, workstations, and minisupercomputers were the growing product lines during the 1980s.

This trend was created by the technology discontinuities of parallelism and networks, and it had severe impacts on IBM's portion of the overall computer market. In 1975, IBM sold 37 percent of the total computer sales in the world market of $30 billion dollars, but in 1990, IBM sold 21 percent of the total world market of $305 billion. Even though IBM sales had grown as the total market grew, IBM's share of the market had declined.

Then, in 1990, IBM had a severe dilemma. It had 70 percent of the mainframe market, which was then rapidly declining toward 25 percent of the total computer market and down. IBM also had 24 percent of the minicomputer market, which has headed for 20 percent of the total computer market and extinction. In contrast, in the growth markets, IBM had only 18 percent of the PC market, 14 percent of the workstation market, 0 percent of the mini-supercomputer market, and 0 percent of the supercomputer market.

IBM was in the unenviable position of then being the dominant producer of the *obsolete* product lines and having only a small position in the *emerging* product lines. To compound the dilemma, the only product line in which IBM could obtain its traditional 20 percent return was the mainframe line. Management had structured its costs and expectations on that level of return and was then in trouble when structural changes and competition in the industry were no longer making possible such a high rate of return. To appreciate IBM's problems, mostly due to management decisions, one should understand that IBM's researchers even invented some of the new technology.

For example, a critical innovation in workstation technology in the early 1980s was the development of a reduced instruction set chip (RISC) microprocessor. The RISC speeded up the processing of the majority of uses of the microprocessor. "In the mid-1970s, IBM researchers pioneered this revolutionary technology for designing faster computers. But the advance wasn't rushed into products, largely because it was seen as a menace to Big Blue's core mainframe business. 'Everything they do is protecting that mainframe market,' says Alan Sims, chief engineer for Cigna Corp, a big IBM customer" (Carey, 1991, p. 118).

The RISC was the key to the new workstation computers of the 1980s, a market that IBM entered late and gained only a 14 percent share (which was ridiculous, since IBM had first invented the RISC technology): "RISC technology wasn't commercialized until scientists at Stanford University and the University of California at Berkeley, funded by the Pentagon's Defense Advanced Research Projects Agency (DARPA), in effect reinvented it. Then upstarts [new firms] like Sun Microsystems, Inc., harnessed the technology—and they now [in 1991] lead in RISC-based workstations" (Carey, 1991, p. 118).

By 1991, IBM no longer controlled key strategic technologies in the larger computer industry. For example, then the key strategic technologies for workstations and parallel-processing computers lay in the microprocessor computer component of which IBM was not a producer and in the system control software. In the late 1980s, IBM did try to gain dominance in the personal computer market with a proprietary system control software called PS2, with a notable lack of success.

Thus in 1991, IBM did not announce any bold new technology strategy. Instead, the CEO then moved to reorganize IBM into separate companies. The old IBM was gone. After the System/360 mainframe innovation, IBM management had long stopped using its technological strengths for competitive business advantages.

The Challenge in Technology Discontinuities

In historical hindsight, it is easy to pick out why and even when a new technology substituted for an old technology or even how over time incremental improvements in part of a technology system resulted in a wholly new technology system—technology discontinuities. Yet the problem in forecasting and planning technology is to anticipate the change. Why is this difficult for discontinuities?

The business context in which any technological innovation is to be implemented affects the probability of its implementation. Initially, a substituting new technology will likely perform a given function less well than an existing technology. Its potential for substitution lies in a natural advantage in the nature of its phenomenal base compared with the older phenomenal base. However, there is never any guarantee that the potentially large natural advantage will ever be realized, because any new technology requires much incremental improvement before its performance surpasses that of the previously existing technology. Moreover, while the incremental improvements to the new technology are being sought, competitors are spurred to seek incremental improvements in the existing technology. Thus the performance goal for which the new technology is aiming is a running target—a race between a new technology and an existing technology. In next-generation technologies, this race causes confusion for a business strategy.

William Abernathy and K. B. Clark (1985) coined the phrase *transilience of technology* to emphasize the pervasive impact of technological change on business operations. Literally, *transilience* means "to pass through," and technological innovation has far-reaching business impacts as technological change "passes through" the different aspects and levels of organizational functioning. In passing through, it can alter both the production and marketing capabilities of a firm.

For example, under production capabilities, changes in technology can alter product design, production systems, skills and knowledge base, materials used, and capital equipment and facilities. Under market capabilities, changes in technology can alter customer bases, customer applications, channels of distribution and service, and customer knowledge and modes of communication (Abernathy and Clark, 1985, p. 3).

Therefore, technological innovation can affect a firm by either preserving or destroying competencies in production and/or in marketing. Abernathy and Clark (1985, p. 8) classified technological innovations according to whether they preserved or destroyed such corporate competencies:

1. *Regular innovations* preserve production competencies and market competencies.

2. *Niche-creation innovations* preserve production competencies but disrupt market competencies.

3. *Revolutionary innovations* obsolete production competencies but preserve market competencies.

4. *Architectural innovations* obsolete production competencies and disrupt market competencies.

Here is the impact of technological discontinuities: Current corporate strategies can successfully deal with regular and niche-creation innovations because the technical skills base of the organization is not affected. (In the latter case, of course, marketing skills must be sharpened and new markets entered.) However, the revolutionary and architectural innovations require true *strategic reorientations.* Revolutionary innovations require reorientation of production competencies, and architectural innovations require reorientations of both production and market competencies.

> *Next-generation technologies are always either* revolutionary *or* architectural *innovations requiring restructuring of production and/or market competencies.*

The Chief Technology Officer (CTO) in a Diversified Firm

As we saw in the IBM illustration, successfully dealing with technology discontinuity requires intervention by a senior management official to *force* a corporation to reformulate technology, product, and business strategies. In the late 1980s, some firms created a new position, a chief technology officer (CTO), to try to deal with this kind of problem in a diversified corporation. The CTO job was to make sure that senior management pays appropriate attention to technology capability.

For example, William Lewis and Lawrence Linden (1990) argued that for a corporation to be a technology leader in its competitive positioning, both a corporate-level research capability was required and a senior manager representing a technology perspective at the corporate level was necessary. They argued the need for a person in top management who could bring in a technology perspective on strategic issues. That same person is also needed to promote the company's technological commitment throughout the firm.

Lewis and Linden noted several companies had instituted a new managerial position for this purpose. "During the past several years, a number of companies have recognized these basic needs and have responded to them by elevating their most senior R&D manager to the recently mined position of 'chief technology officer.'...It represents a significant effort to carve out a leadership role for the spokesperson and orchestrator of a company's technology strategy. The CTO's key tasks are not those of a lab director but, rather, of a technical businessperson deeply involved in shaping and implementing overall corporate strategy" (Lewis and Linden, 1990, p. 64).

What should senior management pay attention to in technology strategy? The CEO team should have an appreciation of the areas of rapidly changing technologies (without necessarily understanding the technical details of the changes), and it is important to understand the potential business impact of such changes. Accordingly, the chief technology officer (CTO) should ensure that senior management is acquainted with and considers the business implications of

- Areas of rapidly changing technology
- Areas of defensive technology
- Level of the firm's technological efforts
- Major competitors' technology strategies and levels of efforts
- Relevant industrial-level university/industry/government research consortia for next-generation technologies

Senior management should be informed when new basic innovations occur and should be given forecasts of potential impacts on markets and competition and estimates of time lines for developments in the new technologies. Senior management also should be informed about the implications of technology change for the firm's manufacturing and service capabilities and should make the appropriate capital investments in improving these capabilities. And senior management should attend to and make appropriate investments in new business opportunities that are arising from corporate research.

It is the role of the CTO to assist senior management in these strategic responsibilities. "For their part, CTOs share this view of the central part of their role. One notes, 'I'm not close enough to the bench to understand the details of our technology programs, so I cannot be a chief scientist. It's not what we need anyway. Rather, I focus on what the CEO needs. I provide input to all the major items that go to the CEO before he addresses them'" (Lewis and Linden, 1990, p. 65).

A chief technology officer facilitates senior management's attention to technology strategy as:

1. *Technological overview:* areas of rapidly changing technology, areas of defensive technology, levels of technological effort, major competitors' technology strategy, and major university/industry/government research consortia

2. *New basic innovations:* significant new inventions, market development, competitive situation, and opportunities and problems

3. *Next-generation technologies:* vision of NGTs, business implications of NGTs, and competitors' activities in NGTs

4. *Manufacturing strategy:* technological changes in manufacturing domains, engineering design system changes, business implications of manufacturing strategies, and competitors' manufacturing strategies

5. *Telecommunications strategy:* technological changes in telecommunications systems, impact on applications, business implications of technical changes, and competitors' telecommunications systems

6. *New business ventures:* basic innovations from new ventures, NGT strategic reorientations, new ventures from manufacturing strategies, and new ventures from telecommunications strategies

7. *Defensive R&D:* existing businesses, market shares and profitability and competitors' strategies; and defensive strategies, technology, market, manufacturing, and telecommunications.

Practical Implications

The lesson to be learned in this chapter is the importance of anticipating and preparing for the infrequent but very important large changes that may occur in a technology system—technology discontinuities.

1. Technology discontinuities are leaps in technical progress when a technology performance parameter advances in multiples over previous capability.

   a. Such leaps require different designs or production processes and hence are called *discontinuities in technical practice.*

   b. Pursuing the science bases for technologies provides a fertile area for improving a technology in large leaps.

   c. It is unlikely that the same productive unit can effectively manage both radical and incremental innovation processes.

   d. The concept of next-generation technologies can be used to reconcile the management of both short-term incremental innovation processes and long-term radical innovation processes in the same firm.

   e. Properly managed, the linear innovation process can work with the cyclic innovation process by feeding into it the technical feasibility of a radical innovation.

   f. The connection between the two innovation processes is the functional prototype of a next generation of a technology system.

   g. At a technology discontinuity, any firm must depend heavily on technical knowledge produced from without.

2. Firms that survived over the long term and through technology discontinuities usually have had executive teams who correctly anticipated structural changes and were prepared with an appropriate strategy, product, and organization and afterwards held a steady course.

a. Anticipating when a discontinuity will occur requires parceling the technology applications into distinct classes of applications and customers and evaluating, on a case-by-case basis, when the performance/cost ratio for a class of applications/customers will clearly justify technology substitution.

b. Next-generation technologies that substitute for current technologies require restructuring of production and/or market competencies.

c. The principal reasons for the difficulty of formulating a technology plan for a technology discontinuity are the technical uncertainties of a new technical vision and the differing perspectives among the different product-group managers and technical staff about that technical vision.

d. To formulate a technology plan for a technology discontinuity, a high-level executive must force and organize the necessary strategic effort.

3. The role of a chief technology officer can facilitate the appropriate involvement of senior management of a diversified firm in new technology strategy.

a. The CTO should facilitate senior management's attention to technology strategy as technological overview, new basic innovations, next-generation technologies, manufacturing strategy, telecommunications strategy, new business ventures, and defensive R&D.

For Further Reflection

1. Identify an industry with a long history (over 75 years). What was the original radical innovation that founded the industry?

2. Subsequently, how many technological discontinuities occurred in the industry?

3. How did major companies in the industry fare through the discontinuities?

4. Choose a company in the most recent discontinuity. What do you think should have been its technology strategy through the discontinuity? What, in fact, was its strategy, if any?

Research Strategy

Research is the principal means for both systematically advancing technologies for incremental innovation and for altering the underlying structures of technology for radical innovation. Research for technological progress is pursued in industrial, university, or government laboratories and is supported either by industry or government.

Research methods underpin research planning. What has been unique in the modern world in terms of technological progress has been the systematic use of scientific methods to perform research toward technological goals. This has been called *scientific technology.*

> Scientific technology *is the use of the methods of scientific research, focused by technical goals, for the purposes of technology progress.*

ILLUSTRATION
The Scientific Search for High-Temperature Superconducting Materials

The history of the search for higher-temperature superconducting materials provides an illustration of how scientific research can be used to improve technology. Figure 14.1 shows the technology S-curve for the temperature performance parameter of superconducting materials. As we noted in Chap. 11, the technology S-curve is an analogy, not a model. Therefore, as an analogy, it can not be used for prediction but only as suggestive for anticipation. However, underlying this analogy are research strategies for improving technological invention which we can illustrate in the case of superconductivity.

Superconductivity, as a natural phenomenon, was discovered in 1908 by Keike Kamerlingh Omnes when he was a professor of physics

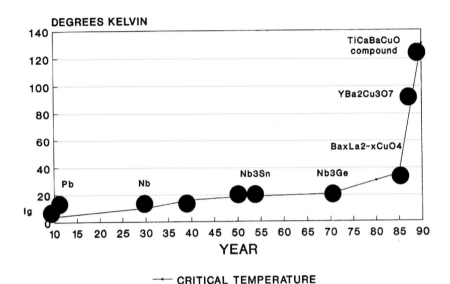

Figure 14.1 Superconducting critical temperatures of different materials.

at the University of Leyden in Holland (McKelvey, 1989). At that time, Omnes was pursuing a scientific research program aimed at (1) achieving new very low temperatures and (2) studying the behavior of materials at such temperatures. Omnes was not motivated by technological goals but by scientific goals for discovering and studying nature. Omnes succeeded in liquefying helium below 4 K.

The temperature of a material is equivalent to the average kinetic energy of the molecules of the material (in motions of vibration or translation). The temperature 0 K is projected to be the temperature at which the vibrations of molecules in a solid stop. Omnes achieved a temperature just 4 degrees above that zero point.

Omnes tried an experiment of conducting an electric current through solidified mercury at this low temperature. He discovered that below 4.2 K, the electrical resistivity of mercury completely disappeared. Figure 14.2 shows the graph on which Omnes plotted his results. The zero electrical resistance in some materials at very low temperatures was called *superconductivity*. Later, in 1913, Omnes received the Nobel Prize for this discovery.

Superconductivity was a surprise because in the passage of electrons through a conductor, some electrical–magnetic force interaction between the traveling electrons and atoms of the conductor usually occurs. These electron-atomic collisions cause the transfer of energy from the electrons to the atoms, slowing the electrons down and stimulating

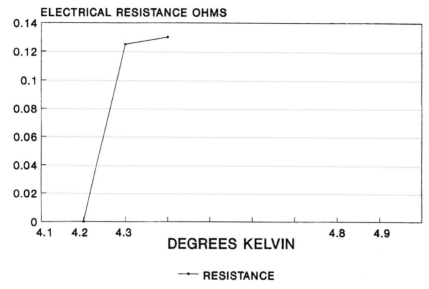

ELECTRICAL RESISTANCE OHMS

Figure 14.2 Electrical resistance of mercury at critical superconducting temperatures.

vibration of the atoms. The macroscopic property of these microscopic interactions is observed as electrical *resistance* in the material, and energy loss from the current is seen as *heating* of the material.

Now why did mercury lose resistivity below a critical low temperature? Why were electrons, even at low temperatures, not colliding with atoms in the solid lattice of the mercury? Omnes could not explain the new phenomenon, but he had discovered it.

However, even if a phenomenon is only discovered and not yet fully understood, technology may use it—as long as the process for manipulating nature has been established. Omnes had established the process by using liquefied helium to cool materials. However, he did not apply for a patent on superconducting materials (as we will see that later discoverers of new superconducting materials did).

The very low temperatures (below 20 K) required for the class of superconducting materials which Omnes discovered can only be attained by the liquid helium equipment, which is complex, cumbersome, and expensive. For economic applications of the property of superconductivity, scientists were aware that higher-temperature superconducting materials were desirable. Therefore, a scientific search for more superconducting materials was continued, first by Omnes and later by others.

World War I interrupted Omnes' research. After 1918, however, Omnes returned to superconductivity—which his research group continued to dominate until his retirement in 1924. Then other researchers entered the physics specialty of superconductivity.

As a result of their experiments, many metals (in addition to mercury) were discovered to display the phenomenon of superconductivity—zinc, cadmium, thallium, indium, gallium, and aluminum. All their transition temperatures, however, were also very low, from 0.85 K for zinc to 3.4 K for indium. In addition, the scientists had discovered that many other metals never display the superconducting phenomenon (such as sodium, potassium, copper, silver, and gold).

Then, around 1929, Walther Meissner of the University of Berlin discovered a new group of superconducting metals with significantly higher transition temperatures—with niobium displaying the highest transition temperature of 9.2 K. During the 1950s through 1970s, many new superconducting materials were discovered that were intermetallic compounds with niobium (such as Nb_2Sn, Nb_3Ga, Nb_3Ge). As a group, these had raised the critical temperature to a maximum of 23.2 K for Nb_3Ge in 1973. (Figure 14.1 shows the historical technology S-curve for that long search from 1910 to 1990.)

Empirical Research Strategies for Technology Push

The preceding illustration shows a general pattern of research strategy which scientific researchers have often used to explore nature, and it consists of a systematic empirical search. To describe this strategy, we must first define more precisely some commonly used terms in scientific method.

A commonly used term in research methodology is that of an *object* as a focus of the research study.

A research object *is a generalized class of existing and individuated things (or processes or forces) grouped according to a shared abstraction.*

A star is an example of a thing as a research object. The nuclear transformations inside the star are examples of phenomenal processes. The strong nuclear and electromagnetic forces involved in stellar dynamics are examples of phenomenal forces.

The empirical aspect of research methodology is to observe nature in a way that uncovers an underlying form of existence—the abstraction.

A classification of a research object next groups together all the observed things (or processes or forces) into memberships in the class, sharing the same underlying form of abstraction. A research object is thus a way of individuating existing things according to underlying shared forms (abstracted and generalized). A research object thus partitions nature, for the researcher, into finite and universal forms.

Objects summarize the abstraction of an underlying essence of the appearance of a general class of existing things.

For example, the air we breathe, although usually invisible in appearance, can be directly experienced as breath or wind. The *object* of these direct experiences is *generalized and abstracted as a gas* (containing freely moving molecules of nitrogen, oxygen, carbon dioxide, and others).

Another important term often used in discussions of research methodology is the term *phenomenon* to indicate the defining properties of objects.

A phenomenon *is a natural property (or process or force) observable in nature defining of an object, and it is regarded as a fundamental form of existence.*

In the example of air, the gas mixture has the phenomenal properties of volume, temperature, and pressure.

The discovery of a new fundamental property (or process or force) has historically often been the basis for scientific recognition and for Nobel Prizes (such as Omnes received).

Phenomenal properties, processes, or forces are complex forms. Researchers often use the term *attributes* for describing these forms.

A physical attribute *is a defining characteristic of a phenomenon.*

In the example of the superconducting phenomenon, its physical attributes consist of: (1) the absence of electrical resistance, (2) the absence of magnetic fields inside the superconducting material, and (3) a transition temperature above which the phenomenon disappears.

Now we can formally define the term *technological parameter,* which we have been using throughout this book.

A technological parameter *is a* functionally relevant *physical attribute of the phenomenon that is required for a generic class of applications.*

We see that a technology parameter selects the physical parameter of an object that is most relevant to the functional transformations in the technology.

With these definitions, we can now describe a general empirical search strategy for technological invention—empiricism in scientific technology:

1. A technologically interesting object is an abstracted class of existing things each of which displays the natural phenomenon for a technological application.

2. The technically "best" object of that class will be the exemplar which displays the highest value of a technology parameter. (For superconducting metals as an *object,* the example of niobium was technically best because niobium had the highest transition temperature.)

3. The first step in an empirical search strategy for technology invention is to explore all members of an object class for the values of the technology parameters of interest.

4. When all members have been researched and their technically relevant physical attributes have been established, then a technically best *examplar* can be determined for a technological application.

Empirical search need not stop here, however, for a second step may be possible. This step involves a deliberate construction of a subclass of objects from the technically best exemplar.

ILLUSTRATION (*Continued*)
Empirical Search for High-Temperature Superconductors

In the case of superconductivity, compounds of niobium were deliberately constructed to produce metallic compounds with a higher transition temperature. Look again at the technology S-curve in Fig. 14.1, observing that the progress from 1950 through 1970 occurred in compounds of niobium (Nb_3Sn in 1953, Nb_3Ge in 1970). We can now explain this progress using the concepts just described.

An example (mercury) of a class of objects (metals) was discovered to display a natural phenomenon (superconductivity). Scientific methodology then suggested to researchers to explore other examples (zinc, cadmium, thallium, etc.) of the same object class (metals). Up to then, the research strategy in superconductivity had a dual motivation, scientific and technological. The scientific motivation was to establish how many objects of the class display the property. The technological motivation, in contrast, centered on finding an example (niobium) in the object class (superconducting metals) that displayed the highest technical performance parameter (critical temperature) required for envisioned technological applications.

Thus one sees that one can sometimes perform the same scientific research from two different motivations—the motivation of the scientist for discovery and explanation and the motivation of the technologist to improve technology. Industrial researchers can be interested in the same research strategy as university researchers, but for different reasons.

Empirical research search can be continued until an object class is exhausted. For superconductivity research, then a technologically best-of-class object (niobium) was established. However, since the technology performance parameter was still not sufficient for applications, systematic technological research was stymied until a new class was discovered. One way to try to find a new class is by constructing a subclass (niobium compounds) from the technically best example (niobium). From 1930 through 1973, the technology-push research concentrated on intermetallic compounds of niobium. These compounds constituted a "constructed" object subclass (niobium intermetallic compounds) of the phenomenon—constructed from the technically best object (niobium) of the object class (superconducting metals).

Technological Constructions of Phenomena

As part of an empirical search strategy in scientific technology, a researcher may try to manipulate, change, alter, or construct variants on the technical best example of the object in order to seek a new examplar with an even higher value of the technically desirable attribute.

> A technological construction *is the deliberate manipulation of a technically best examplar of an object to create a technically best subclass with improved technological parameters.*

ILLUSTRATION (*Continued*)
Superconductivity Multiparameters

By 1965, the new class of superconducting materials based on compounds of niobium had been established, and the compounds were beginning to be used in the first commercial application of superconductivity—as superconducting magnets. For this application, the niobium-compound metals had to be drawn into wire and wound as a solenoid. Large superconducting solenoids having strong magnetic fields then found an application in nuclear magnetic resonance (NMR) instruments for both scientific research and medical diagnostics.

In ordinary electromagnets, an electric current traveling in a coil of wire around a solenoid creates a magnetic field through the solenoid. However, as the current passes through the solenoid, substantial and continuing energy loses occur through electrical resistance heating of the solenoid wire. In a superconducting electromagnet, once the current is turned on, very little energy loss occurs in the electromagnet as long as the temperature is maintained below the superconducting transition temperature.

There is an energy loss in a superconducting electromagnet, but it is in the cooling system for liquid helium temperatures and not in the electrical resistance. At high magnetic fields, the energy loss from keeping the system cold is much, much less than the energy losses would be in the electrical resistance of ordinary electromagnets. Superconducting electromagnets have allowed much higher magnetic fields to be obtained than ordinary electromagnets because of this great difference of energy loss.

In applications, such as NMR imaging, these very high magnetic fields are necessary for instrumental performance. This commercial application of superconductivity meant not only that the transition temperature was an important technical parameter but also that the performance of the superconducting material in high-intensity magnetic fields was an important technical parameter.

Thus several more attributes of the niobium compounds (in addition to the critical temperature) became technically important:

- The ductility of the material (for drawing into wires)

- The strength of the material (for resisting the mechanical forces on coils caused by the magnetic fields)

- The ability of the material to maintain zero resistivity in a large magnetic field

The performance region on a two-parameter (magnetic field and critical temperature) graph for the metallic superconductors niobium-titanium and niobium-tin will show that niobium-tin has a region of higher performance for superconductivity at higher fields (up to 30 T) and at higher temperatures (up to 19 K). The superconducting performance region for niobium-titanium has its highest field up to 20 T and highest temperature up to 10 K. This means that in practice, niobium-tin might be chosen to be used at around 25 T and 10 K.

Multiparameter Technology
Performance Measures

A single performance parameter may summarize the general direction of desired technical progress for a generic set of applications. However, as the application set gets more specific, usually more than one performance parameter is required. As noted previously, technical performance parameters for an application may consist of several parameters—multiparameter performance graphs. In the case of superconductivity, the first general commercial application was in superconducting magnets. This application introduced a second performance parameter, the capability of enduring a magnetic field while still superconducting.

In new technologies, functionality is established first, and then an application is made. When the functionality is first established, a critical performance parameter is evident. In the example of superconductivity, the transition temperature was the critical performance parameter.

When the first application begins, however, it usually turns out that more parameters enter the performance picture. Thereafter, desirable technical performance is indicated as regions in a multiparameter graph. This technique is useful for understanding the interaction of many performance parameters, displaying a multidimensional performance region.

A technical parameter performance graph must display one dimension for each performance parameter, and the regions of performance necessary for a commercial application can then be identified on the graph.

For an application, the desirable regions on a multiparameter performance graph can be used to define technical goals for research strategy.

ILLUSTRATION (*Continued*)
The Theory of Superconductivity
in Metals

As late as 1954, the search for improved niobium compounds had been guided purely by empiricism. First, there was a systematic exploration of all metals for the metal with the highest transition temperature (niobium). Second, there was a systematic exploration of intermetallic compounds of niobium (Nb_3Sn). The reason that the search was purely empirical was that no scientific theory had yet modeled (explained) the superconducting phenomenon.

Then, in 1957, scientists John Bardeen, Leon Cooper, and J. Robert Schrieffer published a theory that explained superconductivity as a quantum mechanical effect of coupled electrons. Later, in 1972, they received the Nobel Prize in physics for explaining superconductivity in metals.

The discovery of a new basic phenomenon of nature has always been worth a Nobel Prize, and explanation of the basic phenomenon has also always been worth a Nobel Prize. You can see in this illustration why in a preceding chapter we defined *science* as the discovery and understanding (or explanation) of nature.

Briefly, let us recall the physical model of a metallic material. Metal atoms in solid form crystallize in regular three-dimensional lattice structures, with even spacing between the atoms. The bivalent bonding of the atoms into a lattice structure occurs by the sharing of outer orbit electrons that circle not around a single atom but around pairs of atoms. This electron sharing produces the molecular bonding of the crystal lattice of the metallic material.

Yet in some metals, the outer orbits have excess electrons that are weakly bonded and therefore are yielded in molecular interactions to nonmetals (valent bonding). Thus, when metal atoms crystallize into a lattice structure, some of these outer electrons in the last shell are shared between them (for the bivalent bonding) and some are free to wander around within the lattice. And (in an analogy to gas molecules wandering freely in space) the wandering electrons are often called an "electron gas" within the space of the lattice structure.

Now in this atom lattice model of metallic crystals, electrical conductivity can be explained as the movement of the "free" electrons through the lattice structure under the motive force of an externally applied electric field. Altogether, these free electrons of the electron gas form an electric current that passes through the metallic conductor.

Then electrical resistance can be explained as the collisions between the free electrons (as they are drawn along by the external force) and the shell-bound electrons of the atoms fixed in the lattice structure. These collisions lose some of the energy that was gained by the free electrons from the external force to the atoms.

As a result of the energy transferred in the collisions to the atoms, these atoms vibrate with the gained energy. At the macroscopic level of the material, these atomic vibrations appear as heat. Thus do electronic collisions of the current create heat as electrical resistivity. The absence of electrical resistance in superconductivity means that the conduction electrons are no longer colliding with the bound outer-shell electrons of the atomic lattice.

How this comes about is exactly the situation that the Bardeen-Cooper-Schrieffer (BCS) theory describes. The BCS theory proposed that a spin-spin coupling of paired conducting electrons enabled these conduction electrons to avoid collisions with the atomic lattice because of the exclusion principle of electron quantum states. The Pauli exclusion principle of electron quantum states is a fundamental law of physics which states that two electrons (or two fermion particles) cannot populate the same quantum state. Dirac proposed an intrinsic spin variable for electrons of the value of 1/2. What this means is that an electron at rest not only displays the property of an electric charge but also displays the property of a magnetic field (such that the poles of the field between any two adjacent electrons can be oriented opposite to each other).

Pauli first proposed this postulate from the empirical observation that electrons around an atom do not all go to the lowest energy state. In nature, all electrons orbiting around the nucleus of an atom form a shell structures of larger and larger orbits around the nucleus of the atom—with only two electrons in any quantum orbit. Thus the exclusion principle simply states a generalization of the empirical observation about atoms that electrons do not all fall into the closest and lowest energy orbit around the nucleus.

Now this is how the BCS theory proposed to explain superconductivity:

> *Two electrons traveling as a quantized wave packet with opposite spin states completely populate all the allowed quantum states of the conducting electrons below the superconductivity critical temperature.*

What this means is that below a critical low temperature in some metals, the freely moving conduction electrons in the metal can pair off as spin-coupled pairs and pass as these couples through the atomic lattice of atoms in the material without any collisions with the bound electrons around the atoms. The Pauli exclusion principle does not allow another electron orbiting an atom in the lattice to interact with spin-coupled pairs of conducting electrons. To do so would require other electrons to participate in the same quantum state as the spin-paired electrons, and this the Pauli exclusion principle does not allow.

In addition to the quantum-mechanical spin coupling of paired superconducting electronic currents, the Bardeen-Cooper-Schrieffer (BCS) theory also explained how two electrons could be forced together close enough (despite their mutually repulsive negative electric charges) to create the spin-coupled state between paired electrons. Their explanation was that superconductivity occurs when an atomic lattice is sufficiently stilled at very low temperatures such that an electron moving between atoms in a lattice can attract the positively ionized atoms of the lattice off their normal sites sufficient to create a region of excess positive charge in that area which in turn can attract a second electron into the region close enough to the first to create the spin coupling between these two electrons.

Use of Scientific Theory in Invention Search Strategy

In the preceding illustration, the BCS theory depicts superconductivity as a result of the quantum physics of electrons able to pair in some atomic lattice materials at low temperatures. This theory thus provided an understanding (an explanation) of what situations are required for the superconducting phenomenon to occur. By understanding the situation, later researchers could then seek to deliberately construct new situations and thereby discover new superconducting materials.

Theory provides an understanding of the structural aspects of a phenomenon. A scientific theory provides a quantitative model of a phenomenon that mathematically both describes the essential properties of the phenomenon and explains the derivative variable properties as the effect of causative primary variables.

> *Scientific theory models a phenomenon as the macroscopic property of an underlying microscopic mechanistic process.*

Recall from Chap. 5 that this was called a *reductionist methodology.* Mechanistic/structural insights from scientific theory are important to guide the construction of natural objects for optimizing the desired technical properties of the phenomenon. Such insights are also important to guide the search for new classes of constructed objects that create the conditions for the phenomenon.

> *The importance of a scientific theory (model) for technological invention is that it can provide guidance for technologically motivated research strategies.*

Researchers can pursue inquiry that may lead to technological invention, guided by systematic empirical strategies cleverly focused by theoretical strategies.

ILLUSTRATION (*Concluded*)
Theory Used in Searching for a New Class of Superconductors

Let us next illustrate how the structural insights of the BCS theory of superconductivity in metals were used to provide "hunches" for creating new classes of superconducting materials. In January of 1986, a major breakthrough in superconducting research occurred. Karl Müller and Johannes Bednorz (scientists at the IBM Research Laboratory in Zurich, Switzerland) discovered that a barium-lanthanum-copper oxide crystal ($Ba_xLa_xCuO_4$) displayed superconductivity with a transition temperature at the much higher temperature of 35 K. In 1987, Müller and Bednorz received a Nobel Prize.

Müller and Bednorz had decided to try a radical approach to achieving a new material class exhibiting superconductivity. They knew that metals and metallic compounds had not provided a path for breaking the low-temperature barrier for over 70 years. They also knew from the BCS theory of superconductivity that an interaction between the conducting electrons and the lattice structure was necessary for forcing electrons to spin couple.

Thus, they thought, why not try some radically new kind of lattice structure that could not be found in metals? In this way, empirical results and theory were brought together to motivate a different kind of search strategy for invention.

Müller and Bednorz decided to try exploring copper-oxide compounds. One of the reasons for this search decision was that they knew that the compound of lanthanum–copper oxide (La_2CuO_4) had a particularly interesting property (Cava, 1990). Lanthanum–copper oxide is an example of a crystal structure in which the smallest geometric figure is a metal atom surrounded by oxygen atoms in a polyhedron configuration. Nine oxygen atoms surround each lanthanum atom as a *coordination polyhedron*. What this means is that two electrons in the outer shell of the lanthanum atom are shared with the surrounding nine oxygen atoms. In lanthanum–copper oxide, the oxygen atoms surround the lanthanum atom in this pattern. Also, the copper atom is surrounded with oxygen atoms in a coordination polyhedron, with six oxygen atoms around every copper atom. However, the lanthanum–copper oxide compound is not a superconductor.

Yet Müller and Bednorz were intrigued by a peculiar property of this crystalline structure. They knew that in this compound the coordination polyhedron of copper was not a perfectly symmetrical octahedron. This is due to the influence of the nearby lanthanum coordination polyhedron. Instead, with the lanthanum atom nearby, the copper polyhedron is distorted, with the two oxygen atoms at opposite corners of the octahedron farther from the copper than the oxygen atoms at the four other corners.

Now the significance of this distortion to Müller and Bednorz was that it suggested that the electrons of the coordination polyhedron interact strongly with the positions of the copper and oxygen atoms in the crystal lattice. This kind of interaction between electrons and the lattice was the key to creating the superconductivity phenomenon in metals (according to theory). Müller and Bednorz's clever idea then was: Why not try modifying the compound so that it became a superconductor? This is another example of trying to manipulate nature to obtain a desired phenomenon.

Again from theory, Müller and Bednorz knew the reason that the lanthanum–copper oxide compound did not conduct electricity. Each copper donates two electrons to the neighboring oxygen atoms and retains nine electrons in its outer shell. Each electron has a magnetic moment due to its spin. Thus, of the nine electrons left to the outer shell, eight will pair up in spin couples, but the odd ninth is left out. This results in an overall magnetic moment for the copper atom in the coordination lattice, which is called *antiferromagnetism*. This antiferromagnetism acts to pin electrons in the crystal lattice, therefore eliminating the property of conductivity (and, of course, any possible superconductivity). For this reason, Müller and Bednorz knew that they had to modify the lanthanum–copper oxide to get rid of the antiferromagnetism and allow conductivity.

Understanding the structure of physical mechanisms facilitates imagination about what to change in that structure in order to obtain properties useful to technology.

One way Müller and Bednorz tried was to add barium atoms to the compound. And for every barium atom added to the formula, they would remove one lanthanum atom. The addition of barium atoms in coordination structures near the copper oxide coordination structures influenced the copper to give up an additional electron to the oxygen atoms, from two to three electrons (or, as it is said, increased the copper valence from +2 to +3). Then the copper atoms were no longer antiferromagnetic. And then the new compound of lanthanum-barium-copper oxide could conduct electricity.

Also, as luck had it, the new compound superconducted. And it superconducted with a transition temperature at 35 K. They had done it! They broke the 20 degree barrier.

By creating a new class of compounds, oxide ceramics, Müller and Bednorz opened up a whole new path for constructing superconducting materials that broke the old low-temperature barrier.

In retrospect, the story sounds straightforward, but in practice, it was a combination of hunches, trial, and error, helped with theoretical guidance. Moreover, since the research was such a long shot, Müller and Bednorz actually worked on this important problem only on a part-time basis for about 4 years before hitting "paydirt."

Empirical Search and Incomplete Theory

An important point of this illustration is how theory, even incomplete theory, can provide help in empirical research strategies for technological progress. While the theory of superconductivity in metals did not apply to ceramics, the structural insights of the theory did apply—the insight that the arrangement of the atomic lattices can affect the squeezing of conduction electrons into spin-coupling pairs. This structural insight suggested to Müller and Bednorz that they should try to create alternate structures to squeeze electron pairs.

Theory facilitates technological invention both in modeling *phenomena for prediction and in providing* structural insights *for empirical search.*

And once a new empirical breakthrough appears, further empirical research is possible even with incomplete theory.

Scientific Method and Technology

Science is an approach to nature that is characterized by method, often loosely called the *scientific method.*

Scientific method rests on the two foundations of experiment and theory.

Experiment involves a direct observation of nature in a manner that can result in abstraction of the properties of nature. Experimentation has often been called *controlled observation.* By the term *control* it is meant that the observations are measurable, repeatable, and reproducible within an expressible error.

Scientific experimentation requires measurement in order to provide relatively fine discrimination. It is this fineness of discrimination that allows one to approach closer and closer to the heart of nature. For example, measures of certain natural constants, such as the speed of light, provide us with the opportunity to test whether different experimental means can provide the same observation of nature and which means more accurately provide the observation.

Repeatability of experimental measurements requires the expression of the conditions of the observation such that within the same conditions the measurements can be taken again and again when the same experiments are redone.

Reproducing experiments requires the expression of the techniques as well as the conditions of the observation so that another observer, a different scientist, can obtain the same results of the experiment. This capability of independent confirmation of scientific experiments is the bedrock upon which rests the claim of *objectivity* in science.

By objectivity, *science claims the results of science to be characteristic of the observed object and not of the observers, the scientists.*

However, any repetition and reproduction of an experiment will result in measurements that are a little off from one another (because of variations in conditions and instruments), and the concept of error requires one to express both the measurements and the confidence limits within which the experimenter claims the measurements to be accurate—the errors of the measurement. Thus experimental observations are repeatable and reproducible within experimental error.

Experimentation involves the concepts of the measurement of a property, the objectivity of the measurement, and the errors of experimentation.

In the measurement, a conceptual abstraction is performed about what is a property of the phenomenon. An expression of the conditions of the abstraction and an estimate of the error of the measurement provide the scientist and other scientists with criteria for testing the replicability and reproducibility (and therefore objectivity) of the observation.

The abstracted property of the experimental observation must be generalized as a general property that all objects of the object class share to varying degrees.

Experimentation provides abstraction, while theory provides generality.

A generalized property is said to be a fundamental variable of the phenomenon.

The first level of theory is a listing of all the fundamental variables of all the phenomena included in a field of observation, and these variables constitute the basic or primitive terms of the theory. The second level of theory is established when two variables of the phenomenon can be correlated and expressed as a generalized law.

Fundamental laws are of the form expressing a dependant variable as a function of independent variables. Fundamental laws occur as laws of first principles independent of context or as phenomenological laws dependent on context. An example of a first-principle law in Newtonian mechanics is the law that a force F applied to an object will result in an acceleration a inversely proportional to the mass m of the object: $F = ma$. An example of a phenomenological law in Newtonian mechanics is the first-order law of springs in which the force F exerted by a distended spring is proportional k to the distance of the distension x: $F = kx$. What is phenomenological about this law is that the proportional constant k must be measured for each class of springs.

Fundamental laws between two variables of a phenomenon apply in whatever condition the phenomenon occurs, but phenomenological laws must be remeasured for different conditions in which the phenomenon occurs.

A scientific theory consists of all the fundamental variables and all the relevant laws correlating variables for the range of phenomena within the observational field.

The value of a scientific theory is that it allows the predictive modeling of an example of a phenomenon. And this is important to technology—the predictive ability of scientific models.

To be able to predict the future physical states of a phenomenon from knowledge of the present or past states of a phenomenon assists in the design of functionally oriented morphologies for technology.

Moreover, the more accurate and complete the scientific models of phenomena which underlie technologies, the more useful are the scientific models for purposes of designing and improving technologies.

Theory plays three important roles in technological invention:

1. To provide taxonomies of natural objects in which empirical search made be made for best of class that displays the highest value of a technological parameter.

2. To provide structural insights for constructing new classes of objects which display the desired phenomenon.

3. To determine when further search is likely to be unfruitful.

Recall that in the technology S-curve, the most important feature was establishing an estimate on the natural limits to performance in the technology.

Scientific phenomena are precursors to technology. All technologies are based on exploiting the possibilities of guiding the selection of the courses of natural events and states of a phenomenon. The morphology of a technology is a deliberate arranging and controlling of the sequencing of states of a phenomenon.

The discovery of a scientific phenomenon is a precondition to the utilization of that phenomenon for a technology—a precursor to technology forecasting.

In addition to discovery, the understanding of the phenomenon (expressed as a scientific model of the phenomenon) is also important to knowing why and how well technological invention can use the phenomenon. Thus scientific modeling also can be a precursor to opening up technical bottlenecks.

Radically new inventions usually occur on the basis of new phenomenal discoveries in science. Yet the particular object in nature in which a phenomenon is discovered has historically frequently not been the object which displays the highest value of the useful parameter of the phenomenon.

The second step in technical progress is usually a further scientific search for the same phenomenon in a larger class of objects. This is an empirical search for the *best of the object class,* which means the natural object having the largest parameter value in the family of objects that all display the phenomenon. In the illustration, this was the metallic object niobium.

After the initial discovery of a new phenomenon, an important research strategy is (1) the continuation of the search for a best-of-class material and (2) the construction of a new subclass of objects from a naturally occurring best-of-class object.

Further constructions within this new subclass may result in a new best of class of a constructed object with a higher-value technological parameter. In the illustration, this was niobium-titanium.

At this point, technologically focused scientific research may slow or even stagnate. However, engineering research may then increase in order to form a useful device or process from the best-of-class object. In the illustration, research was done in order to make niobium-titanium into a strong, ductile wire for winding into coils for a superconducting magnet.

A scientific theory may at some point be created to explain and model the phenomenon. The value of the theory for technology is that the phenomenon is then understood as some structural or mechanistic aspects of the physics or chemistry of the object.

Empiricism thus becomes enlightened by a theoretical insight about structure. A renewed empirical search may thus be begun outside the existing set of object classes and constructed object classes for new kinds of structures that may be manipulated into displaying the desired phenomenon. In the illustration, this was the empirical search extended from metals into ceramics.

Empirical search may be renewed for a class of objects displaying a phenomenon when it is "structurally informed" by theory and may result in a new technological breakthrough. In the illustration, the teams of Müller and Bednorz and Chu and Wu both constructed variations of copper oxide ceramics to create superconductivity, and both teams used the structural insights from the BCS theory of superconductivity in metals.

For scientific technology, even incomplete scientific theory is better than no scientific theory in facilitating empirical search for technological innovation. For example, in the new class of ceramic superconducting materials, research continued to find the best of class and to begin to engineer materials of proper shape and composition for applications. Finally, new theory was being worked on to explain the superconducting phenomenon in ceramics and to provide a guide to the natural limitations for the technology.

Empirical search, experimentation, and theory operate together to provide search strategies for the discovery and construction of technologically important objects. This is the underlying importance of science and engineering research in technological progress.

ILLUSTRATION
Semiconductor Research Corporation

The research-sponsoring activities of the Semiconductor Research Corporation (SRC) in the 1980s provides an illustration of organized research strategy at an industrial level. SRC was formed in 1982 in the United States for industrial cooperation in semiconductor research. Its research activity focused on semiconductor technology, which was then useful to a broad range of products and over a broad range of applications. There were two industries then that were particularly concerned with maintaining the U.S. competitiveness in semiconductor integrated circuits, the chip industry and the computer industry. "The goal of the SRC is to plan and undertake a research program that would broaden the generic technology base of the mainstream semiconductor industry and, in the process, increase the relevantly trained human resources of the industry" (Cavin et al., 1989, p. 1327).

The participating chip and computer firms shared an attitude that the advancement of generic technical knowledge was essential for all. "The major portion of SRC's operating budget is provided by membership dues from U.S. companies that are either suppliers or users of integrated circuits, or vendors of material or equipment to the industry. Agencies of the U.S. government have also contributed support to the SRC since 1986..." (Cavin et al., 1989, p. 1327).

This illustration is focused on a research strategy at a national level. The results of its sponsored research projects were published in the scientific and technical literature. The SRC sponsorship also facilitated the training of new generations of engineers skilled in the technology. SRC, however, was not merely a philanthropic act on the part of industry, for its activities were strategically focused and goal-oriented. Each year SRC conducts many technical meetings and research reviews wherein many industry and government technical personnel participate in developing joint assessments of university research results, needs, and directions. From this, SRC committees define specific technical goals to guide and focus the research program. The company forecasts technical progress over 10 years and plans to hasten the pace of progress by 2 years.

The company's activities were organized into three science areas: microstructure science, manufacturing science, and design science. The company used the term *science* to indicate the strategic underpinnings of semiconductor chip technology.

> *With relationship to technology, relevant and focused science is strategic.*

In the microstructure science area, research was focused on the structures of devices and interconnections at the microphysical scale. Research was intended to create new and improved processes and advanced devices, materials, and equipment. Research included the mathematical modeling of existing and new processes and devices that would be useful to their design.

In the design science area, research was intended to provide new design principles, new computer tools, and new methods and software to facilitate the design of integrated circuits.

In the manufacturing science area, research focused on improving the yield and reliability of manufactured chips. Research was intended to improve computer-integrated manufacturing, manufacturing processes, and the packaging of devices.

Research Strategy and Technology Systems

In the preceding illustration we see a defined technology system, semiconductor chips, as a focus for research strategy. We see also that the areas of research include the structure (or morphology) of the major devices in the system (chips) and the production processes required to produce the devices.

> *In any research strategy for technology progress, technical advance should be planned both for the devices of the technology and for the means of producing the devices.*

Devices require research on the structures of the devices, the materials of the devices, the physical understanding and modeling of the phenomena underlying the devices, and tools and software to improve the design of devices. Processes require research on inventing unit processes in the production transformations, physical understanding and modeling of the transformations in the unit processes, sensing and control of the transformations in the unit processes, tools and software to improve design of unit processes, and integration and control of unit processes into a production system.

Technology Maps

Historically, at any point in time, some technologies are changing rapidly, some not changing rapidly, and some new technologies are being introduced. A survey of the areas of rapidly changing technologies and brand new technologies can provide a list of areas to watch. The sources of research on these topics are industrial laboratories, university researchers, governmental laboratories, and government research-sponsoring agencies.

For clarity, one can classify these new and rapidly changing technologies in systems nomenclature, and I called such a classification a *technology map* (Betz, 1987). Also, one can classify these areas of changing technology both by technology push and by market pull.

> *Research strategy should be informed by technology plans that express not only general technology performance goals but also specific performance goals for specific commercial applications of the technology.*

Practical Implications

The lesson to be learned in this chapter is an understanding of how the methods and tools of science can be used in research for technological progress—scientific technology.

1. Research strategy and methods underlay research planning.
    a. Research methodology consists of approaches to observe nature in a way that uncovers an underlying form of existence.
    b. A research *object* is a generalized class of existing and individuated things (or processes or forces) grouped according to a shared abstraction.

    c. A *phenomenon* is a natural property (or process or force) observable in nature defining of an object and regarded as a fundamental form of existence.

    d. A *physical attribute* is a defining characteristic of a phenomenon.

2. Scientific method rests on two foundations, experiment and theory, with experimentation providing abstraction and theory generality.

    a. *Experiment* involves a direct observation of nature in a manner that can result in the abstraction of properties of nature.

    b. *Theory* is a listing of all the fundamental variables of all the phenomena and laws of relationships between variables.

    c. The *predictive ability* of scientific models is important to technological design.

3. A *technology parameter* is a functionally relevant physical attribute of a phenomenon that is required for a generic class of applications.

    a. A technical parameter performance graph must display one dimension for each performance parameter, and the regions of performance necessary for a commercial application can then be identified on the graph.

    b. For an application, the desirable regions on a multiparameter performance graph can be used to define technical goals for research strategy.

    c. The particular object in nature in which a phenomenon is discovered is not likely to be the object which displays the highest value of the useful technical parameter of the phenomenon.

4. In research strategy for technology, technical advance should be planned both for the devices of the technology and for the production processes.

    a. Devices require research on the structures of the devices, the materials of the devices, the physical understanding and modeling of the phenomena underlying the devices, and tools and software to improve the design of the devices.

    b. Processes require research on inventing unit processes in the production transformations, physical understanding and modeling of the transformations in the unit processes, sensing and control of the transformations in the unit processes, tools and software to improve the design of unit processes, and integration and control of unit processes into a production system.

For Further Reflection

1. Identify a fundamental scientific discovery that provided the phenomenal basis for a radically new technology.

2. Describe the empirical search strategies that improved the performance of the phenomenon for the technology.

3. Identify and describe when and how theory was made for the phenomenon—before or after the technological applications?

15

Planning Research

Earlier we saw that technology planning procedures provide the business context for planning research programs in industry. Now we look into how research projects themselves are planned to attain goals envisioned by technology planning.

Research planning logically accompanies technology planning. It is the step between technology planning and technology implementation. Recall that research should be planned for the technologies that support the current or new businesses of the firm, with the goals of

1. Improving current products (or production processes or services)

2. Creating next generations of technology for new models of products (or production processes or services)

3. Creating radical innovations for new businesses

Research planning occurs at different organizational levels—at the divisional laboratories of strategic business units, at the corporate research laboratories, and at the national research supporting and performing institutions of a nation.

ILLUSTRATION
The Tiny Machines of Microdynamics

As an illustration of research planning at a national level for a radical new technology, let us review the planning for research in the radically new microdynamic machine technology in the mid-1980s. Then the techniques for fabricating electronic semiconductor integrated circuit (IC) chips were being applied to fabricate very, very tiny mechanical devices. In 1988, the National Science Foundation (NSF) of the U.S. government funded workshops to plan a research agenda for the newly emerging technology of microelectromechanical devices.

Federal research-sponsoring agencies frequently assemble expert re-searchers in workshops to formulate a research agenda for a field. The NSF workshop occurred in three meetings: July 1987 in Salt Lake City, Utah; November 1987 in Hyannis Port, Massachusetts; and January 1988 in Princeton, New Jersey. Three program officers then at NSF funded the study: George Hazelrigg, Frank Huband, and Howard Moraff. The workshop was organized and the resulting report edited by three Bell Laboratories researchers: Kaigham Gabriel, John Jarvis, and William Trimmer. Attending the workshop and contributing to the report were 15 other researchers (10 from universities and 5 from AT&T Bell Labs, IBM, and Novasensor, Inc). The summary of the re-port began: "The miniaturization of electronics has produced a far-reaching technological revolution. Now mechanics is poised on the brink of a similar miniaturization, and its own revolution. Researchers are working toward creating microdynamical systems, the microscale derivatives of conventional large-scale electromechanical systems" (Gabriel, Jarvis, and Trimmer, 1988).

Some of the first experimental prototypes of very small machinery were shown at this workshop. One was a simple mechanical mecha-nism of a crank which, as it rotated, pushed and pulled a slot. The size of the slot was **103 millionths of a meter**. A second was a gear-slide combination, and a third was a three-gear train with different gear ra-tios. All these devices were so small that they could be seen *only under a microscope*.

The origin of the research ideas to create such tiny devices came from (1) IC chip–producing techniques in electronics and (2) research to cre-ate integrated sensors. Previously, the sensing device and the elec-tronics (to amplify and process the signals) were fabricated separately. Around 1980, several researchers began experimenting with the con-struction of sensing devices on the same silicon chip on which the elec-tronics would be fabricated—thus integrating the sensing structures and amplifying/processing circuits on the same IC chip.

The workshop report listed the following topics as research areas for pursuing the new technology of microdynamic devices:

1. Device architecture and performance
2. Microactuators
3. Design tools
4. Microcharacterization and metrology
5. Fabrication: silicon micromaching, materials, mask design, and etching
6. Science base

The first topic of research would create new types of device structures, and the second would create new types of tiny actuators required to power such devices. The next two areas proposed the need for new tools to facilitate the design of the new devices and to characterize and measure such tiny devices. The fifth research area would create new means of fabricating such devices. The sixth area would deepen the understanding of the physical processes and materials underlying such devices and their fabrication.

This research agenda indicated the kinds of research that should be pursued to make progress in the new technology. It did not predict exactly which devices would be created nor the dates of their creation. Prediction of this kind was not within the knowledge of the experts gathered at the workshop, since it depended on the emergence of markets and applications for the new technology. Within their knowledge, however, was a collective vision of an exciting new technology and the kinds of research required to create it.

Planning Targeted Basic Research

Sometimes the ideas for new directions in research for a technology grow out of research in previous areas. Also, this early exploratory research often has occurred both in universities, in some industrial corporate laboratories, and in some governmental mission laboratories. The support for such research now often comes from federal programs and industrial firms.

> *The best source of information about the* longer-term *directions of new technologies can be found in research projects that are basic in nature but technologically oriented, called* targeted basic research.

In targeted basic research, government programs support exploratory research in broad areas. In the United States, some of the federal agencies that have significant programs to support basic research include the Department of Agriculture, the Department of Defense, the National Science Foundation, and the National Institutes of Health.

When groups of researchers begin to see that particular exploratory projects are opening up a new technological area, then the federal government in collaboration with industry and university researchers can organize national workshops that identify the new opportunities and propose topics for future research. If government and industry subsequently fund research programs on these topics, then projects funded within these programs can be monitored in order to identify both the direction and the pace of technical advance.

Research for technology cannot be planned until a basic technological invention occurs for the technology system. In the preceding illustration, the inventions for microdynamic machines were already being made in integrated sensor research. Prior to the basic technology invention, research can only be planned through a generally scientific orientation—in the support of a board scientific discipline aimed at discovering, exploring, and understanding nature.

After the basic technology invention, research can be planned by focusing on the generic technology system, production processes, or underlying physical phenomena.

Technology-focused research can be planned for (1) generic technology systems and subsystems for product systems, (2) generic technology systems and subsystems for production systems, and (3) physical phenomena underlying the technology systems and subsystems for product systems and for production systems.

In any of these technology systems, research can be focused on improving the system through inventing improvements in any aspect of the system:

1. Improved system boundary
2. Improved components
3. Improved connections
4. Improved materials
5. Improved power and energy
6. Improved system control

Also, the physical phenomena underlying a system can be focused on any of the system aspects:

1. Phenomena involved in the system boundary
2. Phenomena underlying components
3. Phenomena underlying connections
4. Phenomena underlying materials
5. Phenomena underlying power and energy
6. Phenomena underlying system control

With all these possibilities, one can understand why a research agenda for technologies can be complex and costly in terms of effort.

ILLUSTRATION
Discovery of Nuclear Fission

The origin of nuclear power technology provides an illustration of how research strategy couples with research planning. Nuclear power technology involved a new scientific study that was focused immediately on military applications.

In 1933, two university scientists, Irene Curie and Frederic Joliot, discovered that radioactive decay could be *stimulated* in an element after allowing the element to be bombarded by alpha particles emitted from the natural radioactive decay of another element. An alpha particle is a proton and a neutron bound together by the strong nuclear force.

This was an exciting discovery, that radioactivity could be induced in an element by the absorption of particles. Immediately, other university scientists, including the Italian Enrico Fermi, began investigating the new phenomenon: "...At the beginning of 1934, Enrico Fermi in Rome had started using neutrons from radon-beryllium sources, in lieu of alpha particles, to activate many common elements" (Segre, 1989, p. 38).

Recall that an element is determined by the positive charge of the nucleus, the number of protons in the nucleus. An isotope of an element is determined by its nuclear weight, the total number of protons and neutrons.

Members of Fermi's scientific team included Edwardo Amaldi, Oscar D'Agostino, Franco Rasitte, and Emilio Segre. The experiments of Fermi's team established that the nucleus of an element could be transformed into a different element or a different isotope of the element by the absorption of a neutron and the emission of either protons or neutrons. Fermi's team next bombarded the elements of thorium and uranium with neutrons. They were particularly interested in uranium, because at the time it was the most heavy known naturally occurring element. They hoped to create new elements that were heavier than uranium when a uranium nucleus absorbed neutrons.

After a neutron was absorbed, radioactive reactions were observed by the decay products. However, this induced radioactivity was not much larger than the natural radioactivity of uranium. Thus it was very confusing to understand just what was happening. "In Rome, we immediately found that irradiated uranium showed a complex radioactivity with a mixture of several decay periods. We expected to find in uranium only the previously observed [radioactivity] backgrounds, and so we started looking for an isotope of element 93 [a next element above uranium's proton count of 92] produced by a neutron-absorption and gamma decay reaction on the 238 isotope of uranium, followed by a beta decay" (Segre, 1989, p. 39).

The researchers hoped that the 238 weight of the uranium element would absorb a neutron to transmute to a 239 weight and a neutron in it would decay to a proton by emitting an electron (beta decay). Thus the element of uranium with 92 protons and weight 238 would be transmuted to a new element of 93 protons and weight 239. Thereby, these researchers hoped to be the first to create and discover a new element created by transmutation of another element. This was the old dream of medieval European alchemists—the transmutation of matter.

Scientists are human and can make mistakes: "Here we made a mistake that may seem strange today. We anticipated that element 93 would resemble rhenium [element 75]....Products of the bombardment were indeed similar to rhenium" (Segre, 1989. p. 39). The researchers were looking for a new element 93 in the products from the experiment that would have a chemistry similar to the known element of rhenium. This expectation was a mistake, and because of it, they did not recognize the products of the *fission* that was actually occurring. Some of the fission products, as the researchers were to learn much later, were isotopes of technetium (element 43), which does have a chemistry similar to rhenium.

So nature was fooling them. Of course, scientists never anthropomorphize nature. If they did, however, they would note that sometimes nature appears to have cooperative moods and sometimes uncooperative moods (and sometimes even a wicked sense of humor). In this case, the confusion meant that Fermi and his group just missed being the first scientists to discover the phenomenon of nuclear fission.

Yet the puzzle was soon to be solved and nuclear fission discovered by other scientists. After Fermi's group reported their results to the world in 1934, Otto Hahn and Lise Meitner at the Kaiser Wilhelm Institute began work in 1935 to confirm the Fermi group's results. And at first, they also were not able to discern the fission phenomenon: "Hahn and Meitner, working with a neutron source about as strong as the ones used in Rome and Paris, started by confirming our Rome [the Fermi group's] results. This is somewhat surprising because they applied quite different chemistry. Their early papers are a mixture of error and truth, as complicated as the mixture of fission products resulting from the bombardments" (Segre, 1989, p. 40).

Meanwhile, in 1935 in Paris, Irene Curie, who was earlier the discoverer of radioactivity, was working with Hans von Halban and Preiswerk on bombarding neurons on thorium. They found a decay product that chemically resembled lanthanum. However, they did not realize it was lanthanum of atomic weight 141, which was a fission product. They thought it might be an isotope of actinium, and they also just missed discovering fission. By 1938, therefore, the puzzle still remained, neither the Italian, German, or French teams had discovered nuclear fission.

Hahn (German) and Joliot (French) met in Rome at a Tenth International Chemistry Conference. They discussed Curie's results, and Hahn concluded that something was wrong in the suggestion that the decay product was actinium. Back in Germany, Hahn and Meitner decided to repeat some of Curie's experiments.

However, in March of 1938, Hitler's German army marched into Austria. Since Lise Meitner was an Austrian of Jewish ancestry, she then lost the previous protection of her Austrian citizenship against Hitler's evil policies. In danger of arrest by Hitler's Gestapo, Meitner escaped to Holland.

Hahn remained in Austria and continued work (with the help of Strassmann). They concentrated on identifying the radioactive component Curie had thought was actinium (but was really lanthanum). Finally, by December of 1938, Hahn and Strassmann had concluded that the radioactive decay products Fermi and Curie had thought were radium, actinium, and thorium were really barium, lanthanum, and cesium. Hahn wrote to Meitner about the results: "He and Strassmann...were coming 'again and again to the frightful conclusion that one of the products behaves not like Ra but rather like Ba'" (Stuewer, 1985, p. 50).

After receiving Hahn's letters in December, Meitner traveled to Sweden to spend the holidays with friends and with her nephew, Otto Frisch (who was also a physicist). Earlier, Frisch had been forced to leave Hitler's Germany in 1933, going to Denmark to work in Niels Bohr's physics institute. (Bohr was the scientist who provided the first quantum-mechanical model of the atom.)

Just after Frisch arrived in Sweden to holiday with his aunt, Meitner received a third letter from Hahn on December 28. She showed her nephew all the letters, for in the last letter Hahn had asked, "Would it be possible that uranium [of atomic weight 239] bursts into barium [with weight 138] and masurium [with weight 101 and now called technetium]?" (Stuewer, 1985, p. 50).

Hahn pointed out that the atomic weights added correctly—138 nucleons from barium added with 101 nucleons from masurium totaled the 239 nucleons in uranium. But the atomic numbers did not add right! Uranium had 92 protons, barium 56 protons, and masurium 43 protons. The sum of 56 protons and 43 protons equaled 99 protons *instead* of the 92 of uranium. Hahn had also written: "so some of the neutrons would have to be transformed into protons....Is that energetically possible?" (Stuewer, 1985, p. 50).

Very excited over these speculations, on the following day the aunt and nephew had breakfast together and went for a hike outdoors. Frisch pushed along on skis, and Meitner walked. They talked and talked about physics. Frisch suggested that a liquid-drop model of the nucleus might be used to explain the Hahn and Strassmann results (Stuewer, 1985, p. 50):

> I worked out the way the electric charge of the nucleus would diminish the surface tension and found that it would be down to zero around $Z = 100$ [atomic number of 100 nucleons] and probably small for uranium. Lise Meitner worked out the energies that would be available from the mass defect in such a breakup (the splitting of a liquid-drop like mass of uranium with 239 nucleons into fragments liquid-drop like sizes with of 138 and 101 nucleons)....It turned out that the electric repulsion of the fragments would give them about 200 MeV [million electron volts] of energy and that the mass defect would indeed deliver that energy.

This was their idea: The uranium atom could split into two near-halves, like a drop of water might split into two smaller water drops. Meitner and Frisch's idea was the concept of *nuclear fission*.

Of course, in actual fact, as Frisch and Meitner knew, this was only an analogy. Sometimes in new territories, scientists have to explore using analogies. The atomic nuclei contain nucleons that have both positive electric charges in protons (and therefore repulsive forces between protons) and an attractive strong nuclear force between protons and neutrons that binds them all together. Yet the form of the physics of a nucleus is sort of similar to a liquid-drop model. But there are complications to this simple model, since nuclei have different shapes and have shell structures (Greiner and Sandulescu, 1990). Still, overall, when the number of nucleons is small, it takes a particle of large impacting energy to tear the nucleus apart. However, as the nucleus gets very large, it may be unstable and undergo spontaneous fission or even particles of only small energy can distort the nuclear sphere into fissioning halves. And the uranium nucleus is up there in size where larger nuclei become unstable. This is why elements with much larger nuclei than uranium do not occur naturally.

On January 1, 1939, Meitner wrote to Hahn: "We have read and considered your paper very carefully, [and] perhaps it is indeed energetically possible that such a heavy nucleus bursts" (Stuewer, 1985, p. 50).

Very excited, Frisch returned to Copenhagen that same day to tell Niels Bohr about the bursting of the uranium nucleus. Later, Frisch wrote to his aunt that upon hearing the news Bohr had burst out: "Oh what idiots we all have been? Oh but this is wonderful! This is just as it must be!" (Segre, 1989, p. 42).

R&D Infrastructure

Research planning takes place within an R&D infrastructure, and such structure must be understood if research planning is to be done properly. The R&D infrastructure provides the institutional framework and bases of technical expertise for innovation processes as well as the goals for planning.

In the preceding illustration, we see university-employed scientists in different countries studying a natural phenomenon, radioactivity, and discovering another natural phenomenon, nuclear fission. Their research planning followed a kind of style of exploration, motivated by ubiquity.

Anything that occurs everywhere in nature (ubiquity) is of interest to scientists to discover and understand, and this is the basis of exploratory research planning in science—the existence of unexplored scientific territory.

The organization of research for the pursuit of technological progress is a new and noteworthy aspect of the twentieth century. In previous centuries, science and technology interacted but not in a deliberately organized manner. Historically, in the late eighteenth and early nineteenth centuries, many of the early basic inventions for radically new technologies were made by inventors who were not trained as scientists. Such examples of nineteenth-century inventors include Morse, Marconi, Edison, Diesel, and so on.

However, the three major institutional changes altered this early pattern of basic technological invention. The first institutional change was the reform of the university into a research-based educational institution. The second was that establishment of corporate central research laboratories. The third was the support of research by national governments.

Around the year 1800 in Germany, Wilhelm Humboldt began reforming the medieval universities in Prussia into the form of the modern disciplinary science departmental structure and professional schools of the modern university. The Prussian reforms spread through all the German universities, which provided the model for graduate research–based education emulated in the rest of Europe and in America. Public American universities as research-based institutions (such as the University of Michigan at Ann Arbor, the University of Wisconsin at Madison, and the University of California at Berkeley) were established in the land-grant college system after the American Civil War. In the same period after 1865, the American seminary institutions (such as Harvard, Yale, Princeton) also were reorganized toward research-based graduate programs.

In the last part of the nineteenth century, equally important changes were occurring in the new technology-based industries of electricity and chemicals with the establishment of engineering departments in firms. In the early twentieth century, after 1920, many of these same firms also established science-based corporate research laboratories. Famous examples in American firms are General Electric's Corporate Research Laboratory in New York, AT&T's Bell Laboratories in New Jersey, DuPont's Corporate Research Laboratory in Delaware, Dow's Corporate Research Laboratory in Michigan, General Motor's Technical Laboratory in Michigan, and so on.

After World War II, national governments began supporting basic research in universities and applied research in government laboratories and in defense industries.

By the end of the twentieth century, the result of these institutional changes was to more directly couple scientific advance with technology development. Roughly described, the picture had evolved as follows:

1. Increasingly during this century, the sources of radically new technologies have been directly from new scientific advances.

2. Technology forecasting cannot begin until a new technology has begun, and this occurs in a basic invention.

3. Once the basic invention occurs, the early stages of innovation have often been principally motivated by technology push.

4. Once a set of significant applications begins to grow markets for the new technology, then market pull also becomes a principal source of innovation.

5. The diffusion of a new technology along with growth in markets shows complex patterns of development depending on numerous factors, including cost and product substitution and applications development.

6. Incremental innovations normally dominate product and process developments in existing technologies and are critical to maintaining competitive competence.

7. After the beginning of new industry through a basic innovation, tight coupling between scientific and technological research often facilitates incremental improvements in the technology and periodically major changes in the development of a new technology as next generations of technology.

8. Next generations of technology create competitive discontinuities that can lead to restructuring of entire industries, with the entry of new firms and exit of existing firms.

The goals of research in industrial and governmental organizations are primarily technological, whereas in university organizations they are both scientific and technological. It is on the long-term technological goals that university- and industrial- based researchers can plan research cooperation.

Nature and the Complexities of Technology

In the illustration of the discovery of nuclear fission, we saw that one of the reasons for the difficulties in interpreting the natural phenomenon of fission (and later the great technological barrier) was the small portion of fissionable isotope found in nature. It was only 0.7 percent of the uranium found in nature. This was nature's doing! Sometimes nature has been easy to manipulate for technology and in other cases difficult. The complexities of nature are the reason that research planning is necessary—so that science can discover and understand nature sufficiently for technological manipulation.

In the subtle complexities of nature are the sources of all the engineering problems of technology, and the more complex the natural basis of a technology, the more is research planning required.

Even though a scientific phenomenon has been discovered upon which to base a new technology and a concept for manipulating the phenomenon as a technology has been invented, there is still a long way before a technology may become effective and efficient at a reasonable cost. The steps from scientific feasibility toward an effective technology lie in engineering research—establishing an engineering feasibility for the technology.

Only until engineering feasibility is established can the technology forecasts that are based only on scientific phenomena become certain. Engineering research is the connection between science and technology.

The planning of engineering research focuses on the understanding and design principles for generic technological materials, devices, processes, and systems.

Engineering research deepens and fills in the knowledge base of science necessary to improve an invention that manipulates natural phenomena. Inventions may be made by scientists or engineers or other personnel. But engineering research moves the invention to practical fruition.

ILLUSTRATION (*Continued*)
The Conception of the Atomic Bomb

The differences between the discovery of nuclear fission and its transformation into power technology can illustrate the differences between
planning science research and planning engineering research.

A few days after Frisch had informed Niels Bohr about the discovery
of fission, Bohr left for America to attend a physics conference (carrying with him Frisch's momentous news). Bohr sailed with his wife on
the ship *Drottningholm,* and accompanying them was a colleague, Leon
Rosenfeld, who later recalled the voyage: "As we were boarding the
ship, Bohr told me he had just been handed a note by Frisch, containing his and Lise Meitner's conclusions; we should 'try to understand it.'
We had bad weather through the whole crossing, and Bohr was rather
miserable, all the time on the verge of seasickness. Nevertheless, we
worked very steadfastly and before the American Coast was in sight
Bohr had got a full grasp of the new process and its main implications"
(Stuewer, 1985, p. 52).

Bohr announced the news of nuclear fission in America at the Fifth
Conference on Theoretical Physics in Washington. The first meeting of
that conference was held in the afternoon on January 26, 1939, in room
105 of Building C of George Washington University, where "Today a
plaque outside the lecture room commemorates this historic meeting.
It was there...that the bombshell burst" (Stuewer, 1985, p. 54).

The report from this conference reads: "Certainly the most exciting
and important discussion was that concerning the disintegration of
uranium of mass 239 into two particles each of whose mass is approximately half of the mother atom, with the release of 200,000,000 elec-
tron-volts of energy per disintegration" (Stuewer, 1985, p. 54). This was
an enormous release of energy per atom. Leo Szilard was at that conference, learning about the discovery of nuclear fission. He had emigrated to the United States from Britain, having earlier emigrated
from the Nazi takeover in Germany in 1933. After that conference,
Szilard worried that the Germans would be the first to turn the discovery into an atom bomb.

Szilard thought American action urgent! He drafted a letter to the
President of his new country, Franklin Delano Roosevelt, taking the
letter first to Albert Einstein. Then Albert Einstein was the most famous scientist in America, and Szilard hoped the President might
listen if Einstein alerted him to the potential development of nuclear
weapons. It did! Roosevelt responded by creating the U.S. Manhattan Project, which built the world's first atom bomb. In historical
perspective, Szilard's sense of urgency was warranted. When nuclear
fission was announced, anyone who understood the physics of its im-

mense release of energy could immediately imagine its application as a technology of power or destruction. And German scientists did. The Germans also had began an atomic bomb project at about the same time as the Manhattan Project, as we will next describe.

Technology and Scientific Experimentation with Nature

Recall that invention begins with conception, conceiving how to make something work—the structural form for the function. Sometimes the engineering capability for manipulating a new technology has arisen from the scientific capability for experimenting with the phenomena underlying the technology. However, sometimes the engineering capability for manipulating a new technology has had to be invented in alternate means of manipulating nature after science has discovered and experimented with the underlying phenomena.

For technology, the discovery of nature is insufficient. Technology also requires instrumentation and techniques for experimenting with nature. Experimental instruments and techniques for experimenting with nature have often become the original basis for a new technology (or the basis for production process control). For example, nuclear *fission* is a case where the scientific experimentation provided an engineering approach to the controlled release of energy. In contrast, however, nuclear *fusion* is a case where the scientific experimentation has not provided a practical engineering approach to the controlled release of energy. Recall that *fusion* is the process of combining protons and neutrons into nuclei, whereas *fission* is the splitting of a nucleus. Historically, nuclear fusion, as the nuclear processes of stars, was observed in nature before the observation of nuclear fission, yet it was nuclear fission that was first observed experimentally. Experimental observation is necessary before a phenomenon can be used for technology.

> *Until one can experiment with a natural phenomenon, one cannot manipulate the phenomenon, and without manipulation, technology is not possible.*

Historically, the invention of scientific instrumentation has led to its later use as technological tools.

> *Research planning which includes developing and improving the instrumental means of observing the phenomena underlying technologies has often provided a basis for manipulating nature and for controlling systems in the technology.*

ILLUSTRATION (*Continued*)
The Nuclear Fission Reactor

In nuclear fission power, the key was how to experiment with the fission phenomenon, i.e., how to set up a self-sustaining nuclear chain reaction. The invention of the nuclear reactor was conceived independently by several people. However, the form used spelled the difference between the American success in developing the first atomic bomb and the German failure.

After Hahn learned of Meisner's and Frisch's conclusion about the fissioning of the uranium atom, Hahn told a colleague, Wilhelm Hangle. In the following April of 1939, Hangle delivered a lecture in Germany on "...the possible applications of nuclear energy making use of a uranium-graphite pile" (Walker, 1990, p. 53).

After hearing Hangle's lecture, Georg Joos (one of Hangle's colleagues at the University of Gottingen) transmitted a report of the lecture to the German Ministry of Education—which passed it on to Hitler's Reich Research Council. Also independently, Nikolaus Riehl (who was an industrial physicist at the Auer Company in Berlin and a former student of Hahn and Meitner) brought nuclear fission to the attention of the German Army Ordnance Office. And also independently, Paul Harteck and Wilhelm Groth, physical chemists in Hamburg, wrote to the Army about the possibility of making nuclear explosives. So the idea of an atomic bomb was in the air—once nuclear fission was discovered.

Often a possible application of a scientific phenomenon has been envisioned immediately after the discovery, even by different people.

The German Army responded promptly, holding two workshops on uranium in April and September of 1939. These were followed by two more conferences in September and October. These conferences resulted in the German Army planning research for an atomic bomb. Like the American Manhattan Project, the German Army project also planned to create a sustained chain reaction. The Army Ordnance Office distributed work on nuclear fission among several institutes. "Throughout the war, Gerlach, Bothe, Heisenberg, Harteck, Hahn and Klaus Clusius—all scientists of first rank—were involved in the scientific work and the administration of the German fission project" (Walker, 1990, p. 54).

The problem was demonstration of the technical feasibility of a nuclear reactor machine. The idea of such a machine was envisioned as a mixture of uranium interspersed with parts of a moderator to control the diffusion of neutrons between uranium atoms. However, there were many technical issues. What kind of uranium, and what kind of moderator? What forms were the uranium and moderator to be in, and how were they to be arranged? These issues of *engineering research* of the composition and configuration of the new technology formed the basis of the early research planning for both the American and German projects.

In the German research institutes, several experiments were begun using different moderators, including graphite and heavy water. Walther Bothe tried using graphite as a moderator. Paul Harteck in Hamburg tried using dry ice (carbon dioxide) as a moderator, with uranium powder arranged in layers within a sphere. Later others tried using paraffin as a moderator.

Bothe's early measurements and calculations led him to conclude that graphite would absorb too many neutrons to allow the reactor to work. Subsequent experiments by Wilhelm Hanle showed that this was wrong, because there had been impurities of boron and cadmium in the graphite that were really the strong absorbers of thermal neutrons. Then in August of 1940, Robert Dopel in Leipzig demonstrated that heavy water made an excellent moderator for thermal neutrons. Heisenberg calculated that a uranium machine relying on graphite would require much more uranium and much more moderator than a machine relying on heavy water.

Therefore, the German Army Ordnance, based on Heisenberg's recommendation, decided to construct a nuclear reactor using heavy water as the moderator. This was a critical and wrong *engineering* decision. It was the reason the Germans did not succeed in developing an atomic bomb. The American project, under Fermi, used carbon as a moderator and was successful.

The Germans made their choice mostly because of economics and supply. Several German scientists were already expert in dealing with heavy water and had been instrumental in persuading a Norwegian company to produce heavy water in Norway. After Germany occupied Norway, the Norwegian company, Norsk-Hydro, was ordered by Nazi officials to increase its annual production of heavy water from 20 liters to 1 metric ton. "Thus the Germans came to focus on developing a reactor fueled by natural uranium and moderated by heavy water" (Walker, 1990, p. 54).

Light water is ordinary water, with two hydrogen atoms and one oxygen atom making up the water molecule. Heavy water is a rarer, naturally occurring variant of normal water in which a hydrogen nucleus contains a neutron as well as a proton. Heavy water occurs naturally mixed with normal water in the very dilute concentration of one molecule of heavy water for about every million molecules of normal water.

Meanwhile, in America, Enrico Fermi built the first nuclear fission reactor at the University of Chicago for the Manhattan Project. The reactor went critical in 1942, demonstrating that a self-sustaining nuclear fission reaction was possible. Fermi's reactor used graphite as a moderator. The Italian physicist Fermi had made the right engineering choice. The German physicist Heisenberg had not. Thus America got to the bomb first, although the Germans had discovered the fission phenomenon (Hitler's persecution of the Jews cost the Germans many good scientists).

Fermi's graphite-moderated reactor burned uranium in a matrix form, with alternating rods of uranium and graphite. To turn on and turn off the chain reaction, alternating rods of borium could be inserted between the uranium rods to absorb neutrons. The graphite slowed the neutrons without capturing them. Uranium of isotope 239 only fissioned well with slow neutrons.

The heavy-water moderation was as technically feasibile as was graphite moderation. The difference was in functional feasibility— heavy-water moderation produced too low a rate in the fission chain for a self-sustaining reaction in the application of a nuclear reactor.

Technical Feasibility Prototypes and Functional Prototypes

All technology can be conceptualized at two levels: (1) generic engineering principles and (2) specific engineering alternative types. What we saw in the preceding illustration is that although the scientific feasibility established that nature may be manipulated, the actual technical manipulation can still involve alternatives for an application. There will be alternatives in materials, in processes, and in form.

The planning in the early stages of engineering research should be aimed at systematically exploring all technical alternatives to find a technically optimal *solution.*

The criteria for technical optimality include (1) performance, (2) efficiency, (3) cost, (4) safety, and (5) life cycle. Different combinations of these factors will be optimal for different applications. An application prototype should adopt the best engineering solution for a given application context.

Recall from the radical innovation process that in establishing engineering feasibility, there are two stages of prototyping of technology systems: technical feasibility prototypes and application functional prototypes.

Planning research for engineering feasibility requires an explicit application and the technical requirements for the application.

Previously, we discussed the importance of using experts in planning research, but it is important to use a mixture of experts— scientists, engineers, and applications users. Each will view the research issues differently. The views of a technology as a scientific problem and as an engineering problem are profoundly different. It is the responsibility of the engineering perspective in planning research to establish as early as possible the requirements of an applications user of an innovative technology—even before users may exist.

A technical feasibility prototype demonstrates that the engineering principles do work in at least one configuration. A functional prototype is a selection from the technically feasible alternatives that optimizes the performance criteria for a given application.

Different morphologies of a technology system are expressed as different configurations of components and materials, and in a given application, alternate morphologies perform differently.

ILLUSTRATION (*Concluded*)
A Nuclear Reactor for the Navy

To make the atomic bomb, the U.S. wartime Manhattan Project had constructed enormous facilities in Oak Ridge, Tennessee, and Hanford, Washington, for enriching the quantity of the fissionable isotope ^{235}U in naturally occurring uranium ore and for producing fissionable plutonium. Thus the first reactor power programs in the United States had a choice of using naturally occurring uranium, enhanced uranium, or plutonium.

The generic principles of any fission reactor based on Fermi's invention are that all such reactors produce heat by splitting the nuclei of the fissile isotopes of uranium or plutonium. These materials are shaped into fuel elements such as rods or pellets. Fission occurs when an atom absorbs a neutron and fissions with the release of two neutrons, on average. If at least one of these neutrons is absorbed by a fissionable atom, producing fission again, then a nuclear chain reaction is continued.

In addition to the neutrons, the fission releases energy in the form of gamma rays (light particles, photons of very short wavelength). This light is partially absorbed by a coolant that uses the heat to boil water into steam that is used to drive a steam turbine connected to a an electrical generator. Thus about a third to a half of the heat from the reaction is transformed into electricity. The rate of the fission reaction and thus the rate of heat produced is controlled by a matrix of control materials that absorb the neutrons, preventing their use for fission.

After World War II, four engineering alternatives were conceived as possible functional prototypes:

1. The heavy water reactor
2. The light water reactor

3. The gas-cooled reactor

4. The liquid metal–cooled reactor

Each type is distinguished by different types of coolant, which is essential both to moderation of neutrons and to heat transfer for electrical generation. Because of this, there are also differences in the kind of fissionable fuel the reactors could burn and the performance of the reactors:

The heavy-water reactor could burn naturally occurring uranium ore with its very small percentage of fissionable uranium isotope (^{235}U).

The light-water reactor could not burn naturally occurring uranium ore but requires processed ore that has a substantially increased ratio of fissionable uranium isotope (^{235}U).

The gas-cooled reactor also requires enhanced uranium but operates at higher temperatures, thereby increasing the efficiency of the conversion of heat to electricity.

The liquid metal–cooled reactor could not only burn naturally occurring uranium ore but also produce more fissionable material—it could breed new fuel.

Which reactor type was best? It depended on both the application and the relative importance of the different criteria for optimality. In nuclear fission reactor design, there were two principal applications—nuclear-powered submarines and nuclear-powered electric generating plants.

At the end of the war, a new U.S. law, the Atomic Energy Act of 1946, established a new civilian agency, the Atomic Energy Commission (AEC), to control and promote atomic energy matters. In the meetings of this commission, research planning for the new technology was done. In the summer of 1947, the General Advisory Committee of the AEC held a meeting at the Oak Ridge National Laboratory. "The Committee included Conant, DuBridge, Oppenheimer, Fermi, Teller, Rabi and Seaborg, world renowned scientists and participants emeritus in the Manhattan District, who had been appointed by the President (Truman) to advise the new Atomic Energy Commission on the technical goals of its programs. Temporarily assigned to Oak Ridge at the time was a group of naval engineering officers headed by the then virtually unknown Captain Hyman G. Rickover" (Hazelrigg, 1982, p. 156).

Although all the persons in the room were brilliant, talented, productive people, the committee was composed mostly of scientists. Rickover was not a scientist by training but an engineer. Rickover had wheedled the privilege of listening to the General Advisory Committee's deliberations by getting permission for himself and one of his lieutenants to sit on one side of the room merely as observers. "When eventually the Committee's deliberations turned to the question of developing useful power from the atom, the consensus after considerable discussion was that useful power generation might perhaps be demonstrated in 20 years, and its application for specific purposes might come some time later. Rickover stood up and, interrupting the discussion, proceeded to berate the distinguished scientists for using a scientific approach to an engineering problem. This broke up the meeting" (Hazelrigg, 1982, p. 156).

Just after World War II in the realm of the politics of science and technology in the United States, the scientists were the "top dogs." They held the political stage over engineers. They were not ready to be lectured to about anything—let alone by an engineer and a junior officer.

Oppenheimer was then the President's Science Advisor (and popularly referred to as "father of the atomic bomb" for his technical leadership of the Manhattan Project). Teller also was a scientist (who was later to become popularly called "father of the hydrogen bomb"). Rabi and Seaborg also were scientists and eventually became Noble Prize winners. Only Conant and DuBridge were engineers, but academically based engineers with doctoral degrees, closer to scientists—engineering science.

In contrast, Rickover was a practicing military engineer, holding a master's degree in engineering. He had an engineer's technical dedication, with an urgent technical vision to build an atomic-powered submarine. His vision was different from a scientific vision—different as to the application, its urgency, and the timing of the innovation.

Even 6 months before the Atomic Energy Commission had been formed, the Navy had already decided to build nuclear-powered ships. The Navy's General Board had recommended to then Secretary of the Navy, James V. Forrestal, that a study and development of atomic power for propulsion of naval ships be begun. Then the Bureau of Ships sent a group of officers to Oak Ridge, Tennessee, to observe a project that was intended to develop a power reactor for civilian use. The senior officer of this group had been Captain Rickover (Hewlett and Duncan, 1974; Polmar and Allen, 1982).

Thus by 1948, Rickover was in charge of the Navy program and proceeded to develop the naval nuclear submarine program. Rickover chose the light-water reactor in a pressurized form for the Navy. Rickover's choice of the pressurized light-water reactor was focused on the submarine application for which it was believed to be the best technical choice.

Under Rickover's leadership, the Navy had evaluated the liquid metal (sodium)–cooled reactor as being "tons heavier" than a pressurized light-water reactor for the same production of energy (JCAE, 1964). The Navy also saw a major safety problem in this alternative: If seawater flooded a liquid metal–cooled reactor, the liquid sodium would react with water producing a terrible explosion. In addition, the control equipment for such a reactor would be very complex and difficult. Finally, the induced radioactivity in liquid metal coolants was long-lived, making repair difficult. Therefore, the Navy decided against using a liquid metal reactor for a submarine. The Navy also saw problems with the other two types, heavy water and gas cooled. The heavy-water reactor also was envisioned as just too big and heavy for a submarine. The gas-cooled reactor also was very big, and its safety and control requirements were more risky and complex than a pressurized light-water reactor. Thus, at the time, for a ship with modest power requirements, the weight, size, simplicity, and safety of the pressurized light-water reactor was seen as the best engineering choice.

The nuclear-powered submarine, of course, revolutionized naval warfare. Earlier, in the beginning of World War II, the relatively slow diesel-powered German submarine (the U-boat) had almost strangled the English war effort. After nuclear power, by the 1960s, the submarine became the dominant war boat. Nuclear power enabled submarines to remain indefinitely at sea and underwater at depths and speeds undreamed of before, the limits of a cruise being only the amount of food a crew required.

For example, in a popular magazine of the 1980s, a civilian reported about the experience of cruising on a modern nuclear submarine. "A profound sense of wonder swept over me as Cmdr. Alfred E. Ponessa, USN, passed the tuna salad. We were seated in the wardroom aboard the nuclear attack submarine *Norfolk,* enjoying the expertly prepared food and the convivial discourse, when I glanced up at the instrument repeater on the bulkhead. This little party was taking place somewhere off the coast of Florida along the 100-fathom curve. Speed? Well in excess of 20 knots. Depth? Much deeper than 400 ft....*Norfolk*'s nuclear reactor quietly simmered just a few steps aft—converting water into steam, steam into turbine energy.... Requiring refueling every 13 years, the power plants' continuous, controlled fission generates on-board electricity, drives an electrolysis process that separates hydrogen from water to produce oxygen, operates the lithium-hydroxide scrubbers that cleanse the atmosphere of carbon dioxide, and desalinates and de-mineralizes ocean water" (Cole, 1988, p. 28).

Since most developmental research went first into a military application, the later civilian choice built on it. Thus it turned out that the pressurized light-water reactor became the dominant type of reactor in use, both military and civilian; but for the civilian application, it was probably the wrong engineering choice.

Product Life Cycle

In a new technology, research planning should be not only focused by the technology system but also by the application system. One important difference between the concepts of the technology and applications systems is in the notion of a product life cycle.

The product life cycle *concept in engineering design requires consideration of all the technical requirements for producing, maintaining, supplying, and disposing of a product.*

Technologies relevant to a product include

- The technologies embedded in the design
- The technologies relevant to maintaining and repairing the product in use
- The technologies required to provide the product with resources (such as fuel or other supplies) while in use
- The technologies required to dispose of or recycle the product after use

Research planning for the innovation of a new technology should include all technologies for the product life cycle of the product embodying the principal transformation function for an application.

ILLUSTRATION
Life and Death of the American Nuclear Power Industry

In the second half of the twentieth century, the American nuclear power industry, utilizing the previously described fission technology, ran into commercial troubles because of safety and environmental problems. By 1989, the American nuclear industry was at a dead standstill. "…Beginning in the mid-1970s and accelerating after the accident at Three Mile Island in Pennsylvania, orders for new nuclear plants stopped materializing or were canceled in many cases. The demand for new reactors plummeted sharply, leaving suppliers and architect-engineers competing in a market that was rapidly becoming a ghost of its former self. Thousands of technical workers were laid off, and nuclear firms of all kinds and sizes migrated into the services side of the business" (Zorpette, 1989, p. 50).

Several nuclear accidents frightened the public about nuclear safety and led to a strong movement protesting the placement of new reactors. The most noted occurred first in America at Three Mile Island and later at Chernobyl in Russia. The nuclear industry had not replaced a first-generation design with a second-generation design that was inherently much safer.

The second problem was the environment. Nuclear disposal of long-lived radioactive waste had not been addressed effectively nor solved by the American nuclear regulatory agency. Accordingly, the public also lost confidence that the nuclear industry could handle the life-cycle problem of what to do with waste.

As a result of these safety and environment problems, the new-technology industry ground to a halt in America just after it had begun growing. For example, in 1966, 16 plants had been ordered, of which 12 were built and 4 canceled. In 1971, 21 plants had been ordered, of which 8 were built and 13 canceled. In 1973, 41 plants had been ordered, of which 9 were built and 32 canceled. In 1975, only 4 plants had been ordered, and 4 were canceled. In 1978, only 2 plants were ordered, and 2 were canceled. From 1979 to 1989, there were no new orders for nuclear power plants in the United States (Zorpette, 1989).

> *The next generations of a new technology should not only improve performance and lower costs but also improve the product life cycle issues of safety and the environment.*

The nuclear power industry failed because neither goverment nor industry planned next generations of technology to address safety and environmental issues.

Research Planning for Next Generations of Technology

In research planning, procedures should (1) characterize a technology as a system, (2) identify technical bottlenecks in the way of improving the system, (3) imagine if and how new research instrumentation, instrumental techniques, algorithmic techniques, or theoretical modeling could be focused on the bottlenecks, and (4) imagine how to improve the understanding and manipulation of the phenomena of the bottlenecks or to develop alternative phenomena to substitute new technical manipulations at the bottlenecks.

Recall that progress in technology systems can be either incremental in advancing system performance or discontinuous in advancing system performance by multiples of improvement over current performance (next-generation technology). Also recall that current corporate strategies can deal successfully with regular and niche- creation innovations, since the technical skill base of the organization is not affected. However, revolutionary and architectural innovations are discontinuous in technology and require strategic reorientations.

Also recall that technological discontinuities require proactive strategic reorientations at the CEO level. While such reorientations require a clear technological vision, it is difficult for technical personnel to consensually provide that clarity because of the uncertainties of the new technical vision and the differing perspectives among the different product group managers and the technical staff about that technical vision. In order for a company to develop a radically new product-line technology strategy, the whole company must fight out different product plans, and this is usually possible only when a high-level executive can formulate and implement a new technology strategy for the whole company.

Research planning for the long term should try to envision and research discontinuities as a next generation of the technology.

The first step in research planning for a next generation of a technology system requires a vision. Such a vision of a next-generation technology requires not only envisioning a major improvement in performance but also research imagination:

1. A description of the functionality, performance, and features that are very desirable but cannot be obtained in the current generation of technology

2. Imagination of how current applications could be better accomplished with the new technology and new applications that might be possible that cannot be accomplished with the current technology

3. Some new methodological research advances that might be used in a new generation of the technology

ILLUSTRATION
Research Planning for Biotechnology
Processing in 1990

In 1991 in an Engineering Research Center for Biotechnology Process Engineering at the University of Massachusetts directed by Professor Daniel Wang, university and industrial researchers planned basic research for the technologies of producing biotechnology products grown in mammalian cell cultures. The technology system divides into a bioreactor (which grows the cell cultures) and a recovery system (which recovers the desired proteins produced by the cells).

Within the bioreactor, the technological performance variables for protein production are number of cells, functioning of each cell, quantity of the protein produced by the cells, and quality of the protein produced by the cells.

The products of biotechnology cell production are particular proteins that have commercial use in therapeutic medicine. The quantity of protein is a measure of the productivity. The quality of the protein depends on its being properly constructed and folded. Proteins are long chains of organic molecules that fold back in particular configurations. The usefulness of the protein depends not only on its having the right molecules linked together but also on the molecules folding up into the right pattern. One of the technical bottlenecks in protein production was to get the proper folding.

For the design of bioreactors (for the production of proteins from mammalian cells), the biotechnical engineer needed (1) to design reactors that grow a high density of cells, (2) to formulate the proper medium for their growth, (3) to provide proper surfaces for mammalian cells to attach, (4) to control the operation of the reactor, and (5) to provide for recovery of the proteins that are grown in the cells.

In 1991, bioreactor designs for growing mammalian cells were fiber-bed, fixed-bed, and ceramic-matrix designs. Bioreactor control required sensors for monitoring cell growth and growth medium, expert systems for deciding on control strategy, and actuators for controlling the bioreactor temperature and material flows.

For the design of the recovery processes of the protein product from the bioreactors, the biotechnical engineer needed to control the physics and chemistry of protein structures, including protein-protein interactions that avoided aggregation of proteins or mutations of proteins, protein surface interactions that affect aggregation of proteins or adsorption of proteins to separation materials, and protein stability that affected the durability of the protein through the separation processes.

In 1991, the recovery processes for separating the desired proteins from the output of the bioreactor included chromatographic, filtration, and protein refolding procedures.

With this kind of system analysis of the bioreactor and recovery system and the important technological variables, the MIT Biotechnology Process Engineering Center next identified the research areas necessary to provide improved knowledge for the technology. For the bioreactor portion of the system, they listed the following scientific phenomena as requiring better understanding:

1. *Extracellular biologic events:* nutrition and growth factors of the cells, differentiation factors of the cells, redox oxygen conditions in the cellular processes, and secreted products of the cells

2. *Extracellular physical events:* transport phenomena of materials in the medium, hydrodynamics of fluid/gas interactions of the medium, cell-surface interactions, and cell-cell interactions

3. *Intracellular events and states:* genetic expression of the proteins in the cells, folding of the proteins and secretion from the cells, glycosylation of the proteins, cellular regulation of secretion, and metabolic flows in the cells and their energetics

We can see that this is a list of the biologic and physical phenomena underlying the cellular activities in the bioreactor.

Similarly, the MIT Biotechnology Process Engineering Center listed the scientific phenomena underlying the recovery process of the system that required better understanding:

1. *Protein-protein interactions:* aggregation of proteins into clumps and mutations of protein structure

2. *Protein-surface interactions:* aggregation of proteins through denaturation and adsorption of the proteins to surfaces

3. *Protein stability:* surface interactions, chemical reactions, aggregation in the solvent, and stabilization

Accordingly, the Center organized their research into two areas: (1) engineering and scientific principles in therapeutic protein production and (2) process engineering and science in therapeutic protein purification. In 1990, the research projects in the first research area of protein production included the following:

- *Expression:* transcription factors
- *Protein trafficking and posttranslational modifications:* glycosylation/folding
- *Redox potential*
- *Pathway analysis*
- *Intercellular energetics*
- *Regulation of secretion*
- *Hydrodynamics:* gas sparging
- *High-density bioreactor designs*
- *Substrata morphology* for cell attachment

Also in 1991, the research projects in the second research area of protein separation included the following:

- *Protein adsorption:* chromatography and membrane
- *Protein aggregation*
- *In vivo protein folding*

■ *Protein stability* in processing, storage, and delivery

In this illustration, we see that in research planning of the Center, first, the technology system was described, the critical technology performance variables were identified, and the underlying scientific phenomena were listed. Second, the research projects were formulated for improved understanding of the underlying scientific phenomena that could be used for inventions, design aides, and control procedures of the technology system.

Research Imagination

Research imagination is a synthesis of two sources of ideas: (1) location and sources of technical bottlenecks in the current technology system and (2) research methods for improved understanding of the phenomena underlying the technology system to break technical bottlenecks.

Research methods consist of (1) means of observing the phenomena and (2) means of modeling the phenomena.

Central to observing phenomena are instrumentation and techniques. Central to modeling phenomena are theory and analytical techniques.

What research instrumentation may do on the problem of technical bottlenecks is either to (1) enable the improved observation of the phenomena occurring in the technology system at the bottleneck that constrains overall system performance or (2) provide a new instrumental means for controlling phenomenal processes at the bottleneck of the system.

Historically, almost all research instrumentation invented to observe natural phenomena have later found industrial use as process-control equipment.

The invention of new research instrumentation to observe phenomena in the places of technical bottlenecks in a technology system will likely be usable for production control in the technology system.

A second source of research imagination is found in new ways of *using* instrumentation for observing nature—new instrumental techniques. A new research instrument may be improved by imagining new ways to use that instrument. Observation requires both instruments and ways to use instruments. And both can be complex and subtle. Moreover, combinations of instruments may create new and powerful instrumental techniques.

A famous illustration of a research technique that laid the basis for the biotechnology described in the preceding illustration is how the new technology of recombinant DNA was discovered from the tools of restriction enzymes. Restriction enzymes are natural organic molecules that cut DNA molecules at specific sites. A second fundamental discovery about DNA was that when cut, the DNA molecules would automatically reassemble the two cut pieces—DNA had "sticky" ends, as the discoverers described it. Two scientists, Boyer and Cohen, put these two techniques together and invented the basic recombinant DNA technique.

A third source of research imagination is theory, including theoretical models. Recall that theory provides the semantic tools for abstracting and generalizing instrumental observations of nature. What is abstracted and generalized are the essential forms of phenomena. This representation of general forms provide the basis for modeling phenomena, and modeling enables scientific prediction.

For technology, to be able to predict the temporal aspects of phenomena under the impact of external forces provides the basis for understanding how technological manipulation of nature works. And better understanding usually leads to improved manipulation.

Improved modeling of phenomena underlying the bottlenecks of technology systems is a usually a good bet for creating ideas for improving the technology system.

A fourth source of research imagination is algorithmic procedures. The creation of new mathematics for theories (or heuristic procedures, programmed operations, or learning operations) provides progress in science, and the procedures are usually also useful for technology. Theory is the fundamental semantic language of science, and mathematics is the fundamental syntactical language of science that enables scientific theories and models to be expressed in quantitative terms.

New or improved mathematics relevant to modeling the phenomenon underlying a technology can provide the basis for more sophisticated, complete, and accurate theories and models of the phenomenon.

Also, heuristic, programmed, or learning operations also provide syntactical means of describing, following, and prescribing the temporal evolution of a system. Recall that, the invention of the stored-program computer revolutionized heuristic procedures and added programmed and learning operations to the tools of science and technology.

Mathematical process models and algorithmic procedures find application in technology in industrial control and in design aides.

Industrial production control uses sensing, algorithmic decisions, and activators to provide improved control for industrial processes. In addition, "intelligent" devices also use sensing, algorithmic decisions, and activators to control the operation of processes.

Thus research imagination (through instrumentation, instrumental techniques, theoretical modeling, or algorithmic techniques) can focus on the physical phenomena underlying a technological system to improve bottlenecks in the system or develop substituting technological manipulations.One can focus imagination on any of the system aspects:

1. Phenomena involved in the system boundary

2. Phenomena underlying components

3. Phenomena underlying connections

4. Phenomena underlying materials

5. Phenomena underlying power and energy

6. Phenomena underlying system control

In this way, in any technology system, research can be focused on improving the system through inventing improvements in any aspect of the system:

1. Improved system boundary

2. Improved components

3. Improved connections

4. Improved materials

5. Improved power and energy

6. Improved system control

If enough research approaches taken together were to break enough technical bottlenecks to provide *multiples* of improvement in technology system performance, then the vision of a next-generation technology might be created.

Research Planning for a Next-Generation Technology

A vision of a next-generation technology is a description of the functionality, performance, and desirable features that cannot be obtained in the current generation of technology and includes imagination about how current applications could be better accomplished with the new technology and new applications that might be possible that cannot be accomplished with the current technology. A research strategy leading to a technical feasibility demonstration of next-generation technology is a set of cross-disciplinary research thrusts containing projects that are scheduled for integration into an experimental testbed embodying the functionality of a next-generation technology system.

Because the form of a next-generation technology system is complex, one should try an example of a system on an application in order to see which advanced research ideas will be useful, which will not, and what cannot yet be done.

A research testbed for a next-generation technology system allows the demonstration of advanced ideas to reduce risk by making more clear what is the most useful engineering pathway to the future generations of a technology system.

Once a next-generation technology (NGT) vision is created, the next step is to formulate a long-term research strategy:

1. The current state of the technological system should be described, delineating present limitations on functionality as performance limits and existing features and current costs, along with present applications of the technology.

2. A desirable significant change in performance and features is then planned that would constitute a dramatic leap forward to alter existing markets and create new markets. The goals of the next-generation technology are then expressed as planned performance, features, and cost.

3. The research issues are identified which could lead from the present state of technology toward the vision of the next generation of technology. These issues are then grouped into a set of cross-disciplinary research thrust areas, which together constitute a program of basic science and engineering research for the next generation of the technological system.

4. Initial research projects are proposed in these thrust areas to address the range of research issues. They are then coordinated in time to result in prototype subsystems of the next generation of the technological system. A chart in PERT format is formulated which plans the integration of these prototype subsystems into an experimental testbed for the next-generation technology system.

These steps lay out a targeted basic research program aimed at demonstrating an experimental prototype of the next-generation system.

Industrial researchers are very sophisticated about current technology and, in particular, about its problems and wherein lie the roadblocks to technical progress. However, because of the applied and developmental demands on industrial research, such researchers have limited time and resources to explore ways to leap current technical limitations. On the other hand, academic researchers have time, resources, and students to explore fundamentally new approaches and alternatives that leapfrog technologies.

Together, industry and university researchers can see how to effectively bound a technological system in order to envision a next generation of technology. This boundary is an important judgment combining (1) assessments of technical progress and research directions which together might produce a major advance and (2) judgments over the domain of industrial organization that such an advance might produce a significant competitive advantage.

The following procedure facilitates the appropriate kinds of cooperation for technology transfer on next-generation technologies:

1. Corporate research should strategically plan next-generation technology, and this is best done within an industrial/university/governmental research consortium.

2. Corporate research should work jointly with product-development groups in the business divisions to plan next-generation technology products.

3. Marketing experiments should be set up and conducted jointly with corporate research and business divisions, with trial products using ideas tested in the consortium experimental prototype testbeds.

4. The CEO team should encourage long-term financial planning focused on next-generation technology.

5. Personnel planning and development is required to transition knowledge bases and skill mixes for next-generation technology.

Consensus on vision and testing technical feasibility in a program of technology transfer between corporate research and strategic business units should pose and address the following questions:

1. What will be the boundaries of the next generation of a technological system?

2. What NGT ideas are technically demonstrable and should now be planned into product strategy, and what NGT ideas must still be

technically demonstrated and should not yet be planned into product strategy?

3. What is the pace of technical change, and when should the introduction of products based on NGT be planned?

4. What professional development and training should be planned for product-development groups in order to prepare for NGT products?

For technology transfer to be effective, it is necessary to have strategic business units participate with corporate research in formulating and reaching a consensus on a next-generation technology vision and plan.

Practical Implications

The lessons to be learned from this chapter are procedures for planning research projects focused on technological progress.

1. In research planning, procedures should
   a. Characterize a technology as a system
   b. Identify technical bottlenecks in the way of improving the system
   c. Imagine if and how new research instrumentation, instrumental techniques, algorithmic techniques, or theoretical modeling could be focused on the bottlenecks
   d. Imagine how to improve the understanding and manipulation of the phenomena of the bottlenecks or to develop alternative phenomena to substitute new technical manipulations at the bottlenecks

2. Research projects in a technology plan can be related to the advancement of the technology by displaying a matrix showing the relevance of any project to the structure of the devices, to the production processes, and to the design tools.

3. Planning research for engineering feasibility requires making explicit the application and the technical requirements of the application.
   a. The early stages of engineering research should be aimed at systematically exploring all technical alternatives to find a technically optimal solution.
   b. The criteria for technical optimality include
      (1) Performance
      (2) Efficiency
      (3) Cost
      (4) Safety
      (5) Life cycle

4. In planning research projects, research imagination can focus on the physical phenomena underlying a technological system to improve bottlenecks in the system or to develop substituting technological manipulations through instrumentation, instrumental techniques, theoretical modeling, or algorithmic techniques.

5. A research program for a next-generation technology system requires both a technical vision and the planning of research areas and topics that must be addressed to fulfill that vision.

   a. Initial research projects are proposed in these areas to address the range of research issues, and they are coordinated to result in prototype subsystems of the next generation of the technological system.

6. Research for a next-generation technology system is facilitated by an industry/university/government research consortium.

   a. To accomplish technology transfer in next-generation technology, corporate research and the development groups of relevant business units must jointly manage the long-term research to explore and reduce technology risk.

For Further Reflection

1. Select a current generic technology system, and identify the technical bottlenecks in the system.

2. Identify research advances in instrumentation, instrumental techniques, algorithmic techniques, or theoretical modeling that could improve the understanding of the phenomena underlying the bottlenecks or develop substituting technical manipulations.

3. Create a vision of a next-generation technology system using the research approaches.

4. Identify university and company research groups with published reputations in either the technology systems or the relevant research areas.

5. Design the organization of a industry/university research consortium to create a testbed for that vision.

6. Imagine new product lines and/or production systems embodying the next-generation technology.

References

Abell, D. F. 1980. *Defining the Business: The Starting Point of Strategic Planning.* Englewood Cliffs, N.J.: Prentice-Hall.

Abernathy, William J. 1978. *The Productivity Dilemma: Roadblock to Innovation in the Automobile Industry.* Baltimore: Johns Hopkins University Press.

Abernathy, William J., and K. B. Clark. 1985. "Mapping the Winds of Creative Destruction," *Research Policy,* Vol. 14, No. 1, pp. 2–22.

Abernathy, W. J., and J. M. Utterback. 1978. "Patterns of Industrial Innovation," *Technology Review,* Vol. 80, June-July, pp. 40–47.

Achilladelis, Basil, Albert Schwarzkopf, and Martin Cines. 1990. "The Dynamics of Technological Innovation: The Case of the Chemical Industry," *Research Policy,* Vol. 19, February, pp. 1–34.

Adler, Paul S. 1989. "Managing High-Tech Processes: The Challenge of CAD/CAM," in Mary Ann Von Glinow and Susan Albers Morhman (eds.), *Managing Complexity in High Technology Industries, Systems and People.* New York: Oxford University Press.

Allen, Frederick. 1988. "The Letter that Changed the Way We Fly," *Invention and Technology,* Fall, pp. 6–13.

Anderson, Phillip, and Michael L. Tushman. 1990. "Technological Discontinuities and Dominant Designs: A Cyclical Model of Technological Change," *Administrative Science Quarterly,* Vol. 35, pp. 604–633.

Angier, Natalie. 1991. "Molecular 'Hot Spot' Hints at a Cause of Liver Cancer," *The New York Times,* April 4, pp. A1–21.

Avishai, Bernard, and William Taylor. 1989. "Customers Drive a Technology-Driven Company: An Interview with George Fisher," *Harvard Business Review,* November-December, pp. 107–116.

Ayres, Robert U. 1989. "The Future of Technological Forecasting," *Technology Forecasting and Social Change,* Vol. 36, No. 1-2, August, pp. 49–60.

Ayres, Robert U. 1990. "Technological Transformations and Long Waves," Parts I and II, *Technology Forecasting and Social Change,* Vol. 37, pp. 1–37 and Vol. 37, pp. 111–137.

Baker, Dexter F. 1986. "Role of Innovation in Air Product's Growth," *Research Management,* November-December, pp. 18–27.

Banks, Howard. 1984. "General Electric—Going with the Winners," *Forbes,* March 26, pp. 97–106.

Becker, Robert H., and Laurine M. Speltz. 1983. "Putting the S-Curve to Work," *Research Management,* September-October, pp. 31–33.

Becker, Robert H., and Laurine M. Speltz. 1986. "Making More Explicit Forecasts," *Research Management,* July-August, pp. 21–23.

Bell, James R. 1984. "Patent Guidelines for Research Managers," *IEEE Transactions on Engineering Management,* Vol. EM-31, No. 3, pp. 102–104.

Bessler, David A., and Peter J. Chamberlain. 1987. "On Baysian Composite Forecasting," *Omega,* Vol. 15, No. 1, pp. 43–48.

Betz, Frederick. 1974. "Representational Systems Theory," *Management Science,* May.

Betz, Frederick. 1987. *Managing Technology.* Englewood Cliffs, N.J.: Prentice-Hall.

Bond, David F. 1991. "Risk, Cost Sway Airframe, Engine Choices for ATF," *Aviation and Space Technology,* April 29, pp. 20–29.

Bower, Joseph L., and Thomas M. Hout. 1988. "Fast-Cycle Capability for Competitive Power," *Harvard Business Review,* November-December, pp. 110–118.

Brainerd, J. G., and T. K. Sharpless. 1984. "The ENIAC," *Proceedings of the IEEE,* Vol. 72, No. 9., September, pp. 1202–1205.

Brittain, James E. 1984. Introduction to "Hopper and Mauchly on Computer Programming," *Proceedings of the IEEE,* Vol. 72, No. 9, September, pp. 1213–1214.

Byrne, John A., Deidre A. Depke, and John W. Verity. 1991. "IBM–As Markets and Technology Change, Can Big Blue Remake Its Culture?," *Business Week,* June 17, pp. 25–32.

Cameron, H. M., and J. S. Metcalfe. 1987. "On the Economics of Technological Substitution," *Technology Forecasting and Social Change,* Vol. 32, pp. 147–162.

Campbell, Andrew. 1991. "Strategy and Intuition—A Conversation with Henry Mintzberg," *Long Range Planning,* Vol. 24, No. 2, pp. 108–110.

Carey, John. 1991. "The Research Is First Class. If Only Development Was Too," *Business Week,* December 16, p. 118.

Cava, Robert. 1990. "Superconductors Beyond 1-2-3," *Scientific American,* August, pp. 42–49.

Cavin, Ralph, Larry W. Sumney, and Robert M. Burger. 1989. "The Semiconductor Research Corporation: Cooperative Research," *Proceedings of the IEEE,* Vol. 77, No. 9, September, pp. 1327–1344.

Cerf, Christopher, and Victor Navasky. 1984. *The Experts Speak.* New York: Pantheon Books.

Chandler, Alfred D. 1990. "The Enduring Logic of Industrial Success," *Harvard Business Review,* March-April, pp. 130–140.

Chemical & Engineering News. 1991. "ICI Under Takeover Threat," May 20, p. 8.

Clark, Douglas W. 1990. "Bugs Are Good: A Problem-Oriented Approach to the Management of Design Engineering," *Research-Technology Management,* May-June, pp. 23–27.

Clark, Kim B. 1989. "What Strategy Can Do for Technology," *Harvard Business Review,* November-December, pp. 94–98.

Clark, Kim B., and T. Fujimoto. 1988. "Overlapping Problem Solving in Product Development," *Harvard Business School Working Paper,* April.

Cohen, Stephen S., and John Zysman. 1989. "Manufacturing Innovation and American Industrial Competitiveness," *Science,* Vol. 239, March 4, pp. 1110–1115.

Cole, Tim. 1988. "Science at Sea," *Popular Mechanics,* September, p. 28.

Coombs, Rod, and Albert Richards. 1991. "Technologies Products and Firm's Strategies," *Technology Analysis and Strategic Management,* Vol. 3, No. 1, pp. 77–86.

Cooper, Robert G. 1990. "New Products: What Distinguishes the Winners?" *Research-Technology Management,* November-December, pp. 27–31.

Coover, Harry W. 1986. "Programmed Innovation—Strategy for Success," *Research Management,* November-December, pp. 12–17.

Crow, Kenneth A. 1989. "Ten Steps to Competitive Design for Manufacturability," *The Second International Conference on Design for Manufacturability,* sponsored by CAD/CIM Alert. Washington, D.C.: Management Roundtable.

Cusumano, Michael A. 1985. *The Japanese Automobile Industry: Technology and Management at Nissan and Toyota.* Cambridge, Mass.: Harvard University Press.

Cusumano, Michael A. 1988. "Manufacturing Innovation: Lessons from the Japanese Auto Industry," *Sloan Management Review,* Fall, pp. 29- 39.

Davidow, William H. 1986. *Marketing High Technology.* New York: The Free Press.

Depke, Deidre, Neil Gross, Barbara Buell, and Gary McWilliams. 1991. "Laptops Take Off," *Business Week,* March 18, pp. 118–124.

Didrichsen, Jon. 1972. "The Development of Diversified and Conglomerated Firms in the United States, 1920–1970," *Business History Review,* Vol. 46, Summer, p. 210.

Dimanescu, Dan, and James W. Botkin. 1987. *The New Alliance: America's R&D Consortia.* Cambridge, Mass.: Ballinger Publications.

Drucker, Peter E. 1990. "The Emerging Theory of Manufacturing," *Harvard Business Review,* May-June, pp. 94–102.

Dulkey, N. C., and Q. Helmer. 1963. "An Experimental Application of the Delphi Method to the Use of Experts," *Management Science,* Vol. 9, pp. 458–467.

Economist. 1990. "Reshaping ICI," April 28, pp. 21–23.

Economist. 1991. "When GM's Robots Ran Amok," August 10, pp. 64–65.

Economist. 1991a. "ICI'S Pitch," July 27, pp. 12–13.

Economist. 1991b. "ICI—The Waiting Game," July 27, p. 66.

Economist. 1991c. "Pile 'em High, Sell 'em Cheap," November 2, pp. 59–60.

Eschenbach, T. G., and G. A. Geistauts. 1987. "Strategically Focused Engineering," *IEEE Transactions on Engineering Management,* Vol. EM-34, No. 2, May, pp. 62–70.

Finan, William F., and Jeffrey Frey. 1989. "The Effectiveness of the Japanese Research-Development-Commercialization Cycle," Semiconductor Research Corporation, SRC Technical Report No. T89099.

Fischer, J. C., and R. H. Pry. 1971. "A Simple Substitution Model of Technological Change," *Technology Forecasting and Social Change,* Vol. 3, pp. 75–88; reprinted in Marvin J. Cetron and Christine A. Ralph, *Industrial Applications of Technological Forecasting.* New York: John Wiley & Sons, 1971.

Foray, Dominique, and Arnulf Grubler. 1990. "Morphological Analysis, Diffusion and Lock-Out of Technologies: Ferrous Casting in France and the FRG." *Research Policy,* Vol. 19, pp. 535–550.

Ford, David, and Chris Ryan. 1981. "Taking Technology to Market," *Harvard Business Review,* March-April, pp. 117–126.

Foster, Richard. 1982. "Boosting the Payoff from R&D," *Research Management,* January, pp. 22–27.

Freeman, Christopher. 1974. *The Economics of Industrial Innovation.* New York: Penguin Books.

Gabriel, Kaigham, John Jarvis, and William Trimmer. 1988. *Small Machines, Large Opportunities: A Report on the Emerging Field of Microdynamics.* Washington, D.C.: National Science Foundation.

Gellman, Barry. 1992. "Computer Problem Cited in Crash of F-22 Prototype," *Washington Post,* Thursday, April 30, p. A3.

Godet, Michel. 1983. "Reducing the Blunders in Forecasting," *Futures,* Vol. 15, June 3, pp. 181–192.

Godfrey, A. Blanton, and Peter J. Kolesar. 1988. "Role of Quality in Achieving World Class Competitiveness," in Martin K. Starr (ed.), *Global Competitiveness: Getting the U.S. Back on Track.* New York: W. W. Norton.

Gold, Bela, William S. Peirce, Gerhard Rosegger, and Mark Perlman. 1984. *Technological Progress and Industrial Leadership.* Lexington, Mass.: Lexington Books.

Goldhar, J. D., M, Jelinck, and T. W. Schlie. 1991. "Flexibility and Competitive Advantage–Manufacturing Becomes a Service Business," *International Journal of Technology Management,* May, pp. 243–259.

Gomory, Ralph E., and Roland W. Schmitt. 1988. "Science and Product," *Science,* Vol. 240, May 27, pp. 1131–1204.

Gottwald, Floyd D. 1987. "Diversifying at Ethyl," *Research Management,* May-June, pp. 27–29.

Greiner, Walter, and Aurel Sandulescu. 1990. "New Radioactivities," *Scientific American,* March, pp. 58–67.

Hamilton, David P. 1991. "Can Electronic Property Be Protected?" *Science,* Vol. 253, July 5, p. 23.

Hazelrigg, George A. 1982. "Windows for Innovation: A Story of Two Large-Scale Technologies," a report submitted to the National Science Foundation, Washington, D.C., No. 82-180-1, December 1.

Hewlett, Richard G., and Francis Duncan. 1974. *Nuclear Navy, 1946–1962.* Chicago: University of Chicago Press.

Heppenheimer, T. A. 1990. "How Von Neumann Showed the Way," *Invention and Technology,* Vol. 6, No. 2, Fall, pp. 8–17.

Huth, John E. 1992. "The Search for the Top Quark," *American Scientist,* September, pp. 430–435.

Jantsch, Erich. 1967. *Technological Forecasting in Perspective.* Paris: Organization for Economic Cooperation and Development.

Joint Committee on Atomic Energy (JCAE). 1974. Congress of the United States, Hearings of October 30–31 and November 13, 1963, on Nuclear Propulsion for Naval Surface Vessels. Washington, D.C.: U.S. Government Printing Office.

Kahn, Brian. 1990. "The Software Patent Crisis," *Technology Review,* April, pp. 53–58.

Kell, Douglas, and G. Rickey Welch. 1991. "Perspective: No Turning Back," *The Times Higher Education Supplement,* September 13, p. 15.

Kocaoglu, Dundar F., M. Guven Iyigun, and Chuck Valcesschini. 1990. "New Product Development Cycle," paper presented at the TIMS/ORSA Conference, Nashville, Tennessee.

Lal, V. B., Karmeshu, and S. Kaicker. 1988. "Modeling Innovation Diffusion with Distributed Time Lab," *Technology Forecasting and Social Change,* Vol. 34, pp. 103–113.

Landau, Ralph, and Nathan Rosenberg. 1990. "America's High-Tech Triumph," *Invention and Technology,* Fall, pp. 58–63.

Landford, H. W. 1972. "A Penetration of the Technological Forecasting Jungle," *Technology Forecasting and Social Change,* Vol. 4, pp. 207–225.

Lewis, William W., and Lawrence H. Linden. 1990. "A New Mission for Corporate Technology," *Sloan Management Review,* Summer, pp. 57–65.

Lubar, Steven. 1990. "New, Useful and Nonobvious," *Invention and Technology,* Spring-Summer, pp. 8–16.

Macoby, Michael. 1991, "The Innovative Mind at Work," *IEEE Spectrum,* December, pp. 23–35.

Magaziner, Ira C., and Mark Patinkin. 1989. "Cold Competition: GE Wages the Refrigerator War," *Harvard Business Review,* March- April, pp. 114–124.

Magaziner, Ira C., and Mark Patinkin. 1989. *The Silent War: Inside the Global Business Battles Shaping America's Future.* New York: Random House.

Mahmoud, E., J. Motwani, and G. Rice. 1990. "Forecasting U.S. Exports: An Illustration Using Time Series and Econometric Models," *Omega,* Vol. 18, No. 4, pp. 375–382.

Mansfield, Edwin. 1989. "The Diffusion of Industrial Robots in Japan and the United States," *Research Policy,* Vol. 18, pp. 183–192.

Manufacturing Studies Board. 1991. *The Competitive Edge: Research Priorities for U.S. Manufacturing.* Washington, D.C.: National Research Council.

Marshall, Eliot. 1991. "The Patent Game: Raising the Ante," *Science,* Vol. 253, July 5, pp. 20–24.

Marquis, Donald G. 1969, "The Anatomy of Successful Innovations," *Innovation,* November. Reprinted in M. L. Tushman and W. L. Moore (eds.), *Readings in the Management of Innovation.* Marshfield, Mass.: Pitman, 1982.

Martino, Joseph P. 1983. *Technological Forecasting for Decision Making,* 2d ed. New York: North-Holland.

McKelvey, John P. 1989. "Understanding Superconductivity," *Invention and Technology,* Spring/Summer, pp. 48–57.

Meade, Nigel. 1989. "Technological Substitution: A Framework of Stochastic Models," *Technology Forecasting and Social Change,* Vol. 36, pp. 389–400.

Merrills, Roy. 1989. "How Northern Telecom Competes on Time," *Harvard Business Review,* July-August, pp. 108–114.

Mintzberg, Henry. 1990. "The Design School: Reconsidering the Basic Premises of Strategic Management," *Strategic Management Journal,* Vol. 11, March-April, pp. 171–195.

Morita, Akio, with Edwin M. Reingold and Mitsudo Shimomura. 1986. *Made in Japan.* New York: E. P. Dutton.

Moser, Roger A. 1987. "New Program Development at Ethyl," *Research Management,* May-June, pp. 30–32.

Musashi, Miyamoto. 1982. *Book of Five Rings.* New York: Bantam.

National Research Council. 1991. *Manufacturing Studies Board: Improving Engineering Design.* Washington, D.C.: National Academy Press.

National Research Council Steering Group for Management of Engineering and

Technology. 1991. *Research on the Management of Technology.* Washington, D.C.: National Academy Press.

National Science Board. 1984. *Industry/University Research Cooperation.* Washington, D.C.: National Science Foundation.

Nelson, Richard R. 1990. "U.S. Technological Leadership: Where Did It Come From and Where Did It Go?" *Research Policy,* Vol. 19, pp. 117–132.

Nevins, Allen, and Frank E. Hill. 1954. *Ford: The Times, the Man, the Company.* New York: Charles Scribners.

Nocera, Joseph. 1984. "Death of a Computer: TI's Price War with Commodore Dooms the 99/4a," *InfoWorld,* June 11, pp. 63–65; reprinted from *Texas Monthly,* April 1984.

Obermeier, Klaus K. 1988. "Side by Side," *Byte,* November, pp. 275–283.

Okuda, Nobuo, Masamichi Sugai, and Nobuyuki Goto. 1986. "Semicustom and Custom LSI Technology," *Proceedings of the IEEE,* December, pp. 1636–1645.

Osburn, William F. 1937. *Technological Trends and National Policy.* Washington, D.C.: Research Report, National Research Council.

Peirce, William S. 1985. "Naive Forecasting: The Fiasco of Coal Gasification," *Omega,* Vol. 13, No. 5, pp. 435–441.

Perry, Tekla S., and Paul Wallich. 1985. "Design Case History: The Commodore 64," *IEEE Spectrum,* March, pp. 48–58.

Pirsig, Robert M. 1974. *Zen and the Art of Motorcycle Maintenance.* New York: William Morrow and Co.

Polmar, Norman, and Thomas B. Allen. 1982. *Rickover.* New York: Simon and Schuster.

Porter, Michael E. 1985. *Competitive Advantage: Creating and Sustaining Superior Performance.* New York: The Free Press.

Prahald, C. K., and Gary Hamel. 1990. "The Core Competence of the Corporation," *Harvard Business Review,* May-June, p. 79–91.

Pugh, Emerson. 1984. *Memories that Shaped an Industry: Decisions Leading to IBM System/360.* Cambridge, Mass.: MIT Press.

Quinn, James Brian, Jordan J. Baruch, and Penny C. Paquette. 1988. "Exploiting the Manufacturing-Services Interface," *Sloan Management Review,* Summer, pp. 45–55.

Quinn, James Brian, and Penny C. Paquette. 1988. "Ford: Team Taurus," Amos Tuck School, Dartmouth College, Case Study.

Quinn, James Brian, Thomas L. Doorley, and Penny C. Paquette. 1990. "Technology in Services: Rethinking Strategic Focus," *Sloan Management Review,* Winter, pp. 79–87.

Ray, George F. 1989. "Full Circle: The Diffusion of Technology," *Research Policy,* Vol. 18, pp. 1–18.

Reid, T. R. 1985. "The Chip," *Science 85,* February, pp. 32–41.

Robinson, A. L. 1984. "One Billion Transistors on a Chip?" *Science,* Vol. 223, January 20, pp. 267-268.

Ronvray, Dennis. 1991. "Making Molecules by Numbers," *New Scientist,* March 30, pp. 22–26.

Rosenthal, Stephen R., and Artemis March. 1991. "Speed to Market," Boston University Research Report, School of Management Manufacturing Roundtable, September 13.

Rowe, Gene, George Wright, and Fergus Bolger. 1991. "Delphi: A Reevaluation of Research and Theory," *Technology Forecasting and Social Change,* Vol. 39, pp. 235–251.

Rowen, Robert B. 1990. "Software Project Management Under Incomplete and Ambiuous Specifications," *IEEE Transactions on Engineering Management,* Vol. 37, No. 1, February, pp. 10–21.

Rubenstein, Albert H. 1989. *Managing Technology in the Decentralized Firm.* New York: John Wiley & Sons.

Schwartz, Evan I. 1990. "The Dinosaur That Cost Merrill Lynch a Million a Year," *Business Week,* November 26, p. 122.

Shenhar, Aaron. 1988. "From Low- to High-Tech Project Management," Tel Aviv University, Tel Aviv, Israel, Prepublication copy.

Schmenner, Roger W. 1988. "The Merit of Making Things Fast," *Sloan Management Review,* Fall, pp. 11–17.

Schmitt, Roland W. 1985. "Successful Corporate R&D," *Harvard Business Review,* May-June, pp. 124–129.

Segre, Emilio G. 1989. "The Discovery of Nuclear Fission," *Physics Today,* July, pp. 38–43.

Sherman, Stratford P. 1989. "The Mind of Jack Welch," *Fortune,* March 27, pp. 39–50.

Skrzycki, Cindy. 1991. "NCR Corporation Agrees to AT&T Merger," *Washington Post,* July 5, pp. C1–C4.

Smith, Lee. 1980. "The Lures and Limits of Innovation," *Fortune,* October 30, pp. 84–94.

Solberg, James. 1992. "Why Does It Take So Long, Cost So Much?" Seminar, Purdue University.

Sorenson, Charles E. 1956. *My Forty Years with Ford.* New York: W. W. Norton.

Steele, Lowell W. 1989. *Managing Technology: The Strategic View.* New York: McGraw-Hill.

Strothman, James E. 1990. "The Ancient History of System/360,q" *Invention and Technology,* Winter, pp. 34–40.

Stuewer, Roger H. 1985. "Bringing the News of Fission to America," *Physics Today,* October, pp. 49–56.

Suh, Nam P. 1990. *The Principles of Design.* New York: Oxford University Press.

Taguchi, Genichi, and Don Clausing. 1990. "Robust Quality," *Harvard Business Review,* January-February, pp. 65–75.

Takanaka, Hideo. 1991. "Critical Success Factors in Factory Automation," *Long Range Planning,* Vol. 24, No. 4, pp. 29–35.

Therrien, Lois. 1989. "The Rival Japan Respects," *Business Week,* November 12, pp. 108–118.

Tichy, Noel, and Ram Charan. 1989. "Speed, Simplicity, Self-Confidence: An Interview with Jack Welch," *Harvard Business Review,* September-October, pp. 112–120.

Tushman, Michael, and Philip Anderson. 1986. "Technological Discontinuities and Organizational Environments," *Administrative Science Quarterly,* Vol. 31, pp. 439–465.

Tushman, M., and E. Romanelli. 1985, "Organizational Evolution: A Metamorphosis Model of Convergence and Reorientation," in L. Cummins and B. Saw (eds.), *Research in Organizational Behavior.* Greenwich, Conn.: AI Press.

Tushman, M., E. Romanelli, and B. Virany. 1985. "Executive Succession, Strategic Reorientations, and Organization Evolution," *Technology in Society,* Vol. 7, pp. 297–313.

Twiss, Brian. 1968. *Management of Research and Development.* New York: Longmans.

Uenohara, Michiyuki. 1991. "A Management View of Japanese Corporate R&D," *Research-Technology Management,* November-December, pp. 17–23.

Utterback, J. M., and W. J. Abernathy. 1975. "A Dynamic Model of Product and Process Innovation," *Omega,* Vol. 3, No. 6, pp. 639–646.

Valery, Nicholas. 1991. "Consumer Electronics Survey," *The Economist,* April 13, pp. 3–18.

Veraldi, Lew. 1988. "Team Taurus—A Simultaneous Approach to Product Development," speech given at the Thayer School of Engineering, Dartmouth College.

Verity, John W. 1990. "Rethinking the Computer," *Business Week,* November 26, pp. 116–124.

Verity, John W., Thane Peterson, Deidre Depke, and Evan I. Schwartz. 1991. "The New IBM," *Business Week,* December 16, pp. 112–118.

Vesey, Joseph T. 1991, "Speed-to-Market Distinguishes the New Competitors," *Research-Technology Management,* November-December, pp. 33–38.

Walker, Mark. 1990. "Heisenberg, Goudsmit, and the German Atomic Bomb," *Physics Today,* January, pp. 52–60.

Wheelwright, Steven C., and W. Earl Sasser, Jr. 1989. "The New Product Development Map," *Harvard Business Review,* May-June, pp. 112–125.

Willyard, Charles H., and Cheryl W. McClees. 1987. "Motorola's Technology Roadmap Process," *Research Management,* September-October, pp. 13–19.

Wilson, Pete. 1988. "The CPU Wars," *Byte,* May, pp. 213–234.

Wingate, P. J. 1982. *The Colorful Du Pont Company.* New Jersey: Serendipity Press.

Wise, G. 1980. "A New Role for Professional Scientists in Industry: Industrial Research at General Electric 1900–1916," *Technology and Culture,* Vol. 21, pp. 408–415.

Wolff, Michael. 1984. "William D. Coolidge: Shirt-Sleeves Manager," *IEEE Spectrum,* May, pp. 81–85.

Worlton, Jack, 1988. "Some Patterns of Technological Change in High Performance Computers," *Proceedings Supercomputing '88,* pp. 312–319.

Zorpette, Glenn. 1987. "Breaking the Enemy's Code," *IEEE Spectrum,* September, pp. 47–51.

Zwicky, F. 1948. "The Morphological Analysis and Construction," in *Studies and Essays, Courant Anniversary Volume.* New York: Interscience.

Index

ABOUT THE AUTHOR

Frederick Betz, Ph.D., is a program manager in the Engineering Directorate at the National Science Foundation, participating in the selection and evaluation of NSF-funded engineering research centers and engineering education coalitions. He formerly taught in the business schools of the California State University at Hayward, the State University of New York at Buffalo, and the University of Pennsylvania. Dr. Betz received his Ph.D. in physics from the University of California at Berkeley.